MCBU
Molecular and Cell Biology Updates

Series Editors:

Prof. Dr. Angelo Azzi
Institut für Biochemie
und Molekularbiologie
Bühlstr. 28
CH–3012 Bern
Switzerland

Prof. Dr. Lester Packer
Dept. of Molecular
and Cell Biology
Membrane Bioenergetics Group
251 Life Science Addition
Membrane Bioenergetics Group
Berkeley, CA 94720
USA

Proteases New Perspectives

Edited by V. Turk

Springer Basel AG

Editor

Prof. Vito Turk
Jozef Stefan Institute
Jamova 39
1000 Ljubljana
Slovenia

Library of Congress Cataloging-in-Publication Data

Proteases : new perspectives / edited by V. Turk
p. cm. – (Molecular and cell biology updates)
Includes bibliographical references and index.
ISBN 978-3-0348-9750-1 ISBN 978-3-0348-8737-3 (eBook)
DOI 10.1007/978-3-0348-8737-3
(hardcover : alk. paper)
1. Proteolytic enzymes. I. Turk, Vito. II. Series.
QP609.P78P745 1999
572'.76 – dc21

Deutsche Bibliothek Cataloging-in-Publication Data

Proteases : new perspectives / ed. by V. Turk. - Basel ; Boston ;
Berlin : Birkhäuser, 1999
(Molecular and cell biology updates)
ISBN 978-3-0348-9750-1

Originally published by Birkhäuser Verlag, Basel, Switzerland in 1999
Softcover reprint of the hardcover 1st edition 1999
Printed on acid-free paper produced from chlorine-free pulp. TFC ∞

ISBN 978-3-0348-9750-1

9 8 7 6 5 4 3 2 1

Table of contents

List of contributors VII

Peptidases: a view of classification and nomenclature
Alan J. Barrett 1

Carboxypeptidases
Josep Vendrell and Francesc X. Avilés 13

Signal peptidases
Mark O. Lively and Christopher M. Ashwell 35

Proteasomes
A. Jennifer Rivett and Grant G.F. Mason 45

Cathepsin E and cathepsin D
Kenji Yamamoto 59

Cell-associated metalloproteinases
Gary D. Johnson and Judith S. Bond 73

Proteinases in parasites
Juan José Cazzulo 95

Cell-surface proteases in cancer
Lisa L. Demchik and Bonnie F. Sloane 109

Insect proteinases
Gerald Reeck, Brenda Oppert, Michael Denton, Michael Kanost, James Baker and Karl Kramer 125

Alzheimer's disease and proteinases
Shoichi Ishiura 149

Calpains: structure and function of the calpain super family
Yasuko Ono, Hiroyuki Sorimachi and Koichi Suzuki 159

Arg-gingipain and Lys-gingipain: a novel class of cysteine proteinases
Kenji Yamamoto, Tomoko Kadowaki, Kuniaki Okamoto 175

Proteinases in apoptosis
Masanori Tomioka and Seiichi Kawashima 185

Caspases: cytokine activators and promoters of cell death
Guy S. Salvesen and Henning R. Stennicke 195

Lysosomal cysteine proteinases: Structure and regulation
Vito Turk, Gregor Gunčar and Dušan Turk 211

Subject index 225

List of contributors

Christopher M. Ashwell, Growth Biology Lab, United States Department of Agriculture, Beltsville, MD 20705, USA

Francesc X. Avilés, Departament de Bioquímica i Biologia Molecular, Facultat de Ciències, and Institut de Biologia Fonamental, Universitat Autònoma de Barcelona, E-08193 Bellaterra; e-mail: FX.Aviles@uab.es

James Baker, The Grain Marketing and Production Research Center, Agricultural Research Service, U.S.Department of Agriculture, Manhattan, KS 66502, USA

Alan J. Barrett, MRC Molecular Enzymology Laboratory, The Babraham Instiute, Cambridge CB2 4AT, UK; e-mail: alan.barrett@bbsrc.ac.uk

Judith S. Bond, Department of Biochemistry and Molecular Biology, The Pennsylvania State University, MS H171, Hershey, PA 17033-0850, USA; e-mail: jbond@psu.edu

Juan José Cazzulo, Instituto de Investigaciones Biotecnológicas, Universidad Nacional de General San Martín, Av. General Paz y Albarellos, INTI, C.C.30, 1650 San Martín, Prov. Buenos Aires, Argentina; e-mail: jcazzulo@inti.gov.ar

Lisa L. Demchik, Cancer Biology Program, School of Medicine, Wayne State University, 540 E. Canfield, Detroit, MI 48201, USA; e-mail: ldemchik@med.wayne.edu

Michael Denton, The Department of Biochemistry, Kansas State University, Manhattan, KS 66506, USA

Gregor Gunčar, Department of Biochemistry and Molecular Biology, J. Stefan Institute, Jamova 39, 1000 Ljubljana, Slovenia

Shoichi Ishiura, Department of Life Sciences, Graduate School of Arts and Sciences, The University of Tokyo, 3-8-1 Komaba, Meguro-ku, Tokyo 153-8902, Japan; e-mail: cishiura@komaba.ecc.u-tokyo.ac.jp

Gary D. Johnson, Department of Biochemistry, Parke-Davis Pharmaceutical Research, Warner-Lambert Company, Ann Arbor, MI 48105, USA

Tomoko Kadowaki, Department of Pharmacology, Kyushu University Faculty of Dentistry, Higashi-ku, Fukuoka 812-8582, Japan; e-mail: tomokad@dent.kyushu-u.ac.jp

Michael Kanost, The Department of Biochemistry, Kansas State University, Manhattan, KS 66506, USA

Seiichi Kawashima, Department of Molecular Biology, Tokyo Metropolitan Institute of Medical Science, 3-18-22 Honkomagome, Bunkyo-ku, Tokyo 113-8613, Japan; e-mail: s-kawa@rinschoken.or.jp

Karl Kramer, The Grain Marketing and Production Research Center, Agricultural Research Service, U.S.Department of Agriculture, Manhattan, KS 66502, USA

Mark O. Lively, Department of Biochemistry, Wake Forest University School of Medicine, Medical Center Boulevard, Winston-Salem, North Carolina 27157, USA; e-mail: mlively@wfubmc.edu

Grant G.F. Mason, Department of Biochemistry, University of Bristol, University Walk, Bristol BS8 1TD, UK

Kuniaki Okamoto, Department of Pharmacology, Kyushu University Faculty of Dentistry, Higashi-ku, Fukuoka 812-8582, Japan; e-mail: kokamoto@dent.kyushu-u.ac.jp

Yasuko Ono, Molecular Structure and Function, Institute of Molecular and Cellular Biosciences, University of Tokyo, 1-1-1 Yayoi, Bunkyo-ku, Tokyo 113-0032, Japan

Brenda Oppert, The Grain Marketing and Production Research Center, Agricultural Research Service, U.S.Department of Agriculture, Manhattan, KS 66502, USA

Gerald Reeck, The Department of Biochemistry, Kansas State University, Manhattan, KS 66506, USA; e-mail: greeck@ksu.edu

A. Jennifer Rivett, Department of Biochemistry, University of Bristol, University Walk, Bristol BS8 1TD, UK; e-mail: j.rivett@bristol.ac.uk

Guy S. Salvesen, The Program for Apoptosis and Cell Death Research, The Burnham Institute, 10901 North Torrey Pines Road, La Jolla, California, CA 92037, USA; e-mail: gsalvesen@burnham-inst.org

Bonnie F. Sloane, Cancer Biology Program and Department of Pharmacology, School of Medicine, Wayne State University, 540 E. Canfield, Detroit, MI 48201, USA; e-mail: bsloane@med.wayne.edu

Hiroyuki Sorimachi, Molecular Structure and Function, Institute of Molecular and Cellular Biosciences, University of Tokyo, 1-1-1 Yayoi, Bunkyo-ku, Tokyo 113-0032, Japan

Henning R. Stennicke, The Program for Apoptosis and Cell Death Research, The Burnham Institute, 10901 North Torrey Pines Road, La Jolla, California, CA 92037, USA

Koichi Suzuki, Molecular Structure and Function, Institute of Molecular and Cellular Biosciences, University of Tokyo, 1-1-1 Yayoi, Bunkyo-ku, Tokyo 113-0032, Japan; e-mail: kosuzuki@iam.u-tokyo.ac.jp

Masanori Tomioka, Department of Molecular Biology, Tokyo Metropolitan Institute of Medical Science, 3-18-22 Honkomagome, Bunkyo-ku, Tokyo 113-8613, Japan

Dušan Turk, Department of Biochemistry and Molecular Biology, J. Stefan Institute, Jamova 39, 1000 Ljubljana, Slovenia

Vito Turk, Department of Biochemistry and Molecular Biology, J. Stefan Institute, Jamova 39, 1000 Ljubljana, Slovenia; e-mail: vito.turk@ijs.si

Josep Vendrell, Departament de Bioquímica i Biologia Molecular, Facultat de Ciències, and Institut de Biologia Fonamental, Universitat Autònoma de Barcelona, E-08193 Bellaterra; e-mail: Josep.Vendrell@uab.es

Kenji Yamamoto, Department of Pharmacology, Kyushu University, Faculty of Dentistry, Higashi-ku, Fukuoka 812-8582, Japan; e-mail: kyama@dent.kyushu-u.ac.jp

Proteases: New Perspectives
V. Turk (ed.)

Peptidases: a view of classification and nomenclature

Alan J. Barrett

MRC Molecular Enzymology Laboratory, The Babraham Institute, Cambridge CB2 4AT, UK

Introduction

It is beyond question that the results of research on proteolytic enzymes, or peptidases, are already benefiting mankind in many ways, and there is no doubt that research in this area has the potential to contribute still more in the future. One of the clearest indications of the general recognition of this promise is the vast annual expenditure of the pharmaceutical industry on exploring the involvement of peptidases in human health and disease.

The high and accelerating rate of research on peptidases is being rewarded by a rate of discovery that could not have been imagined just a few years ago. One measure of this is the number of known peptidases. At the present time, we can recognise perhaps 600 distinct peptidases, including over 200 that are expressed in mammals, and new ones are being discovered almost daily. This means that there is a clear need for sound systems for classifying the enzymes and for naming them. Only with such systems in place can the wealth of new information that is becoming available be shared efficiently amongst the many scientists now active in this field of research. Without such systems, there will be needless and expensive duplication of effort, and the rate of discovery, and its consequent benefits to mankind, will be slower. The justification for trying to improve the systems is therefore strictly practical, and most of the questions that arise are best dealt with by asking what will be most useful to the scientists working in the field, not by reference to any abstract theory.

The aims of classification and nomenclature are largely simple and obvious. At the present time, it is natural for us to think of these in terms of the storage and retrieval of data on the World Wide Web (WWW), but an approach that is good for the WWW is also good for paper-based archives. The ideal would be that an individual scientist interested in a particular peptidase would be able quickly and unambiguously to retrieve all of the published information about that enzyme, uncontaminated with irrelevant material. This requires that the enzyme has a unique name or code number that is to be found in all the relevant data records, whether they be in the specialized databases of sequences, higher-level structure or genetic information, or in the wider published literature, embracing biological functions and disease involvements. The scientist should then be able to broaden the search to bring in other peptidases with similar activities, with similar structures or sharing evolutionary origins. The technology for all of this exists, but what can be achieved in practice depends upon the quality of the classification and nomenclature that are in use.

Problems of terminology

A terminological muddle is immediately apparent from the many almost synonymous terms that are in use for the group of enzymes as a whole. Thus, *proteolytic enzyme*, *protease* and *proteinase* are almost-overlapping terms for the whole group of enzymes that we are here terming *peptidases*. These terms originally had slightly different shades of meaning (reviewed elsewhere: [1, 2]), but these differences have largely been lost in current usage. It would be helpful to anyone wanting to access all the data if one term were consistently used. We have argued that the most logical term is *peptidase*, subdivided into *exopeptidase* and *endopeptidase*, and this is what is recommended by the International Union of Biochemistry and Molecular Biology (IUBMB) [3]. There is no need whatever for the other familiar terms to be abandoned, but it would be to the advantage of all if 'peptidase' were also included amongst the indexing keywords assigned to papers and database records relevant to this topic.

The names of the individual peptidases also pose special problems. Most enzymes other than peptidases are conveniently named solely on the basis of the reactions they catalyse, but this is generally not a good approach for peptidases. One reason is that the specificities of peptidases are commonly so complex that even when they can be determined unambiguously, they cannot be described briefly enough to form a convenient name. Also, there are many examples of peptidases that catalyse closely similar reactions, and could in principle be given the same name, but need to be distinguished because they are the products of different genes, expressed under different promoters, located in different cell types or compartments, and serve quite different biological functions. A simple example would be the pancreatic and leukocyte forms of elastase; these obviously need to be treated as distinct peptidases, despite their similar specificities in the test-tube. But once we depart from the criterion of the reaction-catalysed as the defining characteristic of an individual peptidase, we find that we need new principles by which to name them. The need for such new principles has not been widely appreciated, and certainly has not yet been met, so that a chaotic situation has arisen over the naming of peptidases. Resolving this is one of the major challenges that face anyone attempting to facilitate communication amongst peptidase scientists.

A three-level system of classification

A three-layer system has been developed for the classification of peptidases by (i) *catalytic type*, (ii) *molecular structure*, and (iii) *individual peptidases* (Fig. 1). This classification is currently managed by a combination of two partially-overlapping systems, the MEROPS system of peptidase clans and families, and the Enzyme Commission (EC) recommendations on enzyme nomenclature. Both can be found on the WWW (Fig. 2).

Rawlings and Barrett [4] proposed a system of classification of peptidases on the basis of similarities in amino acid sequences. This was further developed through articles in two volumes of *Methods in Enzymology* [5–8], and in 1996 was presented in the form of the MEROPS database on the WWW. The word *MEROPS* has no important meaning, but now seems a suitable trivial name for reference to this system as a whole, whether in printed form or on the WWW. MEROPS is important primarily in the first and second levels of the three-level system.

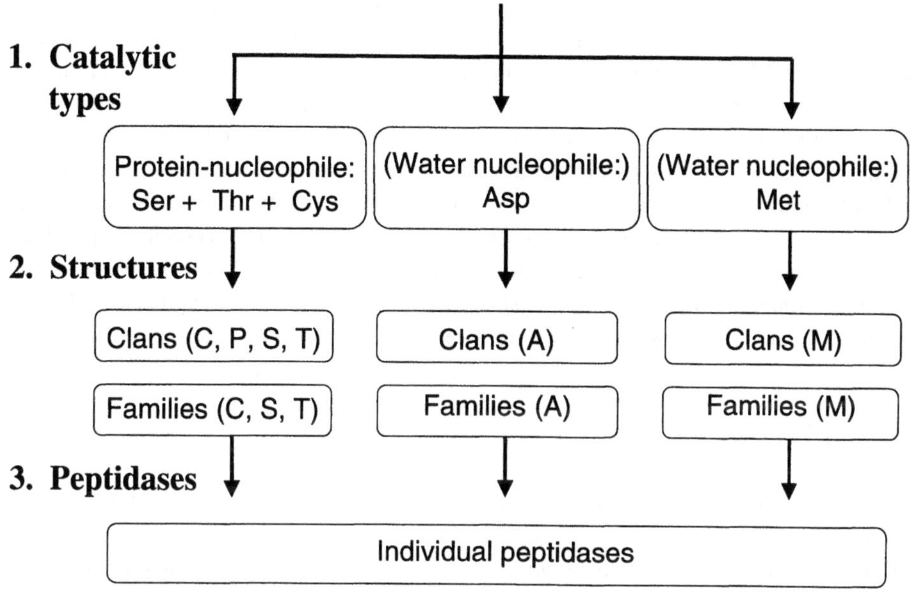

Figure 1. The three-level classification of peptidases.

MEROPS system

http://www.bi.bbsrc.ac.uk/Merops/Merops.htm

IUBMB Enzyme Nomenclature for Peptidases (EC 3.4):

http://www.chem.qmw.ac.uk/iubmb/enzyme/EC34

Figure 2. World Wide Web locations of peptidase classification documents.

The EC system is that of the Nomenclature Committee of IUBMB. The Committee was the successor to the Enzyme Commission [9], and the numbers that it applies to enzymes are still termed *EC numbers*. As is well known, the EC recommendations provide classification and nomenclature for enzymes of all kinds. For the majority of enzymes, the classification is based strictly upon the type of reaction that the enzyme catalyses, and this also leads to a name for the enzyme. As was mentioned above, the reactions catalysed by peptidases are generally too complex to be used in this way, and different peptidases may have similar activities. As a result, the section of the EC recommendations that deals with peptidases (subclass 3.4) is rather different in style from the remainder of the recommendations, but it is this section that will be referred to here as the *EC recommendations*, or simply *EC*. Despite having great difficulty in classifying peptidases, EC plays an important role in their nomenclature, and it is useful in two ways. Firstly, it gives a unique number to each peptidase that is included in the list, and this can be used for unambiguous reference to that enzyme when needed, and secondly, it provides a *Recommended name* for each peptidase. Other names are also listed, so this helps significantly in cutting through the present muddled state of naming of individual peptidases. EC therefore makes its major contribution to the third level of the three-level system.

Level 1: Catalytic type

It has long been recognised that major groups of proteolytic enzymes can usefully be distinguished on the basis of the chemical groups responsible for catalysis. The exact way in which this is done has needed minor adjustments from time to time, but the principle is valuable because it is familiar, and is still working well.

In the EC recommendations, catalytic type is used to subdivide the carboxypeptidases and the endopeptidases (Tab. 1). Catalytic type also forms the highest level of classification for all peptidases in the MEROPS system. At the time of the introduction of the MEROPS system [4], the groupings of *serine*, *cysteine*, *aspartic*, *metallo* and *unknown* catalytic types of peptidase were recognised. As an extension of this, the initial letters S, C, A, M and U have been used in forming the names of clans (by adding a further letter) and families (by adding a number). Recently, the threonine-dependent peptidases of the proteasome group have been recognised [10], and the letter T has been used in the same way.

Using the nature of the amino acid (or metal) primarily responsible for activity as the top level of classification in the MEROPS system was sound only so long as there was no reason to think that a peptidase of one of these types could ever have evolved into one of another. The reason is that this would infringe the hierarchical integrity of the classification, since each clan, in the second layer of the classification, represents a unique evolutionary line (see below). But we now know that a peptidase of one catalytic type has indeed sometimes evolved from one of a different type. The most clear-cut evidence of this came with the demonstration that the protein fold of the cysteine-type picornain of hepatitis A virus (family C3, clan CB) is so close to those of serine peptidases in the trypsin family (family S1, clan SA) that they must have had a common origin [11]. This led to the anomalous situation that we had two clans representing a single protein fold, and a single evolutionary origin, because they differed in catalytic type. Evidently, too much weight was being placed on the exact nature of the amino acid at the catalytic centre. To deal with this and other similar problems, just two major catalytic types are now recognised, and these are termed *protein nucleophile*, combining the older serine, cysteine and threonine types, and *water nucleophile*, which we further divide into aspartic and metallopeptidases. We have no reason to think that a peptidase can cross these boundaries in the

Table 1. Classification of peptidases according to the EC recommendations [3]

Sub-subclass	Kind of peptidase
3.4.11	Aminopeptidases
3.4.13	Dipeptidases
3.4.14	Dipeptidyl-peptidases
3.4.15	Peptidyl-dipeptidases
3.4.16	Serine-type carboxypeptidases
3.4.17	Metallocarboxypeptidases
3.4.18	Cysteine-type carboxypeptidases
3.4.19	Omega peptidases
3.4.21	Serine endopeptidases
3.4.22	Cysteine endopeptidases
3.4.23	Aspartic endopeptidases
3.4.24	Metalloendopeptidases
3.4.99	Endopeptidases of unknown type

course of evolution. The terms protein nucleophile and water nucleophile serve as shorthand for two essentially different types of catalytic mechanism. In the peptidases of serine, threonine and cysteine type, the nucleophilic group that initiates the attack on the peptide bond is an oxygen or a sulfur atom that is part of the protein structure of the peptidase, being in the side chain of an amino acid. As a result of this, a covalent acyl enzyme is formed as an intermediate in catalysis. Typically, this is hydrolysed, but it can also take part in a transfer reaction, in which the more N-terminal of the two products of the peptide bond cleavage is transferred from the acyl enzyme to some acceptor other than water. In contrast, in the water-nucleophile peptidases of aspartic and metallo types, the attacking nucleophile is a water molecule, bound and activated in the catalytic site. The functional groups of the protein that make catalysis possible do not react directly with the substrate, so that this mechanism does not involve the formation of an acyl enzyme, and normally does not lead to transfer reactions. The letter 'P' is used in naming a clan of protein nucleophile peptidases that contains families of more than one catalytic type, so that the original clans SA and CB are now merged as clan PA.

Level 2: Molecular structure

There are strong arguments for using the wealth of data on the amino acid sequences and three-dimensional structures of peptidases in their classification. Crucially important is the fact that simple, automated searches of the sequence databases rapidly return lists of similar peptidases, even in the absence of an ideal nomenclature. The similarities in primary structure tend to reflect shared evolutionary origins, and a wealth of biological meaning can be extracted from this. Accordingly, the classification of peptidases into families is at the heart of the MEROPS system.

Peptidase families

The MEROPS system started with the establishing of peptidase families. All of the amino acid sequences of peptidases that were available in 1993 were searched for statistically significant similarities, so as to group them in families of peptidases that were indisputably homologous. In the course of this exercise, some pairs of sequences were encountered that did show significant relationship, but only in parts of the sequence unlikely to contribute directly to the peptidase activity. The matches arose from the chimeric nature of many protein structures, and were not directly relevant to the classification of peptidases. Accordingly, such relationships were not used in the forming of families [12, 13], and the stipulation was made that only significant relationships *in the part of the proteins responsible for peptidase activity* would justify grouping in a single peptidase family. Application of these methods to the sequences that have been reported since 1993 has led to the growth of most of the families that were established at that time, to the merging of several of the families when 'linking' sequences were discovered, and to the setting up of a number of new families. The total number of families is now about 140.

MEROPS also provides a way of naming the families of peptidases. Until now, there has been no unambiguous way to do this, and a family has generally been referred to by the name of one of its members. For example, one might have spoken of the 'prolyl oligopeptidase family' or the 'dipeptidyl-peptidase IV family'. Not only are these cumbersome names, but it happens that both would be references to the same family, termed *S9* in MEROPS, since both peptidases named are in this family. In the simple system used in MEROPS, the name of each family is constructed from a capital letter representing the catalytic type of the peptidases it contains (S, T, C, A, M or U) followed by a number that is assigned arbitrarily. If the family disappears (usually as a result of being merged with another), the name is not re-used.

Clans

From the first, it was evident that the strict criteria that were being applied in the building of peptidase families solely by reference to amino acid sequences were failing to place together peptidases that were strongly indicated as being related by other forms of evidence, most notably similarities in tertiary structure. It is well established that similarities in protein fold persist in evolution much longer than do close similarities in amino acid sequence, and accordingly, the folds can reveal distant relationships that cannot be seen clearly in the primary structures. Such distantly related groups were termed *clans* [4]. The kinds of evidence that are used in the forming of clans are not easily evaluated by statistical methods, so the assignments are necessarily somewhat subjective, but we can nevertheless make most of them with a good degree of confidence. The total number of clans is now about 30. The clans are named similarly to the families, with a letter indicating the catalytic type of the peptidases contained in the clan, but followed by a capital letter. A clan that contains protein-nucleophile peptidases of more than one catalytic type, such as that containing the trypsin-like serine peptidases as well as the picornain-like cysteine peptidases, is named with a P, making clan PA, in this particular case.

The developments in the MEROPS system since 1993 have been reflected in printed articles (e.g. [5–8]) and in several releases of the WWW version (Fig. 2). A summary of the system as it stands in 1997 can be seen in Table 2.

Table 2. Clans and families of peptidases

a) 'Protein nucleophile': serine, threonine and cysteine peptidases		
Clan	Family	Example
PA	S1	Trypsin
	S2	Streptogrisin A
	S3	Togavirin
	S6	IgA1-Specific serine endopeptidase
	S7	Flavivirin
	S29	Hepatitis C virus NS3 polyprotein peptidase
	S30	Potyvirus P1 proteinase
	S31	Pestivirus polyprotein peptidase p80
	S32	Equine arteritis virus serine endopeptidase
	S35	Apple stem grooving virus protease
	C3	Poliovirus picornain 3C
	C4	Tobacco etch virus NIa endopeptidase
	C24	Feline calicivirus endopeptidase
	C30	Mouse hepatitis coronavirus picornain 3C-like endopeptidase
	C37	Southampton virus processing peptidase
SB	S8	Subtilisin
SC	S9	Prolyl oligopeptidase
	S10	Carboxypeptidase C
	S15	X-Pro dipeptidyl-peptidase
	S28	Pro-X carboxypeptidase
	S33	Prolyl aminopeptidase
	S37	PS-10 peptidase (*Streptomyces lividans*)
SE	S11	D-Ala-D-Ala carboxypeptidase A
	S12	D-Ala-D-Ala carboxypeptidase B
	S13	D-Ala-D-Ala peptidase C
SF	S24	Repressor LexA
	S26	Signal peptidase I
	S41	Tail-specific protease
SH	S21	Assemblin
TA	T1	Proteasome
CA	C1	Papain
	C2	Calpain
	C10	Streptopain
	C12	Deubiquitinating peptidase Yuh1
	C19	Isopeptidase T
CC	C6	Tobacco etch virus HC-proteinase
	C7	Chestnut blight virus p29 endopeptidase
	C8	Chestnut blight virus p48 endopeptidase
	C9	Sindbis virus nsP2 endopeptidase

(continued on next pages)

Table 2. (continued)

a) 'Protein nucleophile': serine, threonine and cysteine peptidases		
Clan	Family	Example
CC	C16	Mouse hepatitis virus endopeptidase
	C21	Turnip yellow mosaic virus endopeptidase
	C23	Blueberry scorch carlavirus endopeptidase
	C27	Rubella rubivirus endopeptidase
	C28	Foot-and-mouth disease virus L proteinase
	C29	Mouse hepatitis coronavirus papain-like endopeptidase 2
	C31	Porcine respiratory and reproductive syndrome arterivirus α
	C32	Equine arteritis virus PCP β endopeptidase
	C36	Beet necrotic yellow vein furovirus papain-like endopeptidase
CD	C14	Caspase-1
CE	C5	Adenovirus endopeptidase
SX	S14	Endopeptidase Clp
	S16	Endopeptidase La
	S18	Omptin
	S19	Chymotrypsin-like protease (*Coccidioides*)
	S34	HflA protease
	S38	Chymotrypsin-like protease (*Treponema denticola*)
CX	C11	Clostripain
	C13	Legumain
	C15	Pyroglutamyl peptidase I
	C25	Gingipain R
	C26	γ-Glutamyl hydrolase
	C33	Equine arterivirus Nsp2 endopeptidase
	C40	Dipeptidyl-peptidase VI
	C41	Hepatitis E cysteine proteinase

b) 'Water nucleophile': aspartic peptidases		
Clan	Family	Example
AA	A1	Pepsin
	A2	HIV 1 retropepsin
	A3	Cauliflower mosaic virus endopeptidase
	A9	Simian foamy virus polyprotein peptidase
	A10	*Schizosaccharomyces* retropepsin-like transposon
	A15	Rice tungro bacilliform virus protease
AB	A6	Nodavirus endopeptidase
AX	A4	Scytalidopepsin B
	A5	Thermopsin
	A7	Pseudomonapepsin
	A8	Signal peptidase II
	A11	*Drosophila* transposon copia peptidase
	A12	Maize transposon bs1 peptidase

(continued on next page)

Table 2. (continued)

c) 'Water nucleophile': metallopeptidases		
Clan	Family	Example
MA	M1	Membrane alanyl aminopeptidase
	M2	Peptidyl-dipeptidase A
	M4	Thermolysin
	M5	Mycolysin
	M9	*Vibrio* collagenase
	M13	Neprilysin
	M30	Hyicolysin
	M36	Fungalysin
MB	M6	Immune inhibitor A
	M7	*Streptomyces* small neutral endopeptidase
	M8	Leishmanolysin
	M10	Interstitial collagenase
	M11	Gametolysin
	M12	Astacin
MC	M14	Carboxypeptidase A
MD	M15	Zinc D-Ala-D-Ala carboxypeptidase
ME	M16	Pitrilysin
MF	M17	Leucyl aminopeptidase
MG	M24	Methionyl aminopeptidase
MH	M18	Aminopeptidase I
	M20	Glutamate carboxypeptidase
	M25	X-His dipeptidase
	M28	Aminopeptidase Y
	M40	Carboxypeptidase (*Sulfolobus sulfataricus*)
	M42	Glutamyl aminopeptidase (*Lactococcus*)
MX	M3	Thimet oligopeptidase
	M19	Membrane dipeptidase
	M22	O-Sialoglycoprotein endopeptidase
	M23	β-Lytic endopeptidase
	M26	IgA-specific metalloendopeptidase
	M27	Tentoxilysin
	M29	Aminopeptidase T
	M32	Carboxypeptidase Taq
	M34	Anthrax lethal factor
	M35	Penicillolysin
	M37	Lysostaphin
	M38	β-Aspartyl dipeptidase
	M41	FtsH endopeptidase
	M45	D-Ala-D-Ala dipeptidase (*Enterococcus*)

(continued on next page)

Table 2. (continued)

d) Unclassified peptidases

Clan	Family	Example
UX	U3	Endopeptidase gpr
	U4	Sporulation sigma E factor processing peptidase
	U6	Murein endopeptidase MepA
	U7	Protease IV
	U9	Prohead proteinase (bacteriophage T4)
	U12	Prepilin type IV signal peptidase
	U26	vanY D-Ala-D-Ala carboxypeptidase
	U27	ATP-dependent protease (*Lactococcus*)
	U28	Aspartyl dipeptidase
	U29	Cardiovirus endopeptidase 2A
	U32	Microbial collagenase (*Porphyromonas*)
	U39	Hepatitis C virus NS2-3 protease
	U40	Protein P5 murein endopeptidase (bacteriophage phi6)
	U43	Infectious pancreatic necrosis birnavirus endopeptidase

The clans and families of peptidases included in the MEROPS database in 1997 are listed, together with one example from each family. The clan PA contains 'protein nucleophile' peptidases of more than one catalytic type, previously assigned to clans SA and CB, as described in the text. The miscellaneous groups of families that cannot yet be assigned to clans are listed as if in 'clans' SX, CX, AX, MX and UX for convenience.

Level 3: Individual peptidases

In any system of classification, the criteria that are used for discriminating between the individual objects that are classified are crucially important, and yet they may be particularly difficult to define. An example that is well known to biologists is the difficulty of defining a distinct species of organism. It is also very difficult to describe what is a distinct and individual peptidase, but some principles emerged during the revision of subsection 3.4 of *Enzyme Nomenclature 1992* [3]. In general, it can be said that two distinct peptidases, as contrasted with forms of a single peptidase, are expected to meet at least one of the following criteria: (a) they have different specificities, (b) they have different sensitivities to inhibitors, (c) they are of different catalytic type, or belong to different peptidase families, or (d) they are encoded by different genes in a single organism. Of course, what is meant by *different* in this context is seldom rigorously defined, and becomes subjective. As was mentioned earlier, the criterion of practical usefulness is an important one in considering systems for classification and nomenclature, and this obviously means usefulness to human beings. This immediately introduces an anthropocentric bias. We may find that smaller differences seem to justify distinguishing peptidases when the enzymes are found in the human body, or in organisms that have an impact on human health or nutrition, than for organisms that are less directly relevant to human welfare.

Conclusions

Working together, the EC and MEROPS systems can provide the basis for a sound classification and nomenclature of peptidases. Despite differences of approach, both EC and MEROPS make real contributions to the overall scheme, although the contributions they make differ between the three levels of the classification. In the upper two levels, catalytic type and molecular structure, the MEROPS system has most to contribute. For classification, catalytic type, clan and family are used together to form a hierarchical system. MEROPS also provides much-needed nomenclature for the clans and families. As the concise and unambiguous MEROPS family names are increasingly introduced into the records of the sequence databases and the EC recommendations, they should provide a very effective way of searching for information about a whole family or clan.

In contrast to MEROPS, EC does not have a great deal to say about the classification of peptidases, since they cannot satisfactorily be classified and named by the reactions they catalyse. As a result, all the entries for peptidases are divided amongst just 13 sub-subclasses (Tab. 1), and homologous peptidases that belong to a single family in MEROPS are sometimes split between several sub-subclasses. The 13 sub-subclasses may be contrasted with the 140 families, grouped in 30 clans, that are provided by MEROPS.

While the EC recommendations may not have much to offer in the higher levels of classification of peptidases, they certainly come into their own at the lowest level, in which individual peptidases are recognized. Almost 300 peptidases are included in the EC recommendations in 1997, and the number is increasing steadily. Not only are the peptidases distinguished as individual enzymes, but a name is recommended for each, and other names that may be encountered in the literature also are listed. Perhaps most importantly, each peptidase is assigned a unique EC number that serves as an unmistakable reference to this enzyme, when the names are not always clear.

Having concluded that together, MEROPS and EC provide a sound basis, it is appropriate to consider what more is needed if we are to achieve the ideal of quick and easy access to all existing knowledge of peptidases that was described at the start of the present chapter. The obvious and major requirement is that both systems be updated regularly to keep pace with the rapid rate of discovery of new peptidases. It will also be essential that the scientists depositing data in the printed literature or the databases take advantage of the availability of unambiguous names and code numbers for the clans, families and individual peptidases by including these names, together with any others they may choose to use, in the public records. This will greatly help other scientists to access that data, so as to use it and cite it, which will unquestionably be to everybody's advantage.

Acknowledgement
My colleague Neil D. Rawlings has made essential contributions at all stages of the development of the MEROPS system, and is curator of the World Wide Web database. I thank him for his advice during the writing of the present chapter.

References

1 Barrett AJ (1986) An introduction to the proteinases. *In*: AJ Barrett, G Salvesen (eds): *Proteinase Inhibitors*. Elsevier Science Publishers, Amsterdam, 3–22
2 Barrett AJ (1998) Proteolytic enzymes: nomenclature and classification. *In*: RJ Beynon, JS Bond (eds): *Proteolytic Enzymes – a Practical Approach*. Oxford University Press, Oxford
3 Nomenclature Committee of the International Union of Biochemistry Molecular Biology (1992) *Enzyme Nomenclature 1992*. Academic Press, Orlando
4 Rawlings ND, Barrett AJ (1993) Evolutionary families of peptidases. *Biochem J* 290: 205–218
5 Rawlings ND, Barrett AJ (1994) Families of serine peptidases. *Meth Enzymol* 244: 19–61
6 Rawlings ND, Barrett AJ (1994) Families of cysteine peptidases. *Meth Enzymol* 244: 461–486
7 Rawlings ND, Barrett AJ (1995) Evolutionary families of metallopeptidases. *Meth Enzymol* 248: 183–228
8 Rawlings ND, Barrett AJ (1995) Families of aspartic peptidases, and those of unknown catalytic mechanism. *Meth Enzymol* 248: 105–120
9 Webb EC (1993) Enzyme nomenclature: a personal perspective. *FASEB J* 7: 1192–1194
10 Seemüller E, Lupas A, Stock D, Löwe J, Huber R, Baumeister W (1995) Proteasome from *Thermoplasma acidophilum*: a threonine protease. *Science* 268: 579–582
11 Allaire M, Chernaia MM, Malcolm BA, James MNG (1994) Picornaviral 3C cysteine proteinases have a fold similar to chymotrypsin-like serine proteinases. *Nature* 369: 72–76
12 Rawlings ND, Barrett AJ (1990) Bone morphogenetic protein 1 is homologous in part with calcium-dependent serine proteinase. *Biochem J* 266: 622–624
13 Rawlings ND, Barrett AJ (1999) *MEROPS*: the peptidase database. *Nucleic Acids Res* 27: 325–331

Proteases: New Perspectives
V. Turk (ed.)

Carboxypeptidases

Josep Vendrell and Francesc X. Avilés

Departament de Bioquímica i Biologia Molecular, Facultat de Ciències, and Institut de Biologia Fonamental, Universitat Autònoma de Barcelona, E-08193 Bellaterra, Spain

Classification of carboxypeptidases: an overview

Carboxypeptidases catalyze the hydrolysis of peptide bonds at the C-terminus of peptides and proteins. This hydrolysis may be a step in the degradation of some substrate molecules or may result in the maturation of others. As for every type of protease, the physiological effect of the hydrolytic action is thus varied and also site- and organism-dependent. Moreover, the carboxypeptidase action may be carried out by at least two different kinds of enzymes with different catalytic mechanisms. In one case, metallocarboxypeptidases possess a tightly bound Zn^{2+} atom which is directly involved in catalysis; on the other hand, the serine-carboxypeptidases contain an active Ser residue at the active centre which belongs to the Ser/His/Asp triad characteristic of serine proteinases.

Classification of proteolytic enzymes has been recognized to be a troublesome issue, since their classification by the catalyzed reaction cannot account for the ample variability in structural and evolutionary relationships (both in terms of primary and tertiary structure), catalytic mechanisms and physiological functions [1]. Rawlings and Barrett proposed the usage of the term 'family' to group peptidases demonstrated to share a significant evolutionary relationship in their primary structures and also introduced the term 'clan' to describe groups of families which share a common ancestor [2]. A hierarchical classification in which the catalytic type is considered first, followed by the evolutionary relationship and the type of reaction catalyzed was subsequently proposed [1]. Since the classification by evolutionary relationships is flexible enough to accommodate the similarities in folding patterns even though the alignment of sequences may sometimes not be conclusive, this approach has been used to elaborate Table 1, where a partial classification of carboxypeptidases is presented.

The data presented in Table 1 are not exhaustive nor does the classification take into account all the possible similarities or differences among carboxypeptidases. The most crowded group is that of family M14 [3, 4], where mammalian carboxypeptidases are grouped. Two subfamilies can be defined based upon sequence homology and overall structure: that of carboxypeptidase A, the structural prototype for most metallocarboxypeptidases, and that of carboxypeptidase H (also named carboxypeptidase E). The degree of amino acid identity is greater than 40% within each subfamily and about 15–20% between subfamilies. Although this classification is used in most references [5–7], the metallocarboxypeptidases in family M14 could also be subdivided upon the basis of their differential involvement in physiological processes. Thus, pancreatic carboxypeptidases A (also called A1), A2 and B typically function only as digestive

Table 1. Classification of carboxypeptidases[a]

Catalytic type	Family		Enzyme	E.C. number	Catalytic feature
Metallocarboxypeptidases					
	M14	Subfamily CPA	Carboxypeptidase A[b]	3.4.17.1	
			Carboxypeptidase A2[b]	3.4.17.15	
			Carboxypeptidase B[b]	3.4.17.2	
			Mast-cell carboxypeptidase A	3.4.17.1	HXXE...H (zbm)[c]
			Carboxypeptidase U	3.4.17.20	
			Carboxypeptidase T[b]	3.4.17.18	
			Carboxypeptidase SG	3.4.17.-	
		Subfamily CPH	Carboxypeptidase H(E)	3.4.17.10	
			Carboxypeptidase M	3.4.17.12	
			Carboxypeptidase N	3.4.17.3	
			Carboxypeptidase D	3.4.17.22	
			Carboxypeptidase Z	3.4.17.-	
	M15		Zinc D-Ala-D-Ala carboxypeptidase[b]	3.4.17.14	H...HXH (zbm)
			Muramoyl-pentapeptide carboxypeptidase	3.4.17.8	
	M32		Taq carboxypeptidase	3.4.17.19	HEXXH (zbm)
	M20		Carboxypeptidase G2 (glutamate CP)[b]	3.4.17.11	An Asp (metal bridging), two Glu and two His bind two zinc ions
			Carboxypeptidase S (Gly-X CP)	3.4.17.4	
	Other families		Alanine carboxypeptidase	3.4.17.6	
			Membrane Pro-X-carboxypeptidase	3.4.17.16	
			Muramoyl-tetrapeptide carboxypeptidase	3.4.17.13	
			Tubulinyl Tyr carboxypeptidase	3.4.17.17	
Serine-carboxypeptidases					
	S10		Carboxypeptidase C[b]	3.4.16.5	
			Carboxypeptidase C[b]	3.4.16.6	Catalytic triad: Ser/His/Asp
			Lysosomal carboxypeptidase	3.4.16.5	
	S28		Lysosomal Pro-X carboxypeptidase	3.4.16.2	

[a]adapted from refs. [1, 3 and 14]
[b]a crystallographic structure is available
[c]zinc-binding motif

enzymes, whereas the rest of mammalian carboxypeptidases, including carboxypeptidase A from mast cells and carboxypeptidase U (or plasma carboxypeptidase B), exert their action in different physiological processes, mainly in non-digestive tissues and fluids, and have been called 'regulatory carboxypeptidases' because of their involvement in more selective processing reactions [5].

Carboxypeptidases of the metallo-type are not homogeneous regarding their zinc-binding motifs and can be classified into distinct groups based upon their zinc binding site [3, 8–10]. The carboxypeptidase A and H subfamilies, which include all of the so-called digestive and regulatory carboxypeptidases, belong to the group having a **HXXE**...**H** binding motif where the three zinc ligands are a histidine, a glutamate located two residues C-terminally and a second histidine located 108–135 amino acids C-terminally from the first histidine. However, the consensus sequences around the zinc ligands permit a further subdivision: all members of the carboxypeptidase H subfamily have consensus sequences XX**H**GX**E**XX...L**H**GGXB (where B stands for bulky amino acids and X stands for any amino acid); members of the carboxypeptidase A subfamily of mammalian origin belong to a group with consensus sequences GB**H**XR**E**WB...B**H**SYSQ; and carboxypeptidases T and SG display the following consensus sequences: XX**H**AR**E**XBT...X**H**TYSE. In contrast to the motif described above and, in fact, to all zinc metalloproteases, members of the zinc D-Ala-D-Ala carboxypeptidase family (M15) display a binding motif of the type **H**...**H**X**H** [11, 12], where the third ligand is located N-terminally from the motif that contains the other two. Finally, Taq carboxypeptidase is an enzyme with no clear homology to any other metallo-peptidase and contains a **HE**XX**H** sequence (the most common zinc-binding motif in zinc metallopeptidases) at positions 276 to 280 of the enzyme [13].

There are at least six families of zinc peptidases with metal ligands which do not belong to the **HEXXH** (or 'zincin' [14]) motif [3, 5, 10, 15]. The 3-D structure of carboxypeptidase G2 (glutamate carboxypeptidase) has been recently solved [16] and shown to be different from the rest of the carboxypeptidases of known structure. The dimeric carboxypeptidase G2 contains a dizinc centre in each subunit similar to that of *Aeromonas* aminopeptidase [17]. Carboxypeptidase S (or Gly-X carboxypeptidase) from *Saccharomyces* has been classified in family M20 of carboxypeptidase G2. Other metallocarboxypeptidases in the table must await the availability of structural data before being classified as members of a known family.

Serine carboxypeptidases are the second major group of carboxypeptidases. They are classified in clan SC [18] and have in common a 'catalytic triad' of Ser, Asp and His with the serine peptidases of clans SA (chymotrypsin) and SB (subtilisin), despite differences in protein fold and in the sequential arrangement of the three amino acids in the triad. Serine carboxypeptidases are normally active at an acidic pH and are widely distributed, although most of the studies have been carried out with proteins from yeast [19, 20]. Protein data banks also contain entries of cysteine carboxypeptidases, but they will not be treated here.

Carboxypeptidase A is a prototype zinc peptidase, and it was the third protein to have its three dimensional structure solved at high resolution [15]. A number of 3-D structures of carboxypeptidases, and also of some of their zymogens, have been solved since then as denoted in Table 1. These studies have led to the observation of similar folding topologies between, for instance, carboxypeptidase A and leucine aminopeptidase [21] or between serine carboxypeptidases and a large family of otherwise unrelated hydrolases [22]. These indications of distant evolutionary relationships between distinct exo- and endopeptidases may open new perspectives in the study of the evolution of the structure and function of these enzymes.

Digestive metallocarboxypeptidases

Background

Pancreatic carboxypeptidase A (A1) was among the first proteases to be discovered [23] and characterized as a zinc enzyme [24]. The characterization and purification of a great number of carboxypeptidases from different sources of mammalian origin [25–28] and non-mammalian origin [29–31], has made these enzymes the best known examples of metallocarboxypeptidases. Although most studies have been carried out with enzymes isolated from animals with a well-developed digestive system, forms which belong to the carboxypeptidase A subfamily have been found in *Streptomyces* [32] and *Thermoactinomyces* [33].

All metallocarboxypeptidases from family M14 (including the digestive and the regulatory enzymes) may in general also be classified upon the basis of their substrate specificities. Carboxypeptidase A-like enzymes have a preference for hydrophobic C-terminal residues, and B-like enzymes exert their action on C-terminal Lys or Arg. However, while most regulatory carboxypeptidases are of the B type (and are thus also called 'basic carboxypeptidases' [34]), the pancreas of different species not only contains carboxypeptidases of types A and B, but also a carboxypeptidase A-like enzyme named carboxypeptidase A2 [28, 35, 36], which shows a preference for C-terminal bulky aromatic residues such as tryptophan [35]. Hence, the notation A1 is now used to describe the formerly-called carboxypeptidase A. The only exception in specificity among regulatory carboxypeptidases is that of mast cell carboxypeptidase A: although it shares a higher degree of homology with carboxypeptidase B than with carboxypeptidase A1 (58% versus 40%), its specificity is towards hydrophobic amino acids [37].

Pancreatic carboxypeptidases are found in the soluble and non-glycosylated form and are synthesized as inactive precursors containing a 94–96 residues activation segment. They are stable as zymogens during storage in the pancreatic granules until secretion occurs. Trypsin-promoted limited proteolysis generates active enzymes with 305–309 residues [6] which contribute to the degradation of dietary proteins in the intestine through their complementary specificities. Allelomorphic forms have been described for carboxypeptidases A and B [28, 38]. A number of methods have been reported for the isolation of digestive carboxypeptidases from both natural sources [28, 39, 40] and recombinant systems [41–43].

Although the normal localization of digestive carboxypeptidases is the pancreas or the intestinal tract, these enzymes can be detected in the serum of patients suffering from acute pancreatitis [44]. The so-called 'pancreas-specific protein' defined as a serum marker for acute pancreatitis and graft rejection [45] was later identified as pancreatic procarboxypeptidase B [46] and also considered a co-marker for chronic renal failure [47]. On the other hand, procarboxypeptidase A has been identified as the ovine pancreatic protein which binds to insulin-like growth factor binding protein-3 [48]. Until recently, the presence of pancreatic carboxypeptidases had not been reported in any other tissues from animals. Normant et al. [49a] reported the presence of carboxypeptidase A1 and A2 gene transcripts in extremely low abundance in rat brain and other tissues. Furthermore, these authors also describe that alternative splicing of the carboxypeptidase A2 gene produces truncated forms of the enzyme with distinct subcellular localizations.

Structure and catalytic mechanism

As commented above, the enzymes of the carboxypeptidase A family, also including those of bacterial origin, are synthesized as proenzymes. The degree of sequence homology between the pancreatic enzymes (A1, A2 and B forms) varies between 42%–64% [6]. Bacterial enzymes share a comparable degree of homology between them but sequence similarity with the digestive enzymes is very low. However, the overall fold of carboxypeptidase T is very similar to carboxypeptidase A, including the active-site region [50]. The overall sequence homology is substantially lower in the pro-segment regions. When comparing, for instance, the pro-segment and the enzyme regions of human procarboxypeptidases A2 and B, the sequence similarity is 42% and 23%, respectively [36].

The only available 3-D structures of carboxypeptidases are those of the pancreatic forms bovine A1 [51], rat A2 [52] and bovine B [53]. Structures of procarboxypeptidases A1 and B from porcine pancreas and A1 from bovine pancreas have also been determined [54–56]. The three enzymes share similar conformational folds and conserve a similar architecture at the active site, with identical catalytic residues and the expected variations in residues at the main specificity-subsites for substrate side-chain binding. The core of the tertiary fold is an eight-strand β-sheet over which eight α-helices pack on both sides to form a globular molecule. The active centre is located between the β-sheet and two helices, and the active centre Zn^{2+} is held by residues located in turns or loops protruding from the secondary structure elements. In the proenzyme structures, the globular region of the pro-segment shields the active site and establishes specific interactions with residues important for substrate recognition [6]. The conformations of the pro-segment moieties of porcine procarboxypeptidases A1 and B are very similar despite the low degree of sequential identity (32%). Other aspects of zymogen structure and activation will be discussed later on.

The refined crystal structure of pancreatic carboxypeptidase A1 is used as the reference for the description of the catalytic mechanism of carboxypeptidases. As for many zinc proteases, the heteroatom is bound by three amino acid residues and a water molecule [57]. Up to now, there is no 3-D information available on non-digestive carboxypeptidases, but the observations on carboxypeptidase A1 can be extrapolated to other metallocarboxypeptidases of family M14 since, despite an overall low degree of sequence similarity, the regions containing the residues which bind zinc or participate in the catalytic mechanism are highly conserved.

Crystallographic and kinetic studies of complexes of bovine carboxypeptidase A1 with naturally occurring inhibitors [58], substrate analogues or 'reaction-coordinate analogues' [59, 60], and kinetic studies with enzymes point-mutated at residues hypothesized to be relevant for substrate binding and catalysis [41, 61–63] have made of this enzyme one of the most thoroughly studied, even used as an example of catalytic mechanism in most textbooks. However, the mechanism has been steeped in controversy for almost 50 years [15]. The acyl pathway hypothesis proposes the existence of a covalent acyl enzyme intermediate whereas the promoted-water pathway considers a simultaneous polarization of the substrate carbonyl group by the zinc ion and a zinc-promoted activation of a water molecule which directly attacks the scissile peptide bond [59]. Arguments are currently in favour of the promoted-water pathway [64], where zinc is penta-coordinated and Glu 270 at the active site is in a position to hydrogen-bond to the intermediate, although this model is still being questioned by a reverse protonation mechanism where the water molecule is activated by the C-terminal carboxylate group of the substrate [65].

The zinc binding residues are His69, Glu72 and His196 (numbering of carboxypeptidase A1). From crystallographic studies a number of amino acids have been defined as important for substrate binding and catalysis and can be classified into several subsites: Asn144, Arg145, Tyr248 in S1'; Arg127 and Glu270 in S1; Arg71, Ser197, Tyr198 and Ser199 in S2; Phe279 in S3 and Glu122, Arg124 and Lys 128 in S4. The terminal carboxylate group of the peptide substrate is fixed by Asn144, Arg145 and Tyr248, while the carbonyl group of the scissile peptide bond becomes positioned near Glu270, Arg127 and zinc. All of those residues are conserved among digestive carboxypeptidases, and differences are restricted to amino acids defining specificity. In carboxypeptidase B, an Asp residue is at position 255 instead of Ile in A1 and A2. This is located at the specificity pocket and allows for the binding of basic residues in the B forms. A further replacement of Ile243 for Gly in the B forms contributes to the creation of a polar environment. Leu203 and Thr268 in A1 are replaced by Met and Ala, respectively, in A2 [52], rendering a binding pocket more capable of accomodating a bulky substrate side-chain.

Substrates and inhibitors

The natural substrates for digestive carboxypeptidases are the peptides and proteins from the diet. The analysis of their activity *in vitro* is performed on synthetic substrates designed to satisfy the specificity of A-like enzymes for hydrophobic amino acids and that of B-like enzymes for basic ones. Most carboxypeptidases also cleave ester substrates, and esterolysis seems to proceed via the same mechanism of proteolysis [59, 66]. Typical substrates for A-like enzymes are benzoyl-glycyl-l-phenylalanine [67], furylacryloyl-l-phenylalanyl-l-phenylalanine [68] and carbobenzoxy-(glycyl)$_{n-l}$-phenylalanine [69]. Similar substrates with Arg or Lys as C-terminal are used for measuring carboxypeptidase B activity. The A forms differ in that A1 preferentially cleaves synthetic substrates with C-terminal Phe, Leu or Tyr, whereas A2 prefers Trp, Phe or Tyr [35]. Azaroformyl peptide surrogates [70] and ^{125}I-labelled substrates [71] have been recently described as being more sensitive and specific.

Metallocarboxypeptidases are all inhibited by chelating agents such as 1,10-phenanthroline. Benzyl succinic acid inhibits A-type enzymes, and compounds like guanidinoethylmercaptosuccinic acid and 2-mercaptomethyl-3-guanidinoethylthiopropanoic acid [72] inhibit B-type enzymes, with K_i in the order of micromolar to nanomolar. Other synthetic carboxypeptidase A inhibitors have been used in order to investigate its catalytic mechanism [6, 59]. Few natural inhibitors of carboxypeptidases have been described. Two exogenous protein inhibitors have been characterized: the 39-residue potato carboxy peptidase inhibitor [73, 74] and a 65-residue worm intestinal inhibitor [75]. Only one mammalian, 223-residue long, endogenous carboxypeptidase inhibitor has been reported thus far [49b]. All of them have K_i at the nanomolar range for carboxypeptidase A1.

Zymogens and activation

The genetic codification and expression of metallocarboxypeptidases as zymogens (proenzymes) has been shown for some of them (but not for all) in particular those belonging to the CPA subfamily, which includes the digestive carboxypeptidases and some enzymes present in tisues other than the pancreas. Most of the members of this subfamily have a 15–22 residues signal peptide and a 92–96 residues pro-segment at the N-terminus of the mature enzyme. An exception to this is carboxypeptidase T (from *Thermoactinomycetes v.*) in which the signal peptide plus the prosegment region is shorter (98 residues) [76]. Metallo-procarboxypeptidases are inactive or show very low peptidolytic and esterolytic activity, and are activated by proteolytic release of the pro-segment.

The sequence homology between the pro-segments of digestive metallo-carboxypeptidases varies from 54% identity in residues when pancreatic A1 and A2 forms are compared to 27% and 22% when these are compared with the pancreatic B form [36]. A similar value is found for the mast cell proenzyme, and a further decrease to below 18% is observed when plasma B or the T form are included in the comparison [76], despite belonging to the same subfamily. Three-dimensional crystal structures are only available for the A1 and B porcine zymogens [54, 55] and a modelled structure has been obtained for the A2 human zymogen [36]. All these structures show a similar topology for their pro-segments: an open sandwich antiparallel α/antiparallel β core (a globular domain formed by three distorted β chains over which two α helices are packed) followed by an extended α helix at the C-terminus which connects to the enzyme moiety.

The globular core of the pro-segment covers the active site of metallocarboxypeptidases, shielding the S2, S3 and S4 subsites [6]. The interaction between both moieties only affects certain residues of S1 (Arg127) and S1' (Tyr248) subsites through a water molecule bridge. An exception to this is Arg145 (belonging to S1') which is directly bound by Asp41 from the pro-segment in the B proenzyme, a link probably responsible for the absolute lack of intrinsic proteolytic activity of this form against peptide substrates in contrast with the measurable (but low) activity of the A1 and A2 forms. A significant difference between the B and A forms is the presence in the former of a short 3_{10} helix in the globular core of the pro-segment, a structure which promotes a more efficient capping of the active site.

Pancreatic metallo-procarboxypeptidases are activated by limited proteolysis in the duodenum, where they arrive after being secreted from the granules of the acinar pancreatic cells. The release of the pro-segments (and thus, activation) is mainly promoted by trypsin in the case of the procaboxypeptidase B and by trypsin complemented with other pancreatic serine proteinases, such as elastase and chymotrypsin, in the case of procarboxypeptidase A [40]. In general, the activation course of these proenzymes is dependent on the capability of the scissed large fragments of the pro-segment to bind and inhibit the active enzyme [6]. This capability seems to be strong in the the A1 forms and null or very weak in the A2 and B forms [40, 77]. The C-terminal proteolytic trimming of the pro-segment reduces or eliminates its inhibitory power. The number of potential proteolytic targets at the connecting region, their accesibility, the extent of helical structure in it and the conformation of the loop placed at its end are important determinants of the proenzyme activation [36, 77]. Some information has also been obtained about the proteolytic activation of metallo-procarboxypeptidases from the digestive system of dogfish [29] and lungfish [30].

Unfolding-refolding studies on the pro-segment of pancreatic metallo-procarboxypeptidases indicate that the globular part of this region acts as an independently folding unit *in vitro*, with a two-state behaviour [78]. The presence of this piece greatly facilitates the heterologous expression of the proenzymes [42, 79]. The pro-segment probably behaves as a co-chaperone in the folding of pancreatic carboxypeptidases *in vivo*.

Knowledge of the structure and functional role of the pro-segment and upon the proteolytic activation of non-pancreatic zymogens is, in contrast, moderate. Modelling studies suggest that the three-dimensional structure of mouse mast cell procarboxypeptidase A is similar to the homologous A1 form from porcine pancreas [80]. In this case, the proteolytic activator is unknown, although it has been proposed that it could be a thiol protease, deduced from *in vivo* activation studies. On the other hand, the three-dimensional structure of procarboxypeptidase U (plasma B) is fully unknown, a fact complicated by its occurrence as a glycosylated form in multiple sites [81] and by its interaction with other blood proteins, such as plasminogen or α2-macroglobulin [81]. Its *in vitro* activation by trypsin, plasmin or thrombin is incomplete and gives rise to an internal cleavage at the enzyme moiety besides the release of the pro-segment [82]. Nothing is known about the structure and activation of procarboxypeptidase T.

Quaternary structure

Pancreatic metallo-procarboxypeptidases A (A1 and A2) occur in different species either as monomers and/or as binary or ternary complexes with zymogens of serine proteinases, such as chymotrypsinogen C (CTGC), chymotrypsinogen B (CTGB) or proproteinase E (PPE) [6, 40]. Thus an A1/PPE binary complex has been reported in human and pig, an A2/CTGC binary complex in rat, an A1/CTGB binary complex in whale, an A1/CTGC binary complex in cattle, and an A1/CTGC/PPE ternary complex in ruminants like cattle, goat and camel. Oligomeric complexes have also been described or suggested for other related non-pancreatic procarboxypeptidases, such as procarboxypeptidase U (plasma B) which binds to plasminogen [81], and for mast cell procarboxypeptidase A which could interact with mast cell chymase mMCP5 [83]

The biological role of those digestive complexes is unknown although, in the case of the digestive zymogens, it has been proposed that they could protect the subunits against inactivation in the acidic conditions of the upper duodenal tract, modulate or coordinate their transformation from zymogens to active forms, potentiate the complementary action of the different components once activated, etc. [6]. The most clear evidence lies in favour of the modulation of the proteolytic activation of the subunits by the complex. Generally, procarboxypeptidases are activated at a slower rate within the complexes than in the isolated forms [84]. In contrast, the activation of the accompaying zymogens is not substantially affected by their presence in the oligomeric structures [84]. X-ray crystallographic studies on the bovine ternary complex indicate that this behaviour is probably related to the quaternary structure in which the pro-segment of the A1 form adopts a central position, linking the carboxypeptidase A1 moiety with the CTGC and PPE subunits [56] (Fig. 1). The burial or limited exposure of the proteolytic sites in the pro-segment of A1 and the full exposure of the corresponding pro-segments of the other two zymogens is thus the determinant for the differential rate of activation of the three subunits [84].

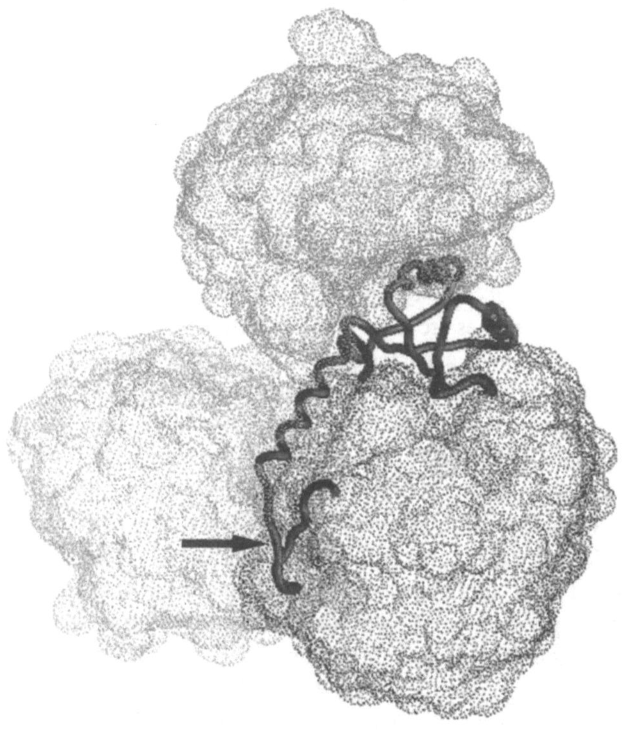

Figure 1. Dot-surface representation of the ternary complex of procarboxypeptidase A1 (front), proproteinase E (top) and chymotrypsinogen C (left) from bovine pancreas [84]. Highlighted with a ribbon representation is the pro-segment of the procarboxypeptidase A1 subunit and a short N-terminal region of the enzyme moiety. The arrow indicates the primary point of proteolytic activation. The pro-segment, covalently linked to carboxypeptidase A1 in the zymogen state, also acts as a non-covalent link with the other two subunits and occupies a central position in the complex.

Regulatory metallocarboxypeptidases

Consideration of the carboxypeptidases that are to be discussed in the following paragraphs under the name 'regulatory carboxypeptidases' does not strictly obey sequence similarities but rather the coincidence in their involvement in physiological functions other than the digestion of alimentary proteins. Following these criteria, mast-cell carboxypeptidase A and plasma carboxypeptidase U (plasma carboxypeptidase B), two enzymes which belong to the carboxypeptidase A subfamily based upon structural homology, will be commented upon here.

Knowledge about the structure and function of regulatory carboxypeptidases has increased dramatically over the last two decades as a consequence of the research upon biologically active peptides and the recognition of the involvement of carboxypeptidases in their processing [34]. The first regulatory carboxypeptidase to be described was carboxypeptidase N, a plasma enzyme found to inactivate bradykinin [85]. Carboxypeptidases M, H (also called E), U and mast cell carboxypeptidase A were described in the 1980 s. Carboxypeptidase M is a membrane-bound enzyme found in various human and animal tissues [86] although it can also be found in soluble form in body fluids [87, 88]. Evidence for a basic carboxypeptidase associated with the production of peptide hormones and neurotransmitters came from studies upon the enzymatic activity of the components of secretory granules [89]. The enzyme was first named 'E' due to its endocrine action and because it was first described as an enkephalin convertase involved in the biosynthesis of enkephalin in bovine adrenal cromaffin granules [90]. The Enzyme Commission renamed it carboxypeptidase H because it is associated with the procesing of many prohormones. Both names are used in the literature.

Carboxypeptidase A from mast cells and plasma carboxypeptidase U (also known as plasma carboxypeptidase B) are the two regulatory carboxypeptidases that belong to the structural subfamily of carboxypeptidase A or pancreatic-like carboxypeptidases. For instance, human mast cell carboxypeptidase A shares a 58% homology with bovine carboxypeptidase B and only about a 15% homology with human carboxypeptidase M. Mast cell carboxypeptidase is found in this type of cell and thus probably in many tissues [91]. It was first isolated from rat peritoneal mast cells [37]. Different approaches including the identification of a carboxypeptidase in human serum distinct from carboxypeptidase N [92], the isolation of a plasminogen-binding protein [81, 93] and the purification of a thrombin-activatable fibrinolysis inhibitor [94] have led to the characterization of the same circulating carboxypeptidase, named U (unstable) because it tends to be very unstable after activation [93].

Carboxypeptidases D and Z are the newest members of this group of enzymes. Carboxypeptidase D was purified from bovine pituitary [95] and shown to be membrane-bound and widely distributed among tissues [96]. The protein is homologous to gp180, a protein from duck liver which binds the duck hepatitis B virus. Like carboxypeptidase H, its pH optimum is 5.5–6.0; in contrast, carboxypeptidase D is a three-domain protein with a molecular mass of approximately 180,000. Carboxypeptidase Z has recently been cloned and expressed in the baculovirus system [7]. The overall arrangement of the catalytic domain and the N-, and C-terminal domains is more similar to carboxypeptidases H, M or N than to carboxypeptidase D. Another protein which shows homology to metallocarboxypeptidases is AEBP1, a transcription repressor recently characterized [97] which lacks essential catalytic residues, but is otherwise able to express carboxypeptidase activity.

Structural properties

While pancreatic carboxypeptidases are synthesized as single-chain non-glycosylated proenzymes of similar molecular mass, a much higher variability is observed among regulatory carboxypeptidases. All of them are of higher molecular mass than the digestive forms, either because they are glycosylated, have extensions at the C-, and/or N- terminus, or both. Most regulatory carboxypeptidases appear in a single polypeptide chain with one exception: carboxypeptidase N is a dimer of heterodimers, with each heterodimer containing one catalytic subunit (50–55 kDa) and one non-catalytic subunit (83 kDa) [98]. Carboxypeptidase N is also an exception in that it is the only case where the catalytic chain is not glycosylated, rendering it highly unstable in the isolated state and thus needing the presence of the glycosylated accompanying subunit to stabilize it and circulate in the blood.

The two plasma carboxypeptidases, N and U, are naturally found in a soluble form. The other regulatory carboxypeptidases have been described to be attached to a macromolecular assembly, although all of them can also be isolated in the soluble form. Carboxypeptidase M is bound to the membrane, probably via a glycosylphosphatidylinositol anchor in its mildly hydrophobic C-terminal region as deduced from its release from the membrane by treatment with phosphatidylinositol-specific phospholipase [99]. The C-terminal amphiphylic α-helical domain of carboxypeptidase H mediates its membrane binding and also sorting of the protein [100]. Carboxypeptidase H binding to the membrane takes place at the acidic pH of the secretory granule interior but not at the neutral extracellular pH reached upon fusion of the secretory granules with the plasma membrane [101]. Mast cell carboxypeptidase A is tightly bound to acidic proteoglycans in the mast cell granules [91], probably through a cluster of basic residues that would form a linking domain [102].

Mast cell carboxypeptidase A and carboxypeptidase U, the two enzymes structurally related to pancreatic carboxypeptidases, are synthesized in the form of zymogens, with activation segments of 94 and 92 residues, respectively [82, 103]. None of the other regulatory carboxypeptidases possess a propeptide that renders the enzyme inactive, although carboxypeptidase H has a short, 14-residue propeptide immediately after the N-terminal signal sequence which is 100% identical between human, rat and mouse [104, 105]. Proteolytic maturation of carboxypeptidase H takes place at a site containing five adjacent arginines [106] and probably occurs in secretory vesicles [105, 107]. Absence of the propeptide does not impair sorting of carboxypeptidase H into the regulated secretory pathway nor its expression, in contrast to pancreatic carboxypeptidases [42]

The enzymes in the carboxypeptidase H subfamily contain a 380–400 residues carboxypeptidase domain which is preceded and followed by N- and C-terminal sequences [7]. The mature enzymes are thus longer than the pancreatic carboxypeptidase-like enzymes. The higher sequence variability among the carboxypeptidase H subfamily is found at the C-terminal sequences, which have a varying length and mechanism of anchoring to supramacromolecular structures [5], and also at the N-terminus. All metallocarboxypeptidases are synthesized with a signal peptide, but only the pancreatic-like enzymes possess a propeptide capable of maintaining the enzymes in the zymogen inactive state until its removal [6]. In this context, the function of the carboxypeptidase H propeptide is not clear. Carboxypeptidase Z is unique in that it contains a 'fz' domain N-terminal to the catalytic domain [7]. This type of domain was defined in frizzled proteins [108] and may affect ligand binding. Carboxypeptidase D, the mammalian homolog of duck gp180 protein [96, 109] is a protein with three homologous carboxypeptidase-like domains. Unlike domains 1 and 2, which contain the putative active-site residues of carboxypeptidases, domain 3 is probably non-functional because of mutations at these points. Figure 2 attempts to summarize the domain structure for the metallocarboxypeptidases which share a catalytic mechanism and belong to family M14.

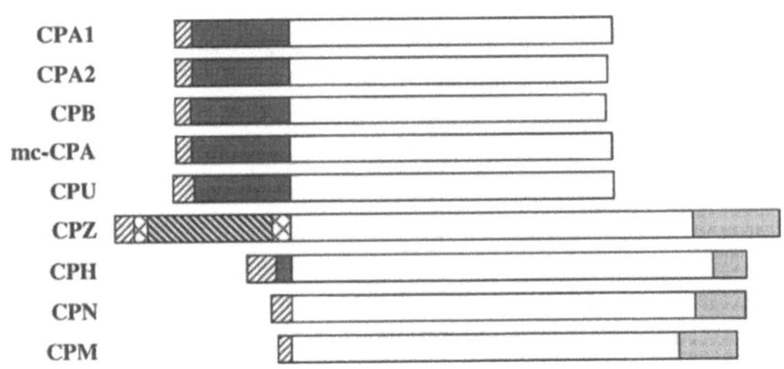

Figure 2. Comparative sizes of digestive and regulatory carboxypeptidases. Shadowed boxes indicate the location of domains in the proteins. Carboxypeptidase D is not included. It contains three carboxypeptidase-like domains followed by a transmembrane domain and a cytosolic tail. Residues important for catalysis are conserved in all enzyme domains shown here. Adapted from Song and Fricker [7]. ▨, Signal peptide; ■, pro-peptide; ▧, 'fz' domain in CPZ; □, enzyme domain; ▦, C-terminal domain; ☒, other.

Regardless of the homology classification of the M14 family of carboxypeptidases into two groups, they all share the same zinc-binding motif (see above). They also show residue conservation at most of the catalytically important residues. Thus, His69, Glu72 and His196 (carboxypeptidase A numbering system) are conserved, as are also Asn144, Arg145 and Tyr248, involved in substrate binding, and Glu270, essential for catalysis. A difference is observed in residue Arg127 in carboxypeptidase A, which is a Lys in forms H, M and N and even more variable in the rest of the H-like enzymes. It has been pointed out [5] that the function of this catalytically important arginine could be performed by Arg136 (carboxypeptidase A numbering), a residue conserved in all of those enzymes. There are also some differences in other residues involved in substrate binding: Arg71 is an Asn in most H-like forms, Ser197 and Tyr198 are always replaced by Gly, and residues Leu203, Ile 243, Ile255 or Thr268, related to specificity, are also variable, although this is to be expected in the latter case given that variability is already observed among pancreatic-like enzymes.

Functional roles

Most regulatory carboxypeptidases (the exception of mast cell carboxypeptidase A has already been mentioned) cleave C-terminal basic residues, with different affinities for arginine or lysine depending on the enzyme [5]. They exert their function on peptides related to a considerable number of relevant physiological processes, both normal and pathological. In the following paragraphs, short overviews of some particularities of regulatory carboxypeptidases will be presented.

Carboxypeptidase N is synthesized in the liver and released in the blood where it is present at high concentrations (30 μg ml^{-1}). It cleaves an ample variety of peptide and protein substrates and is the major blood-borne inactivator of potent peptides such as kinins and anaphylotoxins. It is believed that carboxypeptidase N has a general protective function in blood [5], as shown by a number of clinical studies. To be fully active, carboxypeptidase N needs the presence of the catalytic (50–55 kDa) subunit and the accompanying glycosylated 83 kDa subunit. The latter protects the former from degradation and also from removal from the blood by glomerular filtration [85].

Carboxypeptidase M was first purified and characterized from human placental microvilli [86]. It is a widely distributed membrane-bound ectoenzyme which is found at high levels in placenta and lungs [110] and also in a variety of tissues and cells including blood vessels, brain and peripheral nerves [5]. It was considered to be a marker of differentiation from peripheral blood monocytes to macrophages *in vitro* [111]. The enzyme has an optimal activity at neutral pH and releases Arg faster than Lys. Carboxypeptidase M can be solubilized after release of its hydrophobic anchor to the membrane by treatment with phosphatidylinositol-specific phospholipase C and trypsin [99, 112]. The soluble form of the enzyme has been detected in amniotic fluid, seminal plasma and urine [5].

The localization of carboxypeptidase M in the plasma membrane indicates that it could participate in the control of peptide hormone activity at the cell surface, in the modulation of receptor specificity through the generation of C-terminal-shortened new agonists or antagonists and in extracellular protein processing or degradation. Among other processes, carboxypeptidase M has been implicated in the metabolism of growth factors from its capacity of generating des-Arg53-EGF [113] and the coincidence of distribution of carboxypeptidase M and the epidermal growth factor. The enzyme could also be involved in protective functions at the alveolar surface [110], and in inflammatory and other pathological processes. Generation of des-Arg9-bradykinin, a peptide which stimulates a whole variety of cellular responses through a change in receptor specificity, could also be another function of carboxypeptidase M.

Evidence for the occurrence of carboxypeptidase H initially came from studies upon the processing of polypeptide hormones [114]. It was localized to secretory granules and first defined as proenkephalin convertase [90]. Carboxypeptidase H is present at high levels in brain, pituitary, adrenal medulla and pancreatic islets [89]. The pH optimum of this enzyme is the same as the normal intragranular pH of 5–6 and can be activated by Co^{2+}. A membrane-bound and a soluble form have been found in all the tissues investigated; both are single-chain glycosylated proteins that have the same activity *in vitro* and are highly specific for basic C-terminal amino acid residues, with a preference for Arg. The soluble form of the enzyme has an apparent molecular weight of about 50 kDa, 2–3 kDa lower than the membrane-bound form [115].

Carboxypeptidase H is involved in the biosynthesis of numerous peptide hormones and neurotransmitters such as Met- and Leu-enkephalin, insulin, vasopressin, oxytocin and many others [116], and has also been related with other physiological processes. Recently it was shown that obese diabetic mice with the *fat/fat* mutation bare a defect in proinsulin processing associated with the virtual absence of carboxypeptidase H activity in pancreatic islets and pituitaries, due to a single Ser→Pro mutation [117]. Mutant mice also showed impaired processing of other peptide hormones [118]. Carboxypeptidase H has also been recently identified as a sorting receptor for secretory proteins in pituitary Golgi-enriched and secretory granule membranes [119].

Carboxypeptidases D and Z have been recently incorporated into the increasing list of regulatory carboxypeptidases. Although their precise functional roles are still to be defined, it is believed that the D form may have a broad role in the processing of proteins that transit the regulatory pathway [96] and might be a substitute for carboxypeptidase H in mice with the *fat/fat* mutation. Carboxypeptidase Z is thought to perform an extracellular function [7].

Mast cell carboxypeptidase A is found in the secretory granules of mast cells, forming macromolecular complexes with proteoglycans. The precursor form of the enzyme is very similar to that of the pancreatic zymogens (see Fig. 2), although its activation takes place at a pair of acidic residues [103] and most of the carboxypeptidase stored in the granules appears to be already activated [120]. It has been shown that granule-bound carboxypeptidase A from mast cells is fully active [102] and cleaves peptide substrates in tandem with other proteases. The precise role of mast cell carboxypeptidase is still to be defined, although it may obviously be related to pathological conditions in which mast cells have been implicated, such as allergic response, inflammation and others.

Carboxypeptidase U, a carboxypeptidase B from plasma, is found in blood in the zymogen form [81]. It was initially characterized as a plasminogen-binding protein which becomes unstable upon removal of the glycosylated pro-segment [81]. Indeed, the pro-segment mediates the high-affinity binding of procarboxypeptidase U to plasminogen [82]. Since it has an affinity for basic C-terminal residues, carboxypeptidase U may release Lys at lysine-binding sites for plasminogen on the surface of cells and fibrin and modulate plasminogen activation and the rate of fibrinolysis; it has recently been defined as a thrombin-activatable fibrinolysis inhibitor [94].

Serine carboxypeptidases

Serine carboxypeptidases belong to families S10 and S28 in clan SC of serine peptidases [1]. These enzymes are maximally active at acidic pH and were thus termed acid carboxypeptidases, which may lead to confusion since they belong to the serine peptidase catalytic-type category as demonstrated by their capacity for being inhibited by typical serine proteinase inhibitors [19] and by the determination of the 3-D structures of some representative enzymes [121–125]. They possess a catalytic triad similar to that found in serine endopeptidases of clans SA and

SB although the sequential arrangement of the catalytic residues is particular for that group of enzymes [1]. However, the overall fold of serine carboxypeptidases is completely different from enzymes like chymotrypsin or subtilisin [125] and is common to several hydrolytic enzymes of diverse phylogenetic origin and catalytic function [22]. Serine carboxypeptidases would thus be another example of convergent evolution to a common enzymatic activity [121].

Serine carboxypeptidases are classified into two groups depending upon substrate preferences: carboxypeptidase C-type enzymes prefer hydrophobic amino acids in P1', whereas carboxypeptidase D-type enzymes prefer basic amino acids at this subsite although they can also accept bulky C-terminal hydrophobic residues. The specificity classification can be refined by also considering the preferences for amino acids at the P1 position [20], which are not necessarily linked to those in P1'. Structural information is available for its optimal activity at acidic pH [126, 127].

The wide distribution of serine carboxypeptidases among fungi, higher plants and animals appears to be restricted to eukaryotes. In animals they are found in lysosomes [128] and are vacuolar enzymes in plants and fungi [19]. Serine carboxypeptidases have been implicated in various physiological processes such as general turnover of proteins, maturation of the α-mating pheromone [129] or maturation of angiotensins [130]. They are synthesized as preproenzymes, and the most extensively studied precursor is that of carboxypeptidase Y from yeast (a carboxypeptidase D-type enzyme); the signal peptide is, in this case, 20 amino acids long and the pro region contains 91 residues [131]. This enzyme from *S. cerevisiae* is the one for which the biosynthetic pathway has been studied in most detail. The mature enzymes are glycosylated in different degrees and their molecular masses may thus differ. The enzymes contain 411 to 451 residues and can be proteolytically processed to a two-chain enzyme depending upon their origin [19].

The applications of serine carboxypeptidases are a consequence of their ability to cleave peptide substrates at their C-terminus and have been used for the C-terminal sequencing of proteins. The efficiency of the analysis can be increased by the use of serine carboxypeptidases with different specificities and by its coupling to rapid detection methods [132]. They have also been used to catalyze the synthesis of peptides [19, 20].

Carboxypeptidase Y from yeast and carboxypeptidase II from wheat

These two enzymes are representative of the two general classes of serine carboxypeptidases defined by their substrate specificity. The monomeric yeast enzyme (abbreviated CPD-Y) belongs to class C, and the homodimeric plant enzyme (abbreviated CPD-WII) to class D. Their structures have been determined at high resolution [121, 122] and, although their amino acid sequence identity is only 26%, the topologies of the two folds are identical. The core structure corresponds to the α/β hydrolase fold [22] and in both cases there is a large α-helical insertion domain into the overall fold (residues 180–311 in the wheat enzyme numbering) which surrounds the active-site cavity and may be important in the recognition of substrates. Both enzymes have very different amino acid sequences in the insertion domains, and the three-dimensional structures of these segments are also different. Given its implication in substrate recognition, it was suggested that the insertion domains would evolve much more rapidly than the rest of the molecule to give rise to new functions [122].

The catalytic triads in the active sites of these enzymes superimpose each other with a rmsd of 0.29 Å and are formed by Ser146, Asp338 and His 397. They constitute a geometry type similar to, but not identical with, that of serine endoproteinases [121] although the structure of the 'oxyanion hole', thought to stabilize the tetrahedral intermediate in the hydrolysis reaction, appears in a similar position relative to the catalytic serine side-chain.

The pH and substrate specificities of these enzymes have been studied by use of a combination of enzymatic analysis with different substrates and inhibitors, mutational replacement of key residues, and X-ray crystallography [126, 127]. It has been shown that a hydrogen bond network binds to the substrate C-terminal carboxylate group and stabilizes the transition state [126]. The basis for substrate specificity has also been established [122, 133–135]. Glutamic acid residues 398 and 272 in CPD-WII are replaced by Met and Leu, respectively, in CPD-Y, accounting for the preference of the latter for hydrophobic amino acids. Residues Leu178 and Trp312, in the S1 binding pocket of CPD-Y, also play an important role in the specificity, and their mutational replacement may alter substrate preference [133, 136]; on the other hand, the hybrid nature of the S1 site in CPD-WII may explain the ability of this enzyme to accept both hydrophobic and basic amino acids at position P1 [135].

Propeptides and folding

The zymogen state of serine carboxypeptidases has been studied in detail for yeast carboxypeptidase Y (CPD-Y). Winther and Sorensen [137] showed that the proenzyme can renature efficiently after 6 M guanidinium hydrochloride treatment whereas the mature enzyme cannot, indicating that the propeptide, which is a flexible domain with a high content of secondary structure, but little tertiary structure in the isolated state [138], may act as a 'co-translational chaperone'. It was argued that the N-terminally-linked pro-region would interact productively with the growing polypeptide chain, and some studies have attempted to define those regions in the propeptide which are essential for *in vivo* folding [139, 140]. Those sequences are located in the C-terminal third of the propeptide where some elements have been shown to be essential for its function. However, an overall lack of requirement for sequence conservation may be deduced from the substitution of the original propeptide by another one from a similar carboxypeptidase [140]. Studies on the precursor forms of serine carboxypeptidases have been extended to the effect of mutational analysis and de-glycosylation in intracellular transport and vacuolar sorting. They indicate a low requirement for sequence conservation of the vacuolar sorting signal [141] and the need for glycosylation only in intracellular transport [142].

Lysosomal protective protein

Lysosomal protective protein, a mammalian enzyme of family S10, was previously known as cathepsin A. It is a serine carboxypeptidase whose present name describes its function of forming a high-molecular-weight complex with β-galactosidase and neuraminidase in the lysosome, protecting them against proteolysis. It forms a dimer with sequence similarity to the Kex1 gene product and carboxypeptidase Y from yeast [143]. Its structure has been determined [123] and shown to be similar to that of other serine carboxypeptidases in the core domain. The active site of the enzyme is preformed in the zymogen and blocked by a 'maturation subdomain'. Major conformational changes in this subdomain and removal of the excision peptide are necessary for activation in what has been described as a unique example of an activation mechanism among known proteases [123]. A comparative modelling study on substrate binding in serine carboxypeptidases has shown that the substrate specificity of human protective protein is similar to that of carboxypeptidase Y, with a preference for hydrophobic P1' residues [144].

Acknowledgements
This research was supported by grant BIO95-0848 from the CICYT (Ministerio de Educación y Ciencia, Spain) and by Centre de Referència de Biotecnologia (Generalitat de Catalunya).

References

1 Rawlings ND, Barrett AJ (1994) Classification of peptidases. *Meth Enzymol* 244: 1–15
2 Barrett AJ, Rawlings ND (1993) The many evolutionary families of peptidases. *In*: FX Avilés (ed.): *Innovation in proteases and their inhibitors*. Walter de Gruyter, Berlin, 13–30
3 Rawlings ND, Barrett AJ (1995) Evolutionary families of metallopeptidases. *Meth Enzymol* 248: 183–228
4 Barrett AJ, Woessner JFJ, Rawlings ND (eds) (1998) *Handbook of Proteolytic Enzymes*. Academic Press, London
5 Skidgel RA (1996) Structure and function of mammalian zinc carboxypeptidases. *In*: NM Hooper (ed.): *Zinc Metalloproteases in Health and Disease*. Taylor and Francis, London, 241–283
6 Avilés FX, Vendrell J, Guasch A, Coll M, Huber R (1993) Advances in metallo-procarboxypeptidases. Emerging details on the inhibition mechanism and on the inhibition process. *Eur J Biochem* 211: 381–389
7 Song L, Fricker LD (1997) Cloning and expression of human carboxypeptidase Z, a novel metallocarboxypeptidase. *J Biol Chem* 272: 10543–10550
8 Blundell TM (1994) Metalloproteinase superfamilies and drug design. *Nat Struct Biol* 1:73–75
9 Hooper NM (1994) Families of zinc metalloproteases. *FEBS Lett* 354:1–6
10 Hooper NM (1996) The biological roles of zinc and families of zinc metalloproteases. *In*: Hooper NM (ed.): *Zinc Metalloproteases in Health and Disease*. Taylor and Francis, London, 1–21
11 Joris B, Van Beeumen J, Casagrande F, Gerday C, Frere JM, Ghuysen JM (1983) The complete amino acid sequence of the Zn^{2+}-containing D-ALANYL-D-alanine-cleaving carboxypeptidase of *Streptomyces albus*. *Eur J Biochem* 130: 53–69
12 Dideberg O, Charlier P, Dive G, Joris B, Frere JM, Ghuysen JM (1982) Structure of a Zn^{2+}-containing D-ALANYL-D-alanine-cleaving carboxypeptidase at 2.5 Å resolution. *Nature* 299: 46–47
13 Lee S-H, Taguchi H, Yoshimura E, Minagawa E, Kaminogawa S, Ohta T, Matsuzawa H (1994) Carboxypeptidase Taq, a thermostable zinc enzyme, from *Thermus aquaticus* YT-1: molecular cloning, sequencing, and expression of the encoding gene in *Escherichia coli*. *Biosci Biotechnol Biochem* 58:1490–1495
14 Bode W, Gomis-Rüth FX, Stocker W (1993) Astacins, serralysins, snake venom and matrix metalloproteinases exhibit identical zinc-binding environments (HEXXHXXGXXH and Met-turn) and topologies and should be grouped into a common family, the 'metzincins'. *FEBS Lett* 331: 134–140
15 Lipscomb WN, Sträter N (1996) Recent advances in zinc enzymology. *Chem Rev* 96: 2375–2433
16 Rowsell S, Pauptit RA, Tucker AD, Melton RG, Blow DM, Brick P (1997) Crystal structure of carboxypeptidase G2, a bacterial enzyme with applications in cancer therapy *Structure* 3: 337–347
17 Chevrier B, Schalk C, D'Orchymont H, Rondeau JM, Moras D, Tarnus C (1994) Crystal structure of *Aeromonas proteolytica* aminopeptidase: a prototypical member of the co-catalytic zinc enzyme family. *Structure* 2: 283–291
18 Rawlings ND, Barret AJ (1994) Families of serine peptidases. *Meth Enzymol* 244: 19–60
19 Breddam K (1986) Serine carboxypeptidases: a review. *Carlsberg Res Commun* 51: 83–128

20 Remington SJ, Breddam K (1994) Carboxypeptidases C and D. *Meth Enzymol* 244: 231–248
21 Artymiuk PJ, Grindley HM, Park JE, Rice DW, Willett P (1992) Three-dimensional structural resemblance between leucine aminopeptidase and carboxypeptidase A revealed by graph-theoretical techniques. *FEBS Lett* 303: 48–52
22 Ollis DL, Cheah E, Cygler M, Dykstra B, Frolow F, Fraken S, Harel M, Remington SJ, Silman I, Schrag J, Sussman J, Goldman A (1992) The alpha/beta hydrolase fold. *Protein Eng* 5: 197–211
23 Anson ML (1937) Carboxypeptidase. I. The preparation of crystalline carboxypeptidase. *J Gen Physiol* 20: 663–669
24 Vallee BL, Neurath H (1955) Carboxypeptidase, a zinc metalloenzyme. *J Biol Chem* 217: 253–261
25 Yamasaki M, Brown JR, Cox DJ, Greenshields RN, Wade R, Neurath H (1963) Procarboxypeptidase A-S6. Further studies of its isolation and properties. *Biochemistry* 2: 859–866
26 Puigserver A, Desnuelle P (1977) Reconstitution of bovine procarboxypeptidase A-S6 from the free subunits. *Biochemistry* 16: 2497–4501
27 Kobayashi R, Kobayashi Y, Hirs CHW (1978) Identification of a binary complex of procarboxypeptidase A and a precursor of protease E in porcine pancretic secretion. *J Biol Chem* 253: 5526–5530
28 Pascual R, Burgos FJ, Salvà M, Soriano F, Méndez E, Avilés FX (1989) Purification and properties of five different forms of human procarboxypeptidases. *Eur J Biochem* 179: 609–616
29 Lacko AG, Neurath H (1970) Studies on procarboxypeptidase A and carboxypeptidase A of the spiny pacific dogfish (*Squalus acanthias*). *Biochemistry* 9: 4680–4690
30 Reeck GR, Neurath H (1972) Isolation and characterization of pancreatic procarboxypeptidase B and carboxypeptidase B of the African lungfish. *Biochemistry* 11: 3947–3955
31 Bradley G, Naudé RJ, Muramoto K, Yamauchi F, Oelofsen W (1996) Ostrich (*Strutio camelus*) carboxypeptidase B: purification, kinetic properties and characterization of the pancreatic enzyme. *Int J Biochem Cell Biol* 28: 521–529
32 Narahashi Y (1990) The amino acid sequence of zinc-carboxypeptidase from *Streptomyces griseus*. *J Biochem* 107: 879–886
33 Osterman AL, Grishin NV, Smulevitch SV, Matz MV, Zagnitko OP, Revina LP, Stepanov VM (1992) Primary structure of carboxypeptidase T: delineation of functionally relevant features in Zn-carboxypeptidase family. *Protein Chem* 11: 561–570
34 Skidgel RA (1988) Basic carboxypeptidases: regulators of peptide hormone activity. *Trends Pharmacol Sci* 9: 299–304
35 Gardell SJ, Craick CS, Clauser E, Goldsmith EJ, Stewart C-B, Graf M, Rutter WJ (1988) A novel rat carboxypeptidase, CPA2: characterization, molecular cloning and evolutionary implications on substrate specificity in the carboxypeptidase gene family. *J Biol Chem* 263: 17828–17836
36 Catasús L, Vendrell J, Avilés FX, Carreira S, Puigserver A, Billeter M (1995) The sequence and conformation of human pancreatic procarboxypeptidase A2. *J Biol Chem* 270: 6651–6657
37 Everitt MT, Neurath H (1980) Rat peritoneal mast cell carboxypeptidase: localization, purification and enzymatic properties. *FEBS Lett* 110: 292–296
38 Scheele G (1986) Two-dimensional electrophoresis in the analysis of exocrine pancreatic proteins. *In*: VLW Go et al. (eds): *The exocrine pancreas*. Raven Press, New York, 185–192
39 Vilanova M, Vendrell J, López MT, Cuchillo CM, Avilés FX (1985) Preparative isolation of the two forms of pig pancreatic procarboxypeptidase A and their monomeric carboxypeptidases A. *Biochem J* 22: 605–609
40 Oppezzo O, Ventura S, Bergman T, Vendrell J, Jörnvall H, Avilés FX (1994) Procarboxypeptidase in rat pancreas. Overall characterization and comparison of the activation processes. *Eur J Biochem* 222: 55–63
41 Gardell SJ, Craick CS, Hilvert D, Urdea MS, Rutter WJ (1985) Site-directed mutagenesis shows that tyrosine 248 of carboxypeptidase A does not play a crucial role in catalysis. *Nature* 317: 551–554
42 Phillips MA, Rutter WJ (1996) Role of the prodomain in folding and secretion of rat pancreatic carboxypeptidase A1. *Biochemistry* 35: 6771–6776
43 Laethem RM, Blumenkopf TA, Cory M, Elwell L, Moxham CP, Ray PH, Walton LM, Smith GK (1996) Expression and characterization of human pancreatic preprocarboxypeptidase A1 and preprocarboxypeptidase A2. *Arch Biochem Biophys* 332: 8–18
44 Delk AS, Durie PR, Fletcher TS, Largman C (1985) Radioimmunoassay of active pancreatic enzymes in sera from patients with acute pancreatitis. I. Active carboxypeptidase B. *Clin Chem* 31: 1294–1300
45 Fernstad R, Tyden G, Brattstrom C, Skoldefors H, Carlstrom K, Groth CG, Pousette A (1989) Pancreas-specific protein. New serum marker for graft rejection in pancreas-transplant recipients. *Diabetes* 38: 55–56
46 Yamamoto KK, Pousette A, Chow P, Wilson H, el Shami S, French CK (1992) Isolation of a cDNA encoding a human serum marker for acute pancreatitis. Identification of pancreas-specific protein as pancreatic procarboxypeptidase B. *J Biol Chem* 267: 2575–2581
47 Chen CC, Wang SS, Chen TW, Jap TS, Chen SJ, Jeng FS, Lee SD (1996) Serum procarboxypeptidase B, amylase and lipase in chronic renal failure. *J Gastroenterol Hepatol* 11: 496–499
48 Fowke PJ, Hodgkinson SC (1996) The ovine pancreatic protein which binds to insulin-like growth factor binding protein-3 is procarboxypeptidase A. *Endocrinology* 150: 51–56
49a Normant E, Gros C, Schwartz JC (1995) Carboxypeptidase A isoforms produced by distinct genes or alternative splicing in brain and other extrapancreatic tissues. *J Biol Chem* 270: 20543–20549

49b Normant E, Martres MP, Schwartz JC, Gros C (1995) Purification, cDNA cloning, functional expression, and characterization of a 26-kDa endogenous mammalian carboxypeptidase inhibitor. *Proc Natl Acad Sci USA* 92: 12225–12229
50 Teplyakov A, Polyakov K, Obmolova G, Strokopytov B, Kuranova I, Osterman A, Grishin N, Smulevitch S, Zagnitko O, Galperina O et al. (1992) Crystal structure of carboxypeptidase T from *Thermoactinomyces vulgaris*. *Eur J Biochem* 208: 281–288
51 Rees DC, Lewis M, Lipscomb WN (1983) Refined crystal structure of carboxypeptidase A at 1.54 Å resolution. *J Mol Biol* 168: 367–387
52 Famming Z, Kobe B, Stewart C-B, Rutter WJ, Goldsmith EJ (1991) Structural evolution of an enzyme specificity. The structure of rat carboxypeptidase A2 at 1.9 Å resolution. *J Biol Chem* 266: 24606–24612
53 Schmid MF, Herriott JR (1976) Structure of carboxypeptidase B at 2–8 Å resolution. *J Mol Biol* 103: 175–190
54 Coll M, Guasch A, Avilés FX, Huber R (1991) Three-dimensional structure of porcine procarboxypeptidase B: a structural basis of its inactivity. *EMBO J* 10: 1–9
55 Guasch A, Coll M, Avilés FX, Huber R (1992) Three-dimensional structure of porcine pancreatic procarboxypeptidase A. A comparison of the A and B zymogens and their determinants for inhibition and activation. *J Mol Biol* 224: 141–157
56 Gomis-Rüth FX, Gómez M, Bode W, Huber R, Avilés FX (1995) The three-dimensional structure of the native ternary complex of bovine pancreatic procarboxypeptidase A with proproteinase E and chymotrypsinogen C. *EMBO J* 14: 4387–4394
57 Vallee BL, Auld DS (1990) Zinc coordination, function and structure of zinc enzymes and other proteins. *Biochemistry* 29: 5647–5659
58 Rees DC, Lipscomb WN (1982) Refined crystal structure of the potato inhibitor complex of carboxypeptidase A at 2.5 Å resolution. *J Mol Biol* 160: 475–498
59 Christianson DW, Lipscomb WN (1989) Carboxypeptidase A. *Acc Chem Res* 22: 62–69
60 Kim H, Lipscomb WN (1991) Comparison of the structures of three carboxypeptidase A-phosphonate complexes determined by X-ray crystallography. *Biochemistry* 30: 8171–8180
61 Hilvert D, Gardell SJ, Rutter WJ, Kaiser ET (1986) Evidence against a crucial role for phenolic hydroxyl of Tyr248 in peptide and ester hydrolysis catalized by carboxypeptidase A: comparative studies on the pH dependencies of the native and Phe248 mutant. *J Amer Chem Soc* 108: 5298–5304
62 Gardell SJ, Hilvert D, Barnett J, Kaiser ET, Rutter WJ (1987) use of direct mutagenesis to probe the role of Tyr198 in the catalytic mechanism of carboxypeptidase A. *J Biol Chem* 262: 576–582
63 Phillips MA, Kaplan AP, Rutter WJ, Bartlett PA (1992) Transition-state characterization: a new approach combining inhibitor analogues and variation in enzyme structure. *Biochemistry* 31: 959–963
64 Alvarez-Santos S, González-Lafont A, Lluch JM, Oliva B, Avilés FX (1994) On the water-promoted mechanism of peptide cleavage by carboxypeptidase A. A theoretical study. *Can J Chem* 72: 2077–2083
65 Mock WL, Zhang JZ (1991) Mechanistically significant diastereoselection in the sulfoximine inhibition of carboxypeptidase A. *J Biol Chem* 266: 6393–6400
66 Auld DS, Galdes A, Geoghegan KF, Holmquist B, Martinelli RA, Vallee BL (1984) Cryospectrokinetic characterization of intermediates in biochemical reactions: carboxypeptidase A. *Proc Natl Acad Sci USA* 81: 5041–5045
67 Folk, Schirmer (1963) The porcine pancreatic carboxypeptidase A system. I. Three forms of the active enzyme. *J Biol Chem* 238: 3884–3894
68 Peterson LM, Holmquist B, Bethune JL (1982) *Anal Biochem* 125: 420–426
69 Auld DS, Vallee BL (1970) Kinetics of carboxypeptidase A. II. Inhibitors of the hydrolysis of oligopeptides. *Biochemistry* 9: 602–609
70 Mock WL, Liu Y, Stanford DJ (1996) Arazoformyl peptide surrogates as spectrophotometric kinetic assay substrates for carboxypeptidase A. *Anal Biochem* 239: 218–222
71 Normant E, Schwartz JC, Gros C (1996) A novel 125I]iodinated carboxypeptidase A substrate detects a metallopeptidase activity distinct from carboxypeptidase A in brain. *Neuropeptides* 30: 13–17
72 Plummer TH, Ryan TJ (1981) A potent mercapto bi-product analogue inhibitor for human carboxypeptidase N. *Biochem Biophys Res Commun* 98: 448–454
73 Hass GM, Ryan CA (1981) Carboxypeptidase inhibitor from potatoes. *Meth Enzymol* 80: 778–791
74 Molina M, Avilés FX, Querol E (1994) C-tail valine is a key residue for the stabilization of the complex between potato inhibitor and carboxypeptidase A. *J Biol Chem* 269: 21467–21472
75 Homandberg GA, Litwiller RD, Peanasky RJ (1989) Carboxypeptidase inhibitors from *Ascaris suum*: the primary structure. *Arch Biochem Biophys* 270: 153–161
76 Smulevitch SV, Osterman AL, Galperina OV, Matz MV, Zagnitko OP, Kadyrov RM, Tsaplina IA, Grishin NV, Chestukhina GG, Stepanov VM (1991) Molecular cloning and primary structure of *Thermoactinomyces vulgaris* carboxypeptidase T. A metalloenzyme endowed with dual substrate specificity. *FEBS Lett* 291: 75–78
77 Vendrell J, Cuchillo CM, Avilés FX (1991) The tryptic activation pathway of monomeric procarboxypeptidase A. *J Biol Chem* 265: 6949–6953
Villegas V, Vendrell J, Avilés FX (1995) The activation pathway of procarboxyeptidase B from porcine pancreas: participation of the active enzyme in the proteolytic processing. *Protein Sci* 4: 1792–1800

78 Conejero-Lara F, Sánchez-Ruiz JM, Mateo PL, Burgos FJ, Vendrell J, Avilés FX (1991) Differential scanning calorimetry study of carboxypeptidase B, procarboxypeptidase B and its globular activation domain. *Eur J Biochem* 200: 663–670
Villegas V, Azuaga A, Catasús L, Reverter D, Mateo PL, Avilés FX, Serrano L (1995) Evidence for a two-state transition in the folding process of the activation domain of human procarboxypeptidase A2. *Biochemistry* 34: 15105–15110
79 Villegas S (1994) Caracterización detallada del proceso de activación de la procarboxipeptidasa B mediante el uso de inhibidores. *Ph D Thesis*. Universitat Autònoma de Barcelona
80 Springman EB, Dikov MM, Serafin WE (1995) Mast cell procarboxypeptidase A. Molecular modeling and biochemical characterization of its processing within secretory granules. *J Biol Chem* 270: 1300–1307
81 Eaton DL, Malloy BE, Tsai SP, Henzel W, Drayna D (1991) Isolation, molecular cloning and partial characterization of a novel carboxypeptidase B from human plasma. *J Biol Chem* 266: 21833–21838
Valnickova Z, Thogersen IB, Christensen S, Chu CT, Pizzo SV, Enghild JJ (1996) Activated human plasma carboxypeptidase B is retained in the blood by binding to alpha2-macroglobulin and pregnancy zone protein. *J Biol Chem* 271: 12937–12943
82 Tan AK, Eaton DL (1995) Activation and characterization of procarboxypeptidase B from human plasma. *Biochemistry* 34: 5811–5816
83 Stevens RL, Qui D, McNeil HP, Friend DS, Hunt JE, Austen KF, Zhang J (1996) Transgenic mice that possess a disrupted mast cell protease 5 (mMCP-5) gene cannot store carboxypeptidase A (mMC-CPA) protein in their granules. *FASEB J* 10: A1307–A1307
84 Ventura S, Gomis-Rüth FX, Puigserver A, Avilés FX, Vendrell J (1997) Pancreatic procarboxypeptidases: oligomeric structures and activation processes revisited. *Biol Chem* 378: 161–165
Gomis-Rüth Gomez-Ortiz M, Vendrell J, Ventura S, Bode W, Huber R, Avilés FX (1997) Crystal structure of an oligomer of proteolytic zymogens: detailed conformational analysis of the bovine ternary complex and implications for their activation. *J Mol Biol* 269: 1–20
85 Erdös EG, Sloane EM (1962) An enzyme in human blood plasma that inactivates bradykinin and kallidins. *Biochem Pharmacol* 11: 585–592
Erdös EG (1979) Kininases. *In*: EG Erdös (ed.): *Handbook of experimental pharmacology*, vol 25, Suppl. Springer-Verlag, Heidelberg, 427–448
Plummer TH, Hurwitz MY (1978) Human plasma carboxypeptidase N. Isolation and characterization. *J Biol Chem* 253: 3907–3912
Levin Y, Skidgel RA, Erdös EG (1982) Isolation and characterization of the subunits of human plasma carboxypeptidase N (kininase I). *Proc Natl Acad Sci USA* 79: 4818–4622
86 Skidgel RA, David RMTan F (1989) Human carboxypeptidase M: purification and characterization of a membrane-bound carboxypeptidase that cleaves peptide hormones. *J Biol Chem* 264: 2236–2241
87 Skidgel RA, Deddish PA, Davis RM (1988) Isolation and characterization of a basic carboxypeptidase from human seminal plasma. *Arch Biochem Biophys* 267: 660–667
88 Dragovic T, Schraufnagel DE, Becker RP, Sekosan M, Votta-Velis EG, Erdös EG (1995) Carboxypeptidase M activity is increased in bronchoalveolar lavage in human lung disease. *Amer J Respir Crit Care Med* 152: 760–764
89 Fricker LD (1988) Carboxypeptidase E. *Annu Rev Physiol* 50: 309–321
90 Fricker LDSnyder SH (1982) Enkephalin convertase: purification and characterization of a specific enkephalin-synthesizing carboxypeptidase localized to adrenal cromaffin granules. *Proc Natl Acad Sci USA* 79: 3886–3890
91 Goldstein SM, Wintroub BU (1993) Mast cell proteases. *In*: MA Kaliner, DD Metcalfe (eds): *The mast cell in health and disease.* Marcel Dekker, New York, 343–380
92 Hendriks D, Scharpé S, vanSande M, Lommaert MP (1989) Characterization of a carboxypeptidase in human serum distinct from carboxypeptidase N. *J Clin Chem Clin Biochem* 27: 277–285
93 Wang W, Hendriks DF, Scharpé SS (1994) Carboxypeptidase U, a plasma carboxypeptidase with high affinity for plasminogen. *J Biol Chem* 269: 15937–15944
94 Bajzar L, Manuel R, Nesheim ME (1995) Purification and characterization of TAFI, a thrombin-activable fibrinolysis inhibitor. *J Biol Chem* 270: 14477–14484
95 Song L, Fricker LD (1995) Purification and characterization of carboxypeptidase D, a novel carboxypeptidase E-like enzyme, from bovine pituitary. *J Biol Chem* 270: 25007–25013
96 Song L, Fricker LD (1996) Tissue distribution and characterization of soluble and membrane-bound forms of metallocarboxypeptidase D. *J Biol Chem* 271: 28884–28889
97 He GP, Muise A, Li AW, Ro Hs (1995) A eukaryotic transcriptional repressor with carboxypeptidase activity. *Nature* 378: 92–96
98 Skidgel RA (1995) Human carboxypeptidase N (lysine carboxypeptidase). *Meth Enzymol* 248: 653–663
99 Deddish PA, Skidgel RA, Kriho VB, Li X-Y, Becker RP, Erdös EG (1990) Carboxypeptidase M in Madin-Darby canine kidney cells. *J Biol Chem* 265: 15083–15089
100 Mitra A, Song L, Fricker LD (1994) The C-terminal region of carboxypeptidase E is involved in membrane binding and intracellular routing in AtT-20 cells. *J Biol Chem* 269: 19876–19881
101 Fricker LD, Das B, Angeletti RH (1990) Identification of the pH-dependent membrane anchor of car-

boxypeptidase E (EC 3.4.17.10) *J Biol Chem* 265: 2476–2482
102 Cole KR, Kumar S, Trong HL, Woodbury RG, Walsh KA, Neurath H (1991) Rat mast cell carboxypeptidase: amino acid sequence and evidence of enzyme activity within mast cell granules. *Biochemistry* 30: 648–655
103 Reynolds DS, Stevens RL, Gurley DS, Lane WS, Austen KF, Serafin WE (1989) Isolation and molecular cloning of mast cell carboxypeptidase A. A novel member of the carboxypeptidase gene family. *J Biol Chem* 264: 20094–20099
104 Manser E, Fernández D, Loo L, Goh PY, Monfries C, Hall C, Lim L (1990) Human carboxypeptidase E. Isolation and characterization of the cDNA, sequence conservation, expression and processing *in vitro*. *Biochem J* 267: 517–525
105 Song L, Fricker LD (1997) The pro region is not required for the expression or intracellular routing of carboxypeptidase E. *Biochem J* 323: 265–271
106 Guest PC, Arden SD, Rutherford NG, Hutton JC (1995) The post-translational processing and intracellular sorting of carboxypeptidase H in the islets of Langerhans. *Mol Cell Endocrinol* 113: 99–108
107 Song L, Fricker LD (1995) Processing of procarboxypeptidase E into carboxypeptidase E occurs in secretory vesicles. *J Neurochem* 65: 444–453
108 Rehn M, Pihlajaniem T (1995) Identification of three N-terminal ends of type XVIII collagen chains and tissue-specific differences in the expression of the corresponding transcripts. The longest form contains a novel motif homologous to rat and *Drosophila* frizzled proteins. *J Biol Chem* 270: 4705–4711
109 McGwire GB, Tan F, Michel B, Rehli M, Skidgel RA (1997) Identification of a membrane-bound carboxypeptidase as the mammalian homolog of duck gp180, a hepatitis B virus-binding protein. *Life Sci* 60: 715–724
110 Nagae A, Abe M, Becker RP, Deddish PA, Skidgel RA, Erdös EG (1993) High concentration of carboxypeptidase M in lungs: presence of the enzyme in alveolar type I cells. *Amer J Respir Cell Molec Biol* 9: 221–229
111 Rehli M, Krause SW, Kreutz M, Andreesen R (1995) Carboxypeptidase M is identical to MAX.1 antigen and its expression is associated with monocyte to macrophage differenciation. *J Biol Chem* 270: 15664–15649
112 Skidgel RA, McGwire GB, LIXY (1996) Membrane anchoring and release of carboxypeptidase M: implications for extracellular hydrolysis of peptide hormones. *Immunopharmacology* 32: 48–52
113 McGwire GB, Skidgel RA (1995) Extracellular conversion of epidermal growth factor (EGF) to des-Arg53-EGF by carboxypeptidase M. *J Biol Chem* 270: 17154–17158
114 Docherty K, Steiner DF (1982) Post-translational proteolysis in polypeptide hormone biosynthesis. *Annu Rev Physiol* 44: 625–638
115 Fricker LD, Snyder SH (1983) Purification and characterization of enkephalin convertase, an enkephalin-synthesizing carboxypeptidase. *J Biol Chem* 258: 10950–10955
116 Fricker LD, Adelman JP, Douglass J, Thompson RC, von Strandmann RP, Hutton J (1989) Isolation and seuqnce analysis of cDNA for rat carboxypeptidase E EC 3.4.17.10], a neuropeptide processing enzyme. *Mol Endocrinol* 3: 665–673
117 Naggert JK, Fricker LD, Varlamov O, Nishina PM, Rouille Y, Steiner DF, Carroll RJ, Paigen BJ, Leiter EH (1995) Hyperproisulinaemia in obese *fat/fat* mice associated with a carboxypeptidase E mutation which reduces enzyme activity. *Nature Gen* 10: 135–142
118 Fricker LD, Berman YL, Leiter EH, Devi LA (1996) Carboxypeptidase E activity is deficient in mice with the fat mutation. Effect on peptide processing. *J Biol Chem* 271: 30619–30624
119 Cool DR, Normant E, Shen F, Che HC, Pannell L, Zhuang Y, Loh YP (1997) Carboxypeptidase E is a regulated secretory pathway sorting receptor: genetic obliteration leads to endocrine disorders in Cpe(fat) mice. *Cell* 88: 73–83
120 Dikov MM, Springman EB, Yeola S, Serafin WE (1994) Processing of procarboxypeptidase A and other zymogens in murine mast cells. *J Biol Chem* 269: 25897–25904
121 Liao DI, Breddam K, Sweet RM, Bullock T, Remington SJ (1992) Refined atomic model of wheat serine carboxypeptidase II at 2.2 Å resolution. *Biochemistry* 31: 9796–9812
122 Endrizi JA, Breddam K, Remington SJ (1994) 2.8 Å structure of yeast serine carboxypeptidase. *Biochemistry* 33: 11106–11120
123 Rudenko G, Bonten E, d'Azzo A, Hol WG (1995) Three-dimensional structure of the human 'protective protein': structure of the precursor form suggests a complex activation mechanism. *Structure* 3: 1249–1259
124 Shilton BH, Li Y, Tessier D, Thomas DY, Cycgler M (1996) Crystallization of a soluble form of the Kex1p serine carboxypeptidase from *Saccharomyces cerevisiae*. *Protein Sci* 5: 395–397
125 Liao DI, Remington SJ (1991) Structure of wheat serine carboxypeptides II at 3.5 Å resolution. A new class of serine proteinase. *J Biol Chem* 265: 6528–6531
126 Mortensen UH, Remington SJ, Breddam K (1994) Site-directed mutagenesis on (serine) carboxypeptidase Y. A hydrogen bond network stabilizes the transition state by interaction with the C-terminal carboxylate group of the substrate. *Biochemistry* 33: 508–517
127 Stennicke HR, Mortensen UH, Breddam K (1996) Studies on the hydrolytic properties of (serine) carboxypeptidase Y. *Biochemistry* 35: 7131–7141
128 Tan F, Morris PW, Skidgel RA, Erdös EG (1993) Sequencing and cloning of human prolylcarboxypeptidase (angiotensina C). Similarity to both serine carboxypeptidase and prolylendopeptidase families. *J Biol Chem* 268: 16631–16638

129 Thomas L, Cooper A, Bussey H, Thomas G (1990) Yeast KEX1 protease cleaves a prohormone processing intermediate in mammalian cells. *J Biol Chem* 265: 10821–10824
130 Odya CE, Erdös EG (1981) Human prolylcarboxypeptidase. *Meth Enzymol* 80: 460–466
131 Valls LA, Hunter CP, Rothman JH, Stevens TH (1987) Protein sorting in yeast: the localization determinant of yeast vacuolar carboxypeptidase Y resides in the propeptide. *Cell* 48: 887–897
132 Thiede B, Wittmann-Liebold B, Bienert M, Krause E (1995) MALDI-MS for C-terminal sequence determination of peptides and proteins degraded by carboxypeptidase Y and P. *FEBS Lett* 357: 65–69
133 Olesen K, Mortensen UH, Aasmul-Olsen S, Kielland-Brandt MC, Remington SJ, Breddam K (1994) The activity of carboxypeptidase Y toward substrates with basic P1 amino acid residues is drastically increased by mutational replacement of leucine 178. *Biochemistry* 33: 11121–11126
134 Bullock TL, Branchaud B, Remington SJ (1994) Structure of the complex of L-benzylsuccinate with wheat serine carboxypeptidase II at 2.0-Å resolution. *Biochemistry* 33: 11127–11134
135 Bullock TL, Breddam K, Remington SJ (1996) Peptide aldehyde complexes with wheat serine carboxypeptidase II: implications for the catalytic mechanism and substrate specificity. *J Mol Biol* 255: 714–725
136 Olesen K, Breddam K (1995) Increase in the P1 Lys/leu substrate preference of carboxypeptidase Y by rational design based on known primary and tertiary structures of serine carboxypeptidases. *Biochemistry* 34: 15689–15699
137 Winther JR, Sorensen P (1991) Propeptide of carboxypeptidase Y provides a chaperone-like function as well as inhibition of the enzymatic activity. *Proc Natl Scad Sci USA* 88: 9330–9334
138 Sorensen P, Winther JR, Kaarsholm NC, Poulsen FM (1993) The pro region required for folding of carboxypeptidase Y is a partially folded domain with little regular structural core. *Biochemistry* 32: 12160–12166
139 Ramos C, Winther JR, Kielland-Brandt MC (1994) Requirement of the propeptide for *in vivo* formation of active yeast carboxypeptidase Y. *J Biol Chem* 269: 7006–7012
140 Ramos C, Winther JR (1996) Exchange of regions of the carboxypeptidase Y propeptide. Sequence specificity and function in folding *in vivo*. *Eur J Biochem* 242: 29–35
141 van Voorst F, Kielland-Brandt MC, Winther JR (1996) Mutational analysis of the vacuolar sorting signal of procarboxypeptidase Y in yeast shows a low requirement for sequence conservation. *J Biol Chem* 271: 841–846
142 Winther JR, Sorensen P (1991) Propeptide of carboxypeptidase Y provides a chaperone-like function as well as inhibition of the enzymatic activity. *Proc Natl Acad Sci USA* 88: 9330–9334
143 Jackman HL, Tan FL, Tamei H, Beurling-Harbury C, Li XY, Skidgel RA, Erdös EG (1990) A peptidase in human platelets that deamidates tachykinins. Probable identity with the lysosomal "protective protein". *J Biol Chem* 265: 11265–11272
144 Elsliger MA, Pshezhetsky AV, Vinogradova MV, Svedas VK, Potier M (1996) Comparative modeling of substrate binding in the S1' subsite of serine carboxypeptidases from yeast, wheat, and human. *Biochemistry* 35: 14899–14909

Proteases: New Perspectives
V. Turk (ed.)

Signal peptidases

Mark O. Lively[1] and Christopher M. Ashwell[2]

[1]*Department of Biochemistry, Wake Forest University School of Medicine, Winston-Salem, NC 27157, USA*
[2]*Growth Biology Lab, United States Department of Agriculture, Beltsville, MD 20705, USA*

Introduction

Type I signal peptidases cleave the signal peptides from secretory and membrane-associated proteins that are transported across or into the bacterial cell membrane or the lipid bilayer of the endoplasmic reticulum (ER) [1]. These peptidases are essential for life in most, if not all cells. In prokaryotes, they are single-chain integral membrane proteins that are anchored to the cell membrane by one or two membrane anchors. The eukaryotic enzyme is associated with a multi-subunit membrane protein complex with subunits positioned on both sides of the ER membrane. Current evidence suggests that the prokaryotic and eukaryotic signal peptidases are distantly related enzymes that use an atypical and perhaps very ancient proteolytic mechanism but few details are known about that mechanism. This chapter considers current information about the structure and functions of type I signal peptidases in prokaryotes and in the ER of eukaryotes. Related signal peptidases found in mitochondria [2] and chloroplasts [3] are not considered here.

Secretory protein translocation

In the course of normal cellular function, many proteins must be transported through (translocated across) membranes to reach their proper sites of function within organelles or outside of cells. Integral membrane proteins must also be targeted to their proper membrane sites for integration into the membrane. It has been estimated that 20% of the 1680 proteins encoded by the *Haemophilus influenzae* genome are either integral membrane or translocated proteins [4]. Thirty percent of proteins expressed by eukaryotes are estimated to be membrane and secretory proteins based on the known cellular localizations of identified *Saccharomyces cerevisiae* genes in the complete yeast genome [5]. Thus signal peptidases must play a broad and significant role in protein biosynthesis throughout biology.

Milstein and associates discovered signal peptidase when they found that secreted proteins are synthesized as higher molecular weight precursors [6]. They reported that immunoglobulin light chains are initially synthesized in myeloma cells with a higher molecular weight than the mature, secreted forms of the proteins. They discovered that a peptidase present in the microsomal membrane fraction of the myeloma cells is responsible for the processing of the larger precursor proteins to their mature forms. This is a fundamental mechanism for the majority of secretory proteins and many membrane proteins in all cells. These proteins are synthesized as precursors with signal peptides at their N-termini [7]. Signal peptides are recognized in the cytosol by components of the secretory system and target a protein to the proper membrane for translocation. The role of signal peptidase is to remove the signal from the mature protein after targeting has been accomplished.

Signal peptides – the biological substrate for signal peptidases

The amino acid sequences of signal peptides are highly variable and there is no single consensus sequence that can describe a typical signal peptide. However, the peptides do have common structural features that enable both eukaryotic and prokaryotic sequences to be recognized by the peptidases [8]. The average length of the signal peptide depends on the organism: Gram-positive bacteria, 32 amino acids; Gram-negative bacteria, 25 amino acids; and eukaryotes, 23 amino acids [9].

Three domains can be recognized consistently in signal peptides: the amino-terminal domain, *n*; the hydrophobic domain *h*; and the carboxy-terminal domain, *c* (Fig. 1). The *n*-domain usually contains one or more basic amino acid residues, typically Arg. The *h*-domain, often referred to as the hydrophobic core, averages 10 amino acids in length and contains only apolar residues (although Ser is often found there). The *c*-domain is adjacent to the signal peptide cleavage site where only amino acids with small, uncharged side chains are found at positions P1 and P3. Based on a selected collection of representative signal peptide sequences in the protein sequence databases, a computer method has been created [4, 9] that will predict the site of cleavage in newly identified gene sequences of secretory proteins (http://www.cbs.dtu.dk/services/SignalP/).

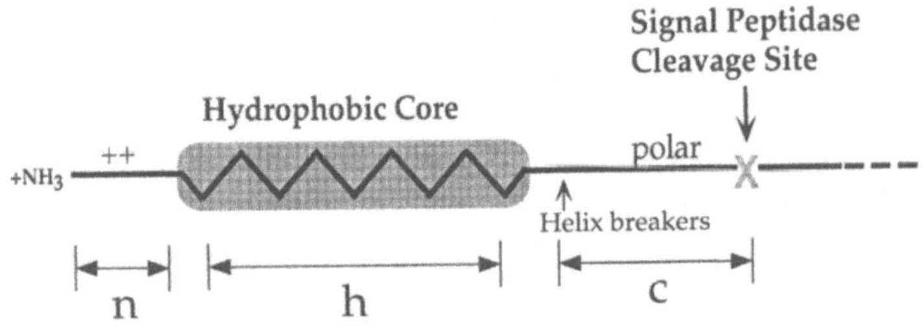

Figure 1. Signal peptide model. Cartoon model representing the major structural elements common to all signal peptides: n-domain, *n*; h-domain, *h*; and c-domain, *c*.

Cellular localization of signal peptidases

In prokaryotes and eukaryotes, the signal peptidase active site is located on the opposite side of the membrane from the site of protein synthesis. Therefore precursor proteins must be transferred from the site of synthesis in the cytoplasm to the signal peptidase active site on the opposite face of the membrane. The removal of the signal peptide is usually a cotranslational event in eukaryotes but cleavage also occurs after the completion of synthesis of some precursor proteins. In bacteria, post-translational cleavage is more common [10, 11]. Bacterial signal peptidases are monomeric integral membrane proteins anchored in the cell membrane by one to three membrane spanning domains, depending on the species, with their globular domains positioned in the periplasm (Fig. 2) [1]. In contrast to prokaryotes, eukaryotic signal peptidases are associated with a complex of four or five membrane proteins [1, 12]. Two subunits of the signal peptidase complex span the ER membrane once such that their N-termini are located in the cytoplasm and their globular domains are located in the lumen of the ER [13]. Two other subunits span the ER membrane twice and have their N-termini and their globular domains disposed to the cytoplasm (Fig. 3) [14]. Present data suggest that only one of the subunits of the

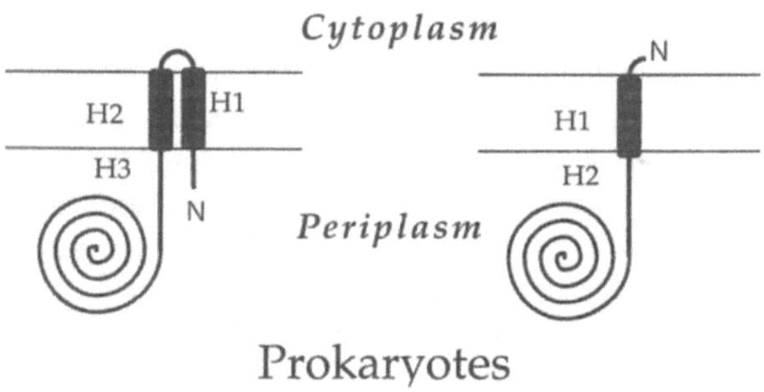

Figure 2. Topological model of prokaryotic signal peptidases. Prokaryotic signal peptidases are integral membrane proteins with one or two membrane spanning domains that anchor the enzyme in the cytoplasmic membrane with the active site domain located in the cytoplasm.

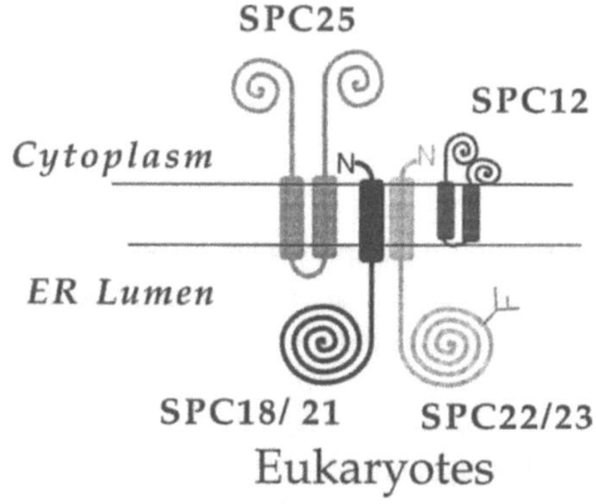

Figure 3. Topological model of eukaryotic signal peptidases. Eukaryotic signal peptidases are associated with a complex of membrane proteins. This model includes one copy of each subunit of the canine signal peptidase complex (SPC) although the precise stoichiometry of subunits found in the SPC in the endoplasmic reticulum is not yet defined.

eukaryotic signal peptidase complex contains the catalytic site while the other individual subunits may have functions in the mechanism of protein translocation into the ER or in substrate recognition for peptidase cleavage.

Prokaryotic signal peptidases

Signal peptidase genes have been isolated from or identified in many different bacterial species: *Escherichia coli* [15]; *Bacillus subtilis* [16, 17]; *Salmonella typhimurium* [18]; *Pseudomonas fluorescens* [19]; *Haemophilus influenzae* [20]; *Rhodobacter capsulatus* [21]; *Bradyrhizobium japonicum* [22]; *Streptococcus pneumoniae* [23]; and *Staphylococcus aureus* [24]. Signal peptidase genes have also been recognized in genomic sequences of archaea: *Methanococcus jannaschii* [25] and *Methanobacterium thermoautotrophicum* [26].

E. coli type I signal peptidase (also named leader peptidase, LEP) is the most studied bacterial signal peptidase and was the first signal peptidase purified [27]. It is a monomeric, single-chain 37 kDa integral membrane protein consisting of two membrane spanning domains, H1 and H2, a third periplasmic hydrophobic domain, H3, and a large globular enzyme domain that is exposed in the periplasm (Fig. 2) [15, 28]. LEP appears to belong to a novel class of serine proteases employing a Ser-Lys catalytic dyad [19, 29–31]. Ser-90 and Lys-145 residues are essential for catalytic activity in LEP [30]. Because LEP is an essential enzyme in *E. coli* [32], it is potentially an effective target for antibiotic development [33]. Expression of a soluble, catalytically active form of leader peptidase, Δ2-75LEP [34, 35], has enabled crystallographic studies of the enzyme's structure [36].

Most bacterial species examined thus far contain only a single chromosomal copy of the signal peptidase gene but two species of *Bacillus*, *B. subtilis* [16] and *B. amyloliquefaciens* [37], have been found to express more than one type I signal peptidase. The complete genomic sequence reveals five Type I signal peptidase genes in *B. subtilis* [17]. It is not known why so many apparently redundant signal peptidase genes are present in these bacteria but the multiple copies may play a central role in the increased secretory capacity of these bacteria.

Eukaryotic signal peptidases

Eukaryotic microsomal signal peptidase complexes are responsible for removal of signal peptides from nascent precursor proteins as they are translocated into the ER [1]. The first microsomal signal peptidase complex was isolated from dog pancreas ER microsomes [38]. The canine microsomal signal peptidase complex (SPC) consists of 5 polypeptides, with relative molecular masses of 25, 22/23, 21, 19, and 12 kDa (Fig. 3). Genes encoding each of these subunits have been cloned: 22/23 kDa glycoprotein, SPC23 [39]; the SPC21 and SPC18 subunits [40, 41]; the 25 kDa subunit, SPC25 [42]; and the 12 kDa subunit, SPC12 [14]. The SPC22/23, SPC18, and SPC21 subunits are type II integral membrane proteins, with the globular domain of each protein lying on the lumenal side of the ER [13]. The SPC12 and SPC25 subunits each have two membrane spanning domains that anchor these proteins in the bilayer such that their globular domains remain in the cytoplasm [14]. SPC18 and SPC21 are homologous isoforms of the same protein and appear to be distantly related to the bacterial signal peptidase family [1]. Based on the apparent conservation of four domains in these proteins and in the bacterial proteins, these subunits are thought to contain the catalytic site (Fig. 4).

	Domain I ▼		Domain II	
E. coli	88	SG.SMMPTLL	127	RGDIVVF
B. subtilis	40	DGDSMYPTLH	68	RGDIVVL
M. jannaschii	16	SD.SMYPIMK	25	RGDLVIV
SEC11	42	SG.SMEPAFQ	51	RGDILFL
SPC18	54	SG.SMEPAFH	63	RGDLLFL
SPC21	66	SG.SMEPAFH	75	RGDLLFL
CONSENSUS		sg-**S**M-P-#-		RGD####

	Domain III ▼		Domain IV ▼	
E. coli	137	EDPKLDYIKRAVGLPGDK	272	GDNRDNSADSR
B. subtilis	75	NGDDVHYVKRIIGLPGDT	145	GDNRRNSMDSR
CONSENSUS-bacteria		------##**K**R##------		GDNR--S-DSr
M. jannaschii	112	KSPTKPVIHRVIDKV.EF	138	GDNNPI.HDPE
SEC11	75	EGKQIPIVHRVLRQHNNH	102	GDNNAG.NDIS
SPC18	88	EGREIPIVHRVLKIH.EK	115	GDNNA..VDDR
SPC21	100	EGRDIPIVHRVIKVH.EK	127	GDNNE..VDDR
CONSENSUS-eukarya		------##**H**R##------		GDNN----D-r

Figure 4. Conserved signal peptidase domains in the Sec11 subunit family. Representative samples of the most similar domains of amino acid sequences from six signal peptidases are shown to highlight amino acids that may be involved in catalysis. The sequences are from *E. coli* [15], *Bacillus subtilis* [17], *Methanococcus jannaschii* [25], *Saccharomyces cerevisiae* [46], canine SPC18 [41], and canine SPC21 [40]. *E. coli* Ser-90 (arrowhead in Domain I) is the active site Ser residue. The role of the highly conserved RGD motif in Domain II is not known. Lys-145 in Domain III plays a key role in the mechanism of *E. coli* signal peptidase. The role of conserved Domain IV is not known. Periods in the amino acid sequences represent gaps inserted to maximize amino acid alignments. Dashes in the consensus sequence represent sites with no sequence consensus. The # symbols represent hydrophobic amino acid residues. The consensus sequences are based on more sequences than are shown here. See [1] for a more comprehensive alignment. (Adapted from [1]).

Signal peptidase has also been purified from the oviducts of laying hens. In apparent contrast to canine SPC, hen oviduct signal peptidase (HOSP) was purified as a complex of only two polypeptides, with relative molecular masses of 22/23 kDa (gp23) and 19 kDa (p19) [43, 44]. The 22/23 kDa polypeptide is glycosylated and, as seen in preparations from dog, migrates during gel electrophoresis as a doublet [43, 44]. A cDNA encoding gp23 revealed that the chicken SPC subunit is 90% identical to SPC22/23 [45]. The purified HOSP complex appears to contain only a single 19 kDa protein band, but genetic evidence shows that chickens also express two forms of the p19 protein that are homologous to SPC18 and SPC21 (Walker and Lively, unpublished results).

The first eukaryotic signal peptidase gene to be cloned was obtained from the yeast *Saccharomyces cerevisiae*. The protein product of the *SEC11* [46] gene, designated Sec11p, has an apparent molecular mass of 18 kDa and is a component of the yeast microsomal signal peptidase complex that contains four subunits [47]. Sec11p is homologous to SPC18, SPC21, and chicken p19 [41] and is an essential gene [46]. The protein product of the SEC3 gene, designated Sec3p, is the glycoprotein component of the yeast signal peptidase complex and it is essential for signal peptidase activity [48, 49]. The *S. cerevisiae* SPC homologs of SPC25 and SPC12 have also been cloned [50]. While the functional role of SPC12 is unknown, it is important for efficient signal peptide cleavage in *S. cerevisiae* [51]. These yeast proteins are highly conserved homologs of the subunits of the microsomal signal peptidase complexes purified from dog pancreas.

Signal peptidase mechanism

The exact subunit stoichiometry of the microsomal signal peptidase complex in the ER is unknown. The functional roles of the individual subunits are also unknown but strong indirect evidence suggests that the proteins of the Sec11 family (SPC18, SPC21, p19, and Sec11p) contain the catalytic site of the eukaryotic signal peptidases [1]. Because signal peptidases are resistant to inhibition by the typical chemical inhibitors of peptidases, it has not been possible to define the catalytic mechanism of the enzyme based on an inhibition profile. However, significant mutagenesis studies with the *E. coli* signal peptidase have provided strong evidence that the bacterial signal peptidases use a catalytic dyad of Ser and Lys to catalyze peptide bond cleavage [29–31]. No other single amino acid residues have been found to be essential for catalysis in the *E. coli* enzyme.

Figure 4 shows an alignment of amino acid sequences from four highly similar domains that are found in all type I signal peptidases [52]. The consensus sequences shown are based on more sequences than are shown in the figure (see [1] for a more complete alignment). Domain I shows a Ser residue that is present in all known sequences from all species. This residue is Ser-90 of *E. coli* signal peptidase and it is essential for catalysis. Based on this fact and the absolute conservation of this amino acid in known signal peptidases, this Ser is presumed to be the active site Ser of the enzyme.

Similarly, Domain III of *E. coli* signal peptidase contains Lys-145 that is required for catalysis [53]. This Lys is absolutely conserved in the bacterial signal peptidases but aligns with a conserved His residue in the archaea and eukaryotic enzymes. If this His residue is involved in catalysis then the archaea and eukaryotes have apparently switched to a different mechanism of catalysis during evolution by substituting Lys for His at this critical position. Proof of this hypothesis must await experimental evidence that proves that this conserved His residue plays a central role in the catalytic mechanism. Interestingly, each of the other conserved signal peptidase domains contains a highly conserved Asp residue providing potential candidates for the

Asp typically found in the catalytic triad of classical serine proteases. There is currently no evidence supporting the involvement of the conserved Asp in catalysis but the conservation of these amino acids in these domains suggests that they do play an important role in the function of these enzymes. Further experiments will be required to clearly define the catalytic mechanism of the type I signal peptidases.

Acknowledgments
This work was supported by grant GM32861 from the US National Institutes of Health.

Note added in proof: An X-ray crystal structure of *E. coli* Δ2-75LEP has been reported: Paetzel M, Dalbey RE, Strynadka NCJ (1998) Crystal structure of a bacterial signal peptidase complex with a β-lactam inhibitor. *Nature* 396: 186 – 190

References

1 Dalbey RE, Lively MO, Bron S, van Dijl JM (1997) The chemistry and enzymology of the type I signal peptidases. *Protein Sci* 6: 1129–138
2 Nunnari J, Fox TD, Walter P (1993) A mitochondrial protease with two catalytic subunits of nonoverlapping specificities. *Science* 262: 1997–2004
3 Shackleton JB, Robinson C (1991) Transport of proteins into chloroplasts. The thylakoidal processing peptidase is a signal-type peptidase with stringent substrate requirements at the -3 and -1 positions. *J Biol Chem* 266: 12152–12156
4 Nielsen H, Engelbrecht J, Brunak S, von Heijne G (1996) Identification of prokaryotic and eukaryotic signal peptides and prediction of cleavage sites. *Protein Eng* 10: 1–6
5 Goffeau A, Barrell BG, Bussey H, Davis RW, Dujon B, Feldman H, Galibert F, Hoheisel JD, Jacq C, Johnston M et al. (1996) Life with 6000 genes. *Science* 274: 546–567
6 Milstein C, Brownlee GG, Harrison TM, Mathews MB (1972) A possible precursor of immunoglobulin light chains. *Nat New Biol* 239: 117–120
7 von Heijne G (1985) Signal sequences. The limits of variation. *J Mol Biol* 184: 99–105
8 von Heijne G (1990) The Signal Peptide. *J Membrane Biol* 115: 195–201
9 Nielsen H, Engelbrecht J, von Heijne G, Brunak S (1996) Defining a similarity threshold for a functional protein sequence pattern: The signal peptide cleavage site. *Protein-Struct Funct Genet* 24: 165–177
10 Oliver DB (1993) SecA protein: autoregulated ATPase catalyzing preprotein insertion and translocation across the *Escherichia coli* inner membrane. *Mol Micro* 7: 159–165
11 Ahn T, Kim H (1996) Differential effect of precursor ribose binding protein of *Escherichia coli* and its signal peptide on the SecA penetration of the lipid bilayer. *J Biol Chem* 271: 12372–12379
12 Müller M (1992) Proteolysis in protein import and export: signal peptide processing in eu- and prokaryotes. *Experientia* 48: 118–129
13 Shelness GS, Lin L, Nicchitta CV (1993) Membrane topology and biogenesis of Eukaryotic signal peptidase. *J Biol Chem* 268: 5201–5208
14 Kalies K-U, Hartmann E (1996) Membrane topology of the 12- and 25-kDa subunits of the mammalian signal peptidase complex. *J Biol Chem* 271: 3925–3929
15 Wolfe PB, Wickner W, Goodman JM (1983) Sequence of the leader peptidase gene of *Escherichia coli* and the orientation of leader peptidase in the bacterial envelope. *J Biol Chem* 258: 12073–12080
16 Tjalsma H, Noback MA, Bron S, Venema G, Yamane K, van Dijl JM (1997) *Bacillus subtilis* contains four closely related type I signal peptidases with overlapping substrate specificities. *J Biol Chem* 272: 25983–25992
17 Kunst F, Ogasawara N, Moszer I, Albertini AM, Alloni G, Azevedo V, Bertero MG, Bessieres P, Bolotin A, Borchert S et al. (1997) The complete genome sequence of the Gram-positive bacterium *Bacillus subtilis*. *Nature* 390: 237–238
18 van Dijl JM, van den Bergh R, Reversma T, Smith H, Bron S, Venema G (1990) Molecular cloning of the *Salmonella typhimurium* lep gene in *Escherichia coli*. *Mol Gen Genet* 223: 233–240
19 Black MT, Munn JG, Allsop AE (1992) On the catalytic mechanism of prokaryotic leader peptidase 1. *Biochem J* 282: 539–543

20 Fleischmann RD, Adams MD, White O, Clayton RA, Kirkness EF, Kerlavage AR, Bult CJ, Tomb J-F, Dougherty BA, Merrick JM et al. (1995) Whole-genome random sequencing and assembly of *Haemophilus influenzae* Rd. *Science* 269: 496–512
21 Klug G, Jager A, Heck C, Rauhut R (1997) Identification, sequence analysis, and expression of the lepB gene for leader peptidase in *Rhodobacter capsulatus*. *Mol Gen Genet* 253: 666–673
22 Müller P, Ahrens K, Keller T, Klaucke A (1995) A TnphoA insertion within the *Bradyrhizobium japonicum* sipS gene, homologous to prokaryotic signal peptidases, results in extensive changes in the expression of PBM-specific nodulins of infected soybean (*Glycine max*) cells. *Mol Microbiol* 18: 831–840
23 Zhang YB, Greenberg B, Lacks SA (1997) Analysis of a *Streptococcus pneumoniae* gene encoding signal peptidase I and overproduction of the enzyme. *Gene* 194: 249–255
24 Cregg KM, Wilding EI, Black MT (1996) Molecular cloning and expression of the spsB gene encoding an essential type I signal peptidase from *Staphylococcus aureus*. *J Bacteriol* 178: 5712–5718
25 Bult CJ, White O, Olsen GJ, Zhou L, Fleischmann RD, Sutton GG, Blake JA, FitzGerald LM, Clayton RA, Gocayne JD et al. (1996) Complete genome sequence of the methanogenic archaeon, *Methanococcus jannaschii*. *Science* 273: 1058–1073
26 Smith DR, Doucette-Stamm LA, Deloughery C, Lee H-M, Dubois J, Aldredge T, Bashirzadeh R, Blakely D, Cook R, Gilbert K et al. (1997) Complete genome sequence of *Methanobacterium thermoautotrophicum* delta H: functional analysis and comparative genomics. *J Bacteriol* 179: 7135–7155
27 Wolfe PB, Silver P, Wickner W (1982) The isolation of homogeneous leader peptidase from a strain of *Escherichia coli* which overproduces the enzyme. *J Biol Chem* 257: 7898–7902
28 Bilgin N, Lee JI, Zhu HY, Dalbey RE, von Heijne G (1990) Mapping of catalytically important domains in *Escherichia coli* leader peptidase. *EMBO J* 9: 2717–2722
29 Black MT (1993) Evidence that the catalytic activity of prokaryote leader peptidase depends upon the operation of a serine-lysine catalytic dyad. *J Bacteriol* 175: 4957–4961
30 Tschantz WR, Sung M, Delgado-Partin VM, Dalbey RE (1993) A serine and a lysine residue implicated in the catalytic mechanism of the *Escherichia coli* leader peptidase. *J Biol Chem* 268: 27349–27354
31 Paetzel M, Dalbey RE (1997) Catalytic hydroxyl/amine dyads within serine proteases. *Trends Biochem Sci* 22: 28–31
32 Date T (1983) Demonstration by a novel genetic technique that leader peptidase is an essential enzyme in *Escherichia coli*. *J Bacteriol* 154: 76–83
33 Allsop AE, Brooks G, Bruton G, Coulton S, Edwards PD, Hatton IK, Kaura AC, McLean SD, Pearson ND, SmaleTC, Southgate R (1995) Penem inhibitors of bacterial signal peptidase. *Bioorg Med Chem Lett* 5: 443–448
34 Kuo DW, Chan HK, Wilson CJ, Griffin PR, Williams H, Knight WB (1993) *Escherichia coli* leader peptidase: production of an active form lacking a requirement for detergent and development of peptide substrates. *Arch Biochem Biophys* 303: 274–280
35 Tschantz WR, Paetzel M, Cao G, Suciu D, Inouye M, Dalbey RE (1995) Characterization of a soluble, catalytically active form of *Escherichia coli* leader peptidase: requirement of detergent or phospholipid for optimal activity. *Biochemistry* 34: 3935–3941
36 Paetzel M, Chernaia M, Strynadka N, Tschantz W, Cao G, Dalbey RE, James MN (1995) Crystallization of a soluble, catalytically active form of *Escherichia coli* leader peptidase. *Protein-Struct Funct Genet* 23: 122–125
37 Hoang V, Hofemeister J (1989) *Bacillus amyloliquefaciens* possesses a second type I signal peptidase with extensive sequence similarity to other *Bacillus* Spases. *Biochim Biophys Acta* 1269: 64–68
38 Evans EA, Gilmore R, Blobel G (1986) Purification of microsomal signal peptidase as a complex. *Proc Natl Acad Sci USA* 83: 581–585
39 Shelness GS, Kanwar YS, Blobel G (1988) cDNA-derived primary structure of the glycoprotein component of canine microsomal signal peptidase complex. *J Biol Chem* 263: 17063–17070
40 Shelness GS, Blobel G (1990) Two subunits of the canine signal peptidase complex are homologous to yeast Sec11 protein. *J Biol Chem* 265: 9512–9519
41 Greenburg G, Shelness GS, Blobel G (1989) A subunit of mammalian signal peptidase is homologous to yeast Sec11 protein. *J Biol Chem* 264: 15762–15765
42 Greenburg G, Blobel G (1994) cDNA-derived primary structure of the 25-kDa subunit of canine microsomal signal peptidase complex. *J Biol Chem* 269: 25354–25358
43 Baker RK, Lively MO (1987) Purification and characterization of chicken oviduct microsomal signal peptidase. *Biochemistry* 26: 8561–8567
44 Lively MO, Newsome AL, Nusier M (1994) Eukaryote microsomal signal peptidases. *Meth Enzymol* 244: 301–314
45 Newsome AL, McLean JW, Lively MO (1992) Molecular cloning of a cDNA encoding the glycoprotein of hen oviduct microsomal signal peptidase. *Biochem J* 282: 447–452
46 Böhni PC, Deshaies RJ, Schekman RW (1988) SEC11 is required for signal peptide processing and yeast cell growth. *J Cell Biol* 106: 1035–1042
47 YaDeau JT, Klein C, Blobel G (1991) Yeast signal peptidase contains a glycoprotein and the *Sec11* gene product. *Proc Natl Acad Sci USA* 88: 517–521

48 Meyer HA, Hartmann E (1997) The yeast SPC22/23 homolog Spc3p is essential for signal peptidase activity. *J Biol Chem* 272: 13159–13164
49 Fang H, Mullins C, Green N (1997) In addition to Sec11, a newly identified gene, SPC3, is essential for signal peptidase activity in yeast endoplasmic reticulum. *J Biol Chem* 272: 13152–13158
50 Mullins C, Meyer H-A, Hartmann E, Green E, Fang H (1996) Structurally related Spc1p and Spc2p of the yeast signal peptidase complex are functionally distinct. *J Biol Chem* 271: 29094–29099
51 Fang H, Panzner S, Mullins C, Hartmann E, Green N (1996) The homologue of the mammalian SPC12 is important for efficient signal peptidase activity in *Saccharomyces cerevisiae*. *J Biol Chem* 271: 16460–16465
52 van Dijl JM, De Jong A, Vehmaanperä J, Venema G, Bron S (1992) Signal peptidase I of *Bacillus subtilis*: Patterns of conserved amino acids in prokaryotic and eukaryotic type I signal peptidases. *EMBO J* 11: 2819–2828
53 Paetzel M, Strynadka NCJ, Tschantz WR, Casareno R, Bullinger PR, Dalbey RE (1997) Use of site-directed chemical modification to study an essential lysine in *Escherichia coli* leader peptidase. *J Biol Chem* 272: 9994–10003

Proteases: New Perspectives
V. Turk (ed.)

Proteasomes

A. Jennifer Rivett and Grant G.F. Mason

Department of Biochemistry, University of Bristol, University Walk, Bristol BS8 1TD, UK

Introduction

Proteasomes are large (700 kDa) multisubunit proteinase complexes which form the catalytic core of 26S proteasomes (approx. 2,000 kDa) and which constitute the major non-lysosomal protein degradation machinery in eukaryotic cells [1]. Related proteinases with similar structures have been purified from some archaebacteria [2] and are also found in eubacteria [3]. 20S proteasomes have a hollow cylindrical structure with catalytic sites located inside the central cavity. They have a novel catalytic mechanism and require protein substrates to be readily unfolded for degradation to small peptides inside the cylindrical structure. Eukaryotic 20S proteasomes are complex particles composed of 14–17 different subunits of 22–34 kDa. They are found in the nucleus and in the cytoplasm and can account for up to 1% of the soluble cellular protein. Regulatory complexes bind to the ends of the cylinder giving rise to 26S proteasomes and other regulated complexes. The 26S proteasome is responsible for the recognition and degradation of proteins involved in cell cycle regulation, transcriptional regulation and signal transduction, in many cases by ubiquitin-dependent mechanisms. Proteasomes also play an important role in the processing of antigens for presentation of peptides by the major histocompatibility complex (MHC) class I pathway.

Subunits

Proteasome subunits are encoded by members of the same gene family with the number of different proteasomal genes depending on the source. The proteasome of *Thermoplasma acidophilum* has two subunits, α (25.8 kDa) and β (22.3 kDa), which are 22% identical at the amino acid level and structurally related [2]. The subunit composition of eukaryotic proteasomes is much more complex. Yeast has 14 different subunits [4], seven of which are alpha type subunits closely related to the *Thermoplasma* alpha subunit and seven subunits more closely related to the *Thermoplasma* beta subunit and therefore described as beta subunits. Many of the beta subunits are synthesized as proproteins which are N-terminally processed during proteasome assembly. All yeast proteasome genes with the exception of Y13 are essential for cell viability. Each proteasome subunit shows a high degree of evolutionary conservation and presumably has the same function in different eukaryotes.

In addition to homologues of the yeast proteasome subunits, higher eukaryotes have three additional subunits, making a total of 17 different subunits which range in size from 22–34 kDa. Post-translational modifications add to the complexity of 2D-PAGE gel patterns which may show as many as 25 spots. Seven of the 17 mammalian proteasome genes encode homologues of the yeast alpha type subunits and ten encode beta type subunits (Fig. 1). The three additional beta type subunits in animal cells are non-essential subunits which are each closely related to one of the other beta subunits (Fig. 1) and which are induced by γ-interferon, a major

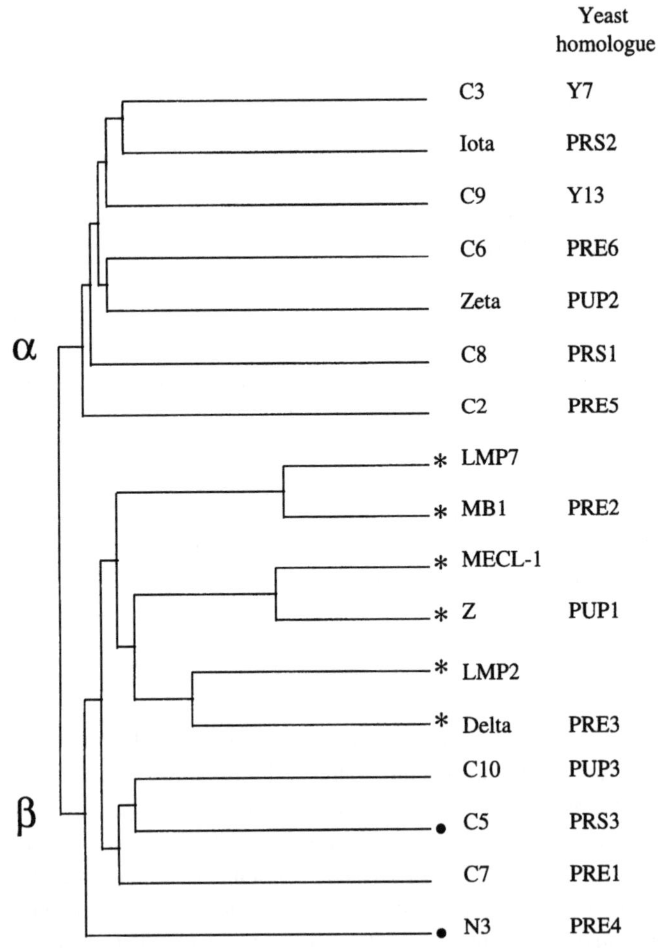

Figure 1. Dendogram of human proteasome subunits showing yeast homologues. Alignment of human proteasome subunit sequences obtained from the data base shows alpha and beta subunit groups. *Subunits containing a catalytically active N-terminal threonine residue demonstrated for yeast proteasome subunits [26]. •Other N-terminally processed beta subunits. The three γ-interferon inducible subunits, LMP2, LMP7 and MECL-1 replace their closely related subunits Delta, MB1 and Z, respectively, in immunoproteasomes which play an important role in antigen processing [6].

immunomodulatory cytokine. The discovery that two of them, LMP2 and LMP7, are encoded within the major histocompatability complex (MHC) class II region close to the two TAP (transporter associated with antigen processing) genes first indicated a role for proteasomes in the processing of antigens for presentation by the MHC class I pathway (see [5, 6] for reviews) and there is now evidence from many different laboratories that they do have this function. The other proteasome genes have different loci and functional analysis of some of these genes has so far failed to explain the regulation of their expression [1].

Structure

Electron microscopy of negatively stained proteasome preparations shows similarities between proteasomes isolated from *Thermoplasma* and several mammalian sources [7, 8]. Two different views of the molecule are observed consistent with a cylindrical structure of approximate dimensions 11 × 16 Å. One is a rectangular side-on view and the other a ring-shaped end-on view with 6–7 subunits visible around the rings. The proteasome isolated from *Thermoplasma acidophilum* has proved to be particularly useful for structural studies including X-ray crystallography [2] because it has only two different types of subunits which are arranged in four heptameric rings, $\alpha7\beta7\beta7\alpha7$. The structure of eukaryotic proteasomes is similar to that of the *Thermoplasma* proteasome but with seven different alpha and seven different beta subunits arranged as a complex dimer with two-fold symmetry. Interestingly, the overall architecture of proteasomes is similar to that of GroEL-like chaperones despite the different evolutionary origin of these molecules and differences in both number and size of the subunits [9].

Determination of the structure of the *Thermoplasma* proteasome [2] with a peptide aldehyde inhibitor (calpain inhibitor I, N-acetyl-Leu-Leu-norleucinal) bound has clearly demonstrated that the beta subunits contain the catalytic sites which are located on the inside of the hollow cylindrical structure. The ends of the cylinder are open but there is restricted access to the interior suggesting that proteins must be unfolded to enter the central cavity where the catalytic sites are located. Surprisingly, the recent X-ray crystallographic analysis of the yeast proteasome [10] has shown that in this case the ends of the cylinder are not open. Instead, there are holes in the sides of the cylinder. The yeast proteasome structure also shows that unlike the *Thermoplasma* proteasome which has 14 catalytic sites (one in each beta subunit) arranged around the central cavity, only three different subunits bind the calpain inhibitor I, consistent with the view that only three of the beta subunits are catalytically active (see below) and there are therefore only six catalytic sites in eukaryotic proteasomes (Fig. 2). The eukaryotic proteasome has the advantage of multiple sites having different specificities but it is not clear why the total number of catalytic sites should be reduced in eukaryotic proteasomes.

Catalytic activities

The proteasome was identified in early studies as having multiple distinct catalytic sites (reviewed in [11, 12]). The multicatalytic endopeptidase activities of proteasomes are usually

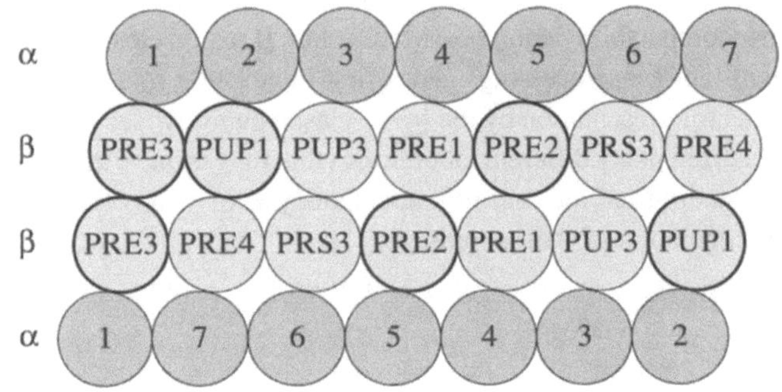

Figure 2. Arrangement of catalytic subunits in the yeast proteasome. A schematic diagram showing the relative position of yeast proteasome subunits in the cylindrical structure of the yeast proteasome [10] with the three catalytic subunits (PRE3, PUP1 AND PRE2) [26] circled in bold. The subunits are arranged as a complex dimer in the $\alpha 7\beta 7\beta 7\alpha 7$ structure.

assayed with a variety of synthetic peptide substrates and the distinction between activities was made on the basis of their inhibitor sensitivity. Eukaryotic proteasomes contain three principal activities described as trypsin-like, chymotrypsin-like and peptidylglutamyl peptide hydrolase activity. Additional activities including a branched chain amino acid preferring, a small neutral amino acid preferring, a second petidylglutamyl peptide hydrolase and an acidic chymotrypsin-like activity have also been described [13–15]. The most sensitive assays use peptidyl-amidomethylcoumarins such as suc-Leu-Leu-Val-Tyr-7-amido-4-methylcoumarin for the chymotrypsin-like activities. Although assignment of some distinct catalytic activities to specific subunits of the complex (see below) and the determination of the structure of the yeast proteasome have helped to understand the number and arrangement of the catalytic sites in eukaryotic proteasomes, it is still not entirely clear exactly how many different types of activity there are. The situation is complicated by the apparent overlap in specificity of some sites, by close interaction between different beta subunits and by the flexibility of the molecule [8, 10]. Moreover, in animal cells there is further complexity due to the heterogeneity and changes in specificity introduced by the γ-interferon-inducible subunits [6, 16].

Inhibitors

In early studies (reviewed in [11, 12]) proteasomes were found to be generally rather poorly inhibited by reagents developed to inhibit serine and cysteine proteases, with the exception of 3,4-dichloroisocoumarin, which was found to be effective on some but not all of the distinct peptidase activities [13, 14]. A variety of peptide aldehydes, peptidyl chloromethanes, and peptidyl boronic acids have subsequently been developed for use as proteasome inhibitors and further characterization of the inhibition of different activities of mammalian proteasomes has confirmed suspected differences in the reactivity of proteasomes containing different amounts of the variable catalytic subunits LMP2, LMP7 and MECL-1 subunits ([16, 17] and see below).

Great care must be taken when interpreting data obtained using protease inhibitors with cultured cells while attempting to estimate the contribution of proteasomes to the degradation of individual proteins as many protease inhibitors are not specific. For example, the peptide aldehyde Z-leucinyl-leucinyl-leucinal (MG132) [18] and calpain inhibitor I are effective inhibitors of proteasomes but also inhibit other proteases. Non-peptidic cell permeable inhibitors of proteasomes, lactacystin [19] and aclacinomycin A [20] may be more specific. The first characterization of the inhibition by lactacystin showed it to be a relatively specific reagent for inhibiting proteasomes [19], and more recent studies with Z-Leu-Leu-Leu vinyl sulphone have demonstrated that the proteasome is the major target for this inhibitor in several different cell types [21]. Such inhibitors are valuable for assigning functions of proteasomes in animal cells and to aid identification of proteins degraded by the ubiquitin-proteasome pathway (see below).

Catalytic mechanism

Determination of the structure of the *Thermoplasma* proteasome provided the first insight into the catalytic mechanism of proteasomes. The position of a peptide aldehyde inhibitor, calpain inhibitor I, implicated the N-terminal threonine residue of the beta subunit as the catalytic nucleophile [2] and it is presumed that the reaction proceeds through an acyl enzyme intermediate involving attachment to the side chain hydroxyl group of Thr 1. The proteasome was the first example of a threonine protease and this novel catalytic mechanism explains its unusual inhibitor characteristics. The importance of Thr 1 has been confirmed by site-directed mutagenesis of the *Thermoplasma* beta subunit [22]. A role for Lys 33, Glu 17 and Asp 166 has been suggested and the recently determined structure of the yeast proteasome [10] shows a water molecule which may be involved in the deacylation step.

Interestingly, the recently described HsIV/HsIU (ClpQ/Y) protease fom *E. coli* has a similar catalytic mechanism and a similar structure but is made up of two rings of six subunits in each [23]. The structure of proteasomes is also similar to that of other Ntn (N-terminal nucleophile) hydrolases which include glutamine PRPP amidotransferase, penicillin acylase, and aspartylglucosaminidase [24]. In some cases an N-terminal serine acts as the catalytic nucleophile and replacement of the Thr1 of the *Thermoplasma* beta proteasome subunit with Ser also produces active enzyme [22, 25]. It is therefore assumed that the absolute conservation of threonine in position 1 in all proteasome catalytic subunits is due to a requirement for threonine for correct N-terminal processing of these subunits (see below).

Identification of catalytic components

Early studies with yeast mutants defective in proteasome activities identified subunits Pre1 and Pre2 as playing a role in chymotrypsin-like activity and Pre3 and Pre4 as subunits responsible for peptidylglutamyl peptide hydrolase activity [26]. Of these, only Pre2 and Pre3 have a catalytic threonine residue. However, the Pre1 and Pre4 subunits are now known to be the nearest neighbours of catalytic subunits and could therefore easily influence their activity (Fig. 2).

Results of recent studies involving site directed mutagenesis of yeast proteasome subunits have shown that the trypsin-like, chymotrypsin-like and peptidylglutamyl peptide hydrolase activities can be ascribed, principally, to the subunits Pup1, Pre2 and Pre3, respectively. A relationship between trypsin-like and chymotrypsin-like activities has been noted in both yeast and mammalian proteasomes [10, 17, 26]. In the mammalian proteasomes, MB1, the homologue of yeast Pre2 (Fig. 1), was first identified as a catalytic subunit involved in chymotrypsin-like activity using radiolabelled lactacystin [19]. Subsequent experiments have shown that this inhibitor can in fact modify all the catalytic subunits as can Z-Leu-Leu-Leu-vinylsulphone [21]. Homologous subunits are believed to have the same functions in yeast and animal proteasomes (Fig. 1).

Subunit processing

Proteasome beta subunits, but not alpha subunits, are mostly synthesized as proproteins which are cleaved during the assembly of subunits to form active proteasomes. The length of the propeptide varies for different subunits from just a few amino acids to more than 60. The *Thermoplasma* proteasome beta subunit has a propeptide of eight amino acids in length which is not required for the assembly of active proteasomes. Processing is required to expose the N-terminal catalytic threonine residue. Only threonine is fully functional in both processing and proteolysis [25] but the Thr 1 → Cys mutant is also active in processing, which is believed to be autocatalytic but may be intermolecular [25].

The processing of the three yeast proteasome catalytic beta subunits, like that of the *Thermoplasma* beta subunit, involves cleavage between the threonine at position 1 and the last glycine of the pro-sequence and is autocatalytic [26], possibly requiring the nucleophilic water molecule [10]. Cleavage of Pre4, which does not itself contain a catalytic threonine residue, is catalysed by the closest catalytic site [26]. In mammalian cells eight of the beta subunits have propeptides and three of these are the non-essential γ-interferon inducible subunits.

Proteasome subunit assembly

Assembly of active proteasomes seems to involve formation of the alpha subunit rings followed by addition and processing of beta subunits. From studies with the *Thermoplasma* proteasome subunits expressed in *E. coli* it is clear that alpha subunits but not beta subunits are able to form rings on their own [27]. Presumably beta subunits are then added and processed, although with the *Thermoplasma* proteasome the propeptide is not essential for assembly. Beta propeptides are required for the correct assembly of the more complex eukaryotic proteasome [28] where assembly of proteasomes is thought to occur via half molecules [28, 29]. Isolated 16S precursor complexes which are imagined to be intermediates in the assembly process are capable of undergoing some processing of the beta subunits *in vitro* [29]. Different beta type subunits may have different kinetics of processing with some precursor forms being more obvious than others. The N3 subunit for example appears only as the mature form in intact proteasomes [30], an observation which may be explained by the fact that N3 (the Pre4 homologue) is itself not catalytic [26]. The heat shock cognate protein Hsc73 has been proposed to play some role in the assembly of mammalian proteasomes [29] as well as in their degradation by lysosomes [31].

The 26S proteasome

The eukaryotic 20S proteasome is able to associate with 19S regulatory complexes (otherwise referred to as ATPase complexes or PA700) which bind at each end of the cylinder to form the 26S proteasome. The 26S proteasome catalyses ATP-dependent protein degradation [32, 33]. The 19S regulatory complexes contain approximately 20 proteins varying from 25–110 kDa (Tab. 1). Six of these subunits are ATPases, while the other non-ATPase subunits are structurally unrelated to each other and most of them are unrelated to other known proteins [33]. The regulatory complexes stimulate 20S proteasome activities but do not themselves possess proteolytic activity. The regulatory subunits are presumably involved in the recognition and unfolding of proteins for degradation within the catalytic core of the proteinase and include the subunit S5a (MBP1) which has been found to bind multi-ubiquitin chains but which is apparently not essential for all ubiquitin-dependent degradation [34]. Several of the regulatory subunits are not essential for cell viability [33]. Another subunit contains an isopeptidase activity which is believed to be responsible for the removal of the ubiquitin chains when ubiquitinated substrate proteins are degraded [35]. It is not clear that PA700 together with 20S proteasome is equiva-

Table 1. Known regulatory subunits of the 26S proteasome

Subunit	Mammalian	Yeast
ATPase subunits		
S4	p56	YTA5/MTS2/YHS4
S6	TBP-7/p48	YTA2
	TBP-1	YTA1
S7	MSS1	YTA3/CIM5
S8	TRIP1/p45	SUG1/CIM3
S10b	p42	SUG2/PCS1
Non-ATPase subunits		
S1	p112	SEN3
S2	TRAP2/p97/55.11	NAS1/HRD2
S3	p91A/p58	SUN2
	p55	NAS5
S5a	Mcb1/MBP1	SUN1
S5b	p50.5	
	p44.5	NAS4
S10	p44/HUMORF07	
	p40.5	
S12	Mov34/p40	NAS3
S14	p31	NIN1/MTS3
	p28	
	p27	NAS2

lent to the 26S proteasome. An additional complex, a PA700-dependent modulator of proteasome activity (300 kDa), has recently been purified and found to contain two 26S proteasome ATPase subunits, TBP1 and p42, and a novel 27 kDa protein [36]. Some of the regulatory subunits may also have other functions. Several of the ATPase subunits have been described as transcriptional regulators but this may be due to their role in the proteolysis of transcription factors rather than any additional direct function. SUG1, which is one of the ATPases has recently been reported to have DNA helicase activity [37].

A γ-interferon-inducible activator

Another proteasome activator complex (approximately 180 kDa) referred to as PA28, REG or the 11S regulator has been identified [38, 39] (Fig. 3). This is a ring-shaped hexameric complex which binds to the ends of the 20S proteasome cylinder and causes up to 100-fold activation of peptidase activities but apparently has little effect on protein degradation. The hexamer contains two different types of subunit, α and β. The two subunits are approximately 50% identical and have similar molecular masses (30 kDa). The C-terminal peptide of the α subunits is important for REG binding to proteasomes. The α subunits on their own can form a heptameric complex which stimulates proteasome activities although not as much as the α β heterohex-

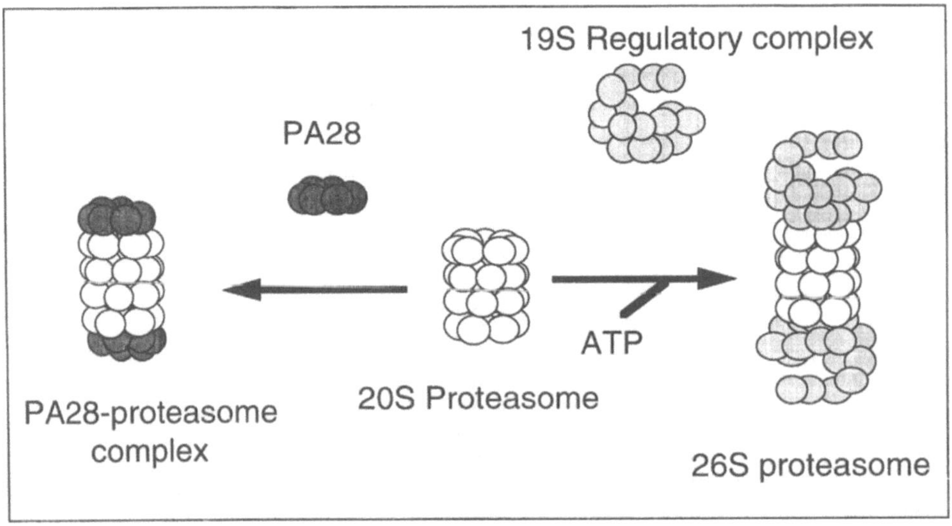

Figure 3. Regulatory complexes of mammalian proteasomes. 20S proteasomes form the catalytic core of 26S proteasomes and PA28-proteasome complexes. 26S proteasomes have a 19S regulatory complex (ATPase complex, PA700) associated with each end of the 20S proteasome cylinder. The 19S regulatory complexes are composed of approximately 20 different polypeptides [33] including six ATPase subunits, a ubiquitin chain binding subunit and an isopeptidase. The size of the subunits which are mostly unrelated to each other varies from 25–110 kDa. The hexameric PA28 (REG or 11S regulator) complex is made up of two different but related types of subunit of around 30 kDa.

amer which is the natural form [38]. Another related protein, Ki antigen [39] or REGγ, binds more weakly than the αβ complex [38]. Like the non-essential proteasome subunits, REG is not found in yeast, is γ-interferon-inducible and has been implicated in the antigen processing pathway (see below).

Subcellular distribution of proteasomes

Results of localization studies with many different cell types using immuno-histochemical, immunofluorescence and immunogold electron microscopy techniques have shown that proteasomes are present in both the nucleus and the cytoplasm of eukaryotic cells and that some of the cytosolic proteasomes are closely associated with the endoplasmic reticulum (ER) (reviewed in [40]). Immunogold electron microscopy with polyclonal anti-proteasome antibodies [41], as well as studies with subcellular fractions isolated from rat liver [42] have shown ER-associated proteasomes to be located on the outside of the smooth ER and the cis-Golgi. Their association with the ER probably reflects functions in antigen processing and in ER-associated protein degradation (see below). Regulatory complexes of 26S proteasomes are also found in both the nucleus and cytoplasm [43]. In cultured cells there are changes in proteasome localization during progression of the cell cycle [44, 45] and in mitosis they have a perichromosomal distribution [45].

It is possible that phosphorylation of proteasome subunits may play a role in determining their distribution. Phosphorylation of the alpha subunit C3 has recently been implicated in transport of proteasomes to the nucleus [46] but the phosphorylation of this subunit is not observed in all cell types [47]. A number of other subunits also have functional nuclear localization signals [48]. Phosphorylation of proteasome alpha subunits C8 and C9 may play a role in the assembly of the 26S proteasome [47].

Functions

The functions of eukaryotic proteasomes have been analysed using genetic and biochemical methods both in yeast and animal cells. Yeast mutants defective in proteasome activities or in ATPase subunits of the 26S proteasome have proved useful as have the more selective of the proteasome inhibitors in animal cells. Proteasomes are now known to be involved in the activation and turnover of transcription factors, in cell cycle control and in antigen processing for presentation by the MHC class I pathway. They are responsible for the degradation of many short-lived proteins and also for some ER-associated protein degradation.

Degradation of regulatory proteins

Yeast mutants in ATPase subunits of the 26S proteasome cause cells to arrest in mitosis [49, 50] and it is now clear that proteolysis by proteasomes plays a major role in regulating the cell cycle. Regulatory proteins such as cyclins and transcription regulators including NFκB, IκB, STAT proteins and the tumour suppressors p53 and retinoblastoma protein are all known substrates of the ubiquitin-proteasome pathway (reviewed in [51, 52]). In this pathway proteins are targetted for degradation by attachment of a multi-ubiquitin chain in a process which involves the proteins E1, E2 and E3, ubiquitin, and ATP. Interestingly, signal-induced degradation of IκB requires ubiquitination but basal turnover was recently found to occur by a ubiquitin-independent pathway [53]. Ornithine decarboxylase, the enzyme which catalyses the rate limiting step in polyamine biosynthesis, is degraded by the 26S proteasome in an ATP-dependent but ubiquitin-independent process [54].

Role of immunoproteasomes in antigen processing

Three of the ten possible beta subunits in animal cells are non-essential subunits. These three extra proteasome subunits (LMP2, LMP7 and MECL-1) are induced by γ-interferon and can replace three other closely related catalytic subunits (delta, MB1, and Z) in 20S proteasomes to give rise to immunoproteasomes ([55, 56] and see [6] for review). The γ-interferon inducible subunits are incorporated into newly synthesized proteasomes rather than directly replacing subunits in existing complexes [55]. From many studies by a number of different groups it is now clear that the presence of the γ-interferon inducible subunits alters the peptidase actvities of proteasomes and also that immunoproteasomes play an important role in antigen processing, but are not absolutely essential (reviewed in [5, 6]). Cells lacking the non-essential subunits are defective in the presentation of viral antigens [57–59].

The γ-interferon inducible regulator protein complex, called PA28 or REG, has also been shown to facilitate processing of viral antigens for presentation by the MHC class I pathway [60] and to lead to the formation of dominant MHC ligands. Peptides displayed at the cell surface in association with MHC class I molecules target cells for destruction by cytotoxic T lymphocytes and one mechanism by which some viruses may evade immune detection could be by the production of proteins (eg HIV tat protein) which interfere with proteasome function [61].

ER-associated protein degradation

Recent studies on ER-associated protein degradation (reviewed in [62]) have implicated proteasomes as the degradation machinery involved. ER-associated protein degradation of some resident ER proteins such as Sec61p [63] and HMGCoA reductase [64], as well as of some proteins that fail to be secreted or to fold properly such as the cystic fibrosis transmembrane regulator [65] and MHC class I molecules [66], involves translocation of the protein out of the ER for degradation by proteasomes on the cytosolic side of the membrane. In some cases this involves ubiquitination by ubiquitin-conjugating enzymes which have been found in the ER membrane [64].

References

1 Coux O, Tanaka K, Goldberg AL (1996) Structure and functions of the 20S and 26S proteasomes. *Annu Rev Biochem* 65: 801–847

2 Löwe J, Stock D, Jap B, Zwickl P, Baumeister W, Huber R (1995) Crystal structure of the 20S proteasome from the archaeon *T. acidophilum* at 3.4 Å resolution. *Science* 268: 533–539

3 Tamura T, Nagy I, Lupas A, Lottspeich F, Cejka Z, Schoofs G, Tanaka K, Mot RD, Baumeister W (1995) The first characterization of a eubacterial proteasome: the 20S complex of *Rhodococcus* sp. strain NI86/21. *Curr Biol* 5: 766–774

4 Hilt W, Heinemeyer W, Wolf DH (1993) Studies on the yeast proteasome uncover its basic structural features and multiple *in vivo* functions. *Enzyme Protein* 47: 189–201

5 York IA, Rock KL (1996) Antigen processing and presentation by the class I major histocompatibility complex. *Annu Rev Immunol* 14: 369–397

6 Tanaka K, Tanahashi N, Tsurumi C, Shimbara N (1997) Proteasomes and antigen processing. *Adv Immunol* 64: 1–38

7 Peters JM (1994) Proteasomes: protein degradation machines of the cell. *Trends Biochem Sci* 19: 377–382

8 Djaballah H, Rowe AJ, Harding SE, Rivett AJ (1993) The multicatalytic proteinase complex (proteasome): structure and conformational changes associated with changes in proteolytic activity. *Biochem J* 292: 857–862

9 Weissman JS, Sigler PB, Horwich AL (1995) From the cradle to the grave: ring complexes in the life of a protein. *Protein Sci* 268: 523–524

10 Groll M, Ditzel L, Löwe J, Stock D, Bochtler M, Bartunik HD, Huber R (1997) Structure of 20S proteasome from yeast at 2.4 Å resolution. *Nature* 386: 463–471

11 Rivett AJ (1989) The multicatalytic proteinase of mammalian cells. *Arch Biochem Biophys* 268: 1–8

12 Orlowski M (1990) The multicatalytic proteinase complex, a major extralysosomal proteolytic system. *Biochemistry* 29: 10289–10297

13 Rivett AJ, Savory PJ, Djaballah H (1994) Multicatalytic endopeptidase complex (proteasome). *Meth Enzymol* 244: 331–350

14 Orlowski M, Cardozo C, Michaud C (1993) Evidence for the presence of five distinct proteolyic components in the pituitary multicatalytic proteinase complex. Properties of two components cleaving bonds on the carboxyl side of branched chain and small neutral amino acids. *Biochemistry* 32: 1563–1572

15 Cardozo C, Chen WE, Wilk S (1996) Cleavage of Pro-X and Glu-X bonds catalyzed by the branched chain amino acid preferring activity of the bovine pituitary multicatalytic proteinase complex (20S proteasome). *Arch Biochem Biophys* 334: 113–120

16 Elutieri AM, Kohanski RA, Cardozo C, Orlowski M (1997) Bovine spleen multicatalytic proteinase complex (proteasome). Replacement of X,Y and Z subunits by LMP1, LMP2 and MECL1 and changes in properties and specificity. *J Biol Chem* 272: 11824–11831

17 Reidlinger J, Pike AM, Savory PJ, Murray RZ, Rivett AJ (1997) Catalytic properties of 26S and 20S proteasomes and radiolabelling of MB1, LMP7 and C7 subunits associated with trypsin-like and chymotrypsin-like activities. *J Biol Chem* 272: 24899–24905

18 Rock KL, Gram C, Rothstein L, Clark K, Stein R, Dick L, Hwang D, Goldberg AL (1994) Inhibitors of the proteasome block the degradation of most cell proteins and the generation of peptides presented on MHC class I molecules. *Cell* 78: 761–771

19 Fenteany G, Standaert RF, Lane WS, Choi S, Corey EJ, Schreiber SL (1995) Inhibition of proteasome activities and subunit-specific amino-acid terminal threonine modification by lactacystin. *Science* 268: 726–731

20 Figueiredo-Pereira ME, Chen WE, Li J, Johdo O (1996) The antitumor drug aclacinomycin A, which inhibits the degradation of ubiquitinated proteins, shows selectivity for the chymotrypsin-like activity of the bovine pituitary 20 S proteasome. *J Biol Chem* 271: 16455–16459

21 Bogyo M, McMaster JS, Gaczynska M, Tortorella D, Goldberg AL, Ploegh H (1997) Covalent modification of the active site threonine of proteasomal β subunits and the *Escherichia coli* homolog HslV by a new class of inhibitors. *Proc Natl Acad Sci USA* 94: 6629–6634

22 Seemüller E, Lupas A, Stock D, Löwe J, Huber R, Baumeister W (1995) Proteasome from *Thermoplasma acidophilum*: a threonine protease. *Science* 268: 579–582

23 Kessel M, Wu WF, Gottesman S, Kocsis E, Steven AC, Maurizi MR (1996) Six-fold rotational symmetry of ClpQ, the *E. coli* homolog of the 20S proteasome, and its ATP-dependent activator, ClpY. *FEBS Lett* 398: 274–278

24 Brannigan JA, Dodson G, Duggleby HJ, Moody PCE, Smith JL, Tomchick DR, Murzin AG (1995) A protein catalytic framework with an N-terminal nucleophile is capable of self-activation. *Nature* 378: 416–419

25 Seemüller E, Lupas A, Baumeister W (1996) Autocatalytic processing of the 20S proteasome. *Nature* 382: 468–470

26 Heinemeyer W, Fischer M, Krimmer T, Stachou U, Wolf DH (1997) The active sites of the eukaryotic 20S proteasome and their involvement in subunit precursor processing. *J Biol Chem* 272: 25200–25209

27 Zwickl P, Kleinz J, Baumeister W (1994) Critical elements in proteasome assembly. *Nat Struct Biol* 1:

765–770
28 Chen P, Hochstrasser M (1996) Autocatalytic subunit processing couples active site formation in the 20S proteasome to completion of assembly. *Cell* 86: 961–972
29 Schmidtke G, Schmidt M, Kloetzel PM (1997) Maturation of mammalian 20S proteasome: purification and characterization of 13S and 16S proteasome precursor complexes. *J Mol Biol* 268: 95–106
30 Thomson S, Rivett AJ (1996) Processing of a mammalian proteasome beta-type subunit, N3. *Biochem J* 315: 733–738
31 Cuervo AM, Palmer A, Rivett AJ, Knecht E (1995) Degradation of proteasomes by lysosomes in rat liver. *Eur J Biochem* 227: 792–800
32 Hoffman L, Rechsteiner M (1996) Nucleotidase activities of the 26S proteasome and its regulatory complex. *J Biol Chem* 271: 32538–32545
33 Tanaka K, Tsurumi C (1997) The 26S proteasome: subunits and functions. *Mol Biol Rep* 24: 3–11
34 Kominami K-I, Okura N, Kawamura M, DeMartino GN, Slaughter CA, Shimbara N, Chung CH, Fujimuro M, Yokosawa H, Shimizu Y et al. (1997) Yeast counterparts of subunits S5a and p58 (S3) of the human 26S proteasome are encoded by two multicopy suppressors of *nin1-1*. *Mol Biol Cell* 8: 171–187
35 Lam YA, Xu W, DeMartino GN, Cohen RE (1997) Editing of ubiquitin conjugates by an isopeptidase in the 26S proteasome. *Nature* 385: 737–740
36 DeMartino GN, Proske RJ, Moomaw CR, Strong AA, Song X, Hisamatsu H, Tanaka K, Slaughter CA (1996) Identification, purification, and characterization of a PA700-dependent activator of the proteasome. *J Biol Chem* 271: 3112–3118
37 Fraser RA, Rossignol M, Heard DJ, Egly J-M, Chambon P (1997) SUG1, a putative transcriptional mediator and subunit of the PA700 proteasome regulatory complex, is a DNA helicase. *J Biol Chem* 272: 7122–7126
38 Realini C, Jensen CC, Zhang Z-G, Johnston SC, Knowlton JR, Hill CP, Rechsteiner M (1997) Characterization of recombinant REGα, REGβ, and REGγ proteasome activators. *J Biol Chem* 272: 25483–25492
39 Ahn K, Tanahashi N, Akiyama K, Hismatsu H, Noda C, Tanaka K, Chung C, Shimbara N, Willy P, Mott J, Slaughter C, DeMartino G (1995) Primary structures of two homologous subunits of PA28, a gamma-interferon-inducible protein activator of the 20S proteasome. *FEBS Lett* 366: 37–42
40 Rivett AJ, Knecht E (1993) Proteasome location. *Curr Biol* 3: 127–129
41 Rivett AJ, Palmer A, Knecht E (1992) Electron microscopic localization of the multicatalytic proteinase in rat liver and in cultured cells. *J Histochem Cytochem* 40: 1165–1172
42 Palmer A, Rivett AJ, Thomson S, Hendil KB, Butcher GW, Fuertes G, Knecht E (1996) Subpopulations of proteasomes in rat liver nuclei, microsomes and cytosol. *Biochem J* 316: 401–407
43 Peters JM, Franke WW, Kleinschmidt JA (1994) Distinct 19S and 20S subcomplexes of the 26S proteasome and their distribution in the nucleus and cytoplasm. *J Biol Chem* 269: 7709–7718
44 Amsterdam AF, Pitzer W, Baumeister W (1993) Changes in intracellular localization of proteasomes in immortalized ovarian granulosa cells during mitosis is associated with a role in cell cycle control. *Proc Natl Acad Sci USA* 90: 99–103
45 Palmer A, Mason GGF, Paramio J, Knecht E, Rivett AJ (1994) Changes in proteasome localization during the cell cycle. *Eur J Cell Biol* 64: 163–175
46 Benedict CM, Clawson GA (1996) Nuclear multicatalytic proteinase subunit RRC3 is important for growth regulation in hepatocytes. *Biochemistry* 35: 11612–11621
47 Mason GGF, Hendil KB, Rivett AJ (1996) Phosphorylation of proteasomes in animal cells: identification of phosphorylated subunits and effect of phosphorylation on proteolytic activity. *Eur J Biochem* 238: 453–462
48 Nederlof PM, Wang H-R, Baumeister W (1995) Nuclear localization signals of human and *Thermoplasma* proteasomal α subunits are functional *in vitro*. *Proc Natl Acad Sci USA* 92: 12060–12064
49 Ghislain M, Udvardy A, Mann C (1993) *S. cerevisiae* 26S protease mutants arrest cell division in G2/metaphase. *Nature* 366 358–362
50 Gordon C, McGurk G, Dillon P, Rosen C, Hastie ND (1993) Defective mitosis due to a mutation in the gene for a fission yeast 26S protease subunit. *Nature* 366 3655–3657
51 Ciechanover A (1994) The ubiquitin-proteasome proteolytic pathway. *Cell* 79: 13–21
52 Hochstrasser M (1997) Ubiquitin-dependent protein degradation. *Annu Rev Genet* 30: 405–439
53 Krappmamm D, Wulczyn FG, Scheidereit C (1996) Different mechanisms control signal-induced degradation and basal turnover of the NF-κB inhibitor IκBα *in vivo*. *EMBO J* 15: 6716–6726
54 Murakami Y, Matsufuji S, Kameji T, Hayashi S, Igarashi K, Tamura T, Tanaka K, Ichihara A (1992) Ornithine decarboxylase is degraded by the 26S proteasome without ubiquitination. *Nature* 360: 597–599
55 Früh K, Gossen M, Wang K, Bujard H, Peterson PA, Yang Y (1994) Displacement of housekeeping proteasome subunits by MHC-encoded LMPs: a newly discovered mechanism for modulating the multicatalytic proteinase complex. *EMBO J* 13: 3236–3244
56 Belich MP, Glynne RJ, Senger G, Sheer D, Trowsdale J (1994) Proteasome components with reciprocal expression to that of the MHC-encoded LMP proteins. *Curr Biol* 4: 769–776
57 Van Kaer L, Ashton-Richardt PG, Eichelberger M, Gaczynska M, Nagashima K, Rock KL, Goldberg AL, Doherty PC, Tonegawa S (1994) Altered peptidase and viral-specific T cell response in LMP2 mutant mice. *Immunity* 1: 533–541
58 Sibille C, Gould K, Willard-Gallo K, Thomson S, Rivett AJ, Powis S, Butcher GW, DeBaetselier P (1995)

LMP2 proteasome subunit required for efficient class I antigen-processing in a T-cell lymphoma. *Curr Biol* 5: 923–930
59 Fehling HJ, Swat W, Laplace C, Kuhn R, Rajewsky K, Muller U, von Boehmer H (1994) MHC class I expression in mice lacking the proteasome subunit LMP-7. *Science* 265: 1234–1237
60 Groettrup M, Soza A, Eggers M, Kuehn L, Dick TP, Schild H, Rammensee HG, Kosinowski UH, Kloetzel PM (1996) Role for the proteasome regulator PA28alpha in antigen presentation. *Nature* 381: 166–168
61 Seeger M, Ferrell K, Rainer F, Dubiel W (1997) HIV-1 Tat inhibits the 20S proteasome and its 11S regulator-mediated activation. *J Biol Chem* 272: 8145–8148
62 Brodsky JL, McCracken AA (1997) ER-associated and proteasome-mediated protein degradation: how two topologically restricted events came together. *Trends Cell Biol* 7: 151–156
63 Biederer T, Volkswein C, Sommer T (1996) Degradation of subunits of the Sec61p complex, an integral component of the ER membrane, by the ubiquitin-proteasome pathway. *EMBO J* 15: 2069–2076
64 Hampton RY, Gardner RG, Rine J (1996) Role of 26S proteasome and HRD genes in the degradation of 3-hydroxy-3-methylglytaryl-CoA reductase, an integral endoplasmic reticulum membrane protein. *Mol Biol Cell* 7: 2029–2044
65 Ward C, Omura S, Kopito RR (1995) Degradation of CFTR by the ubiquitin-proteasome pathway. *Cell* 83: 121–127
66 Wiertz EJHJ, Tortorella D, Bogyo M, Yu J, Mothes W, Jones TR, Rapoport TA, Ploegh HL (1996) Sec61-mediated transfer of a membrane protein from the endoplasmic reticulum to the proteasome for destruction. *Nature* 384: 432–438

Proteases: New Perspectives
V. Turk (ed.)

Cathepsin E and cathepsin D

Kenji Yamamoto

Department Pharmacology, Kyushu University Faculty of Dentistry, Higashi-ku, Fukuoka 812-8582, Japan

Introduction

Aspartic proteinases are produced by a number of cells and tissues. These enzymes share a high degree of similarity which involves primary structures, and most of them are active predominantly in the acidic pH range. Eukaryotic aspartic proteinases (i.e. the pepsin family) are usually bilobed molecules with the active-site cleft located between the lobes, and each lobe contains one aspartic acid residue essential for catalytic activity [1, 2]. On the other hand, retroviral aspartic proteinases (i.e. the retropepsin family) are active as a homodimer which comprises small identical molecules analogous to a single lobe of the pepsin family members [1, 3, 4]. Mammalian aspartic proteinases are usually divided into two groups. Many comprise a secretory group consisting of proteins that fulfil their physiological function in extracellular spaces (e.g. pepsin, gastricsin, chymosin and renin) and the other a non-secretory group consisting of proteins that effect their function primarily within the cell of origin [e.g. cathepsin D (CD) and cathepsin E (CE)]. Most aspartic proteinases are synthesized on the membrane-bound ribosomes of the endoplasmic reticulum (ER) as preproenzymes containing a signal peptide required for ER translocation and a propeptide for control of enzyme activity during transport. Proteolytic removal of the propeptide results in the activation of the enzyme.

CD represents a major portion of the proteolytic activity in the lysosomal compartment. The enzyme is usually localized in lysosomes of various tissues, although the level of expression varies with different cell types [5, 6]. CD is also detected in endosomes of certain cell types, such as macrophages [7] and hepatocytes [8]. About 90% of the total CD activity in lysosomes is soluble [6, 7], whereas about 20% of the activity is membrane-associated in endosomes [7]. Lysosomal CD is believed to play an important physiological role in proteolysis for intra- and extracellular proteins, whereas endosomal CD is postulated to play a role in proteolytic processing of foreign antigens, invariant chain, and prohormones [7, 9–12]. The overexpression and excretion of CD has been implicated in pathological processes such as inflammation [5], tumor progression [13–19] and the release of β-amyloid peptides from the amyloid precursor protein (APP) in Alzheimer's disease [20, 21]. CD is known to be synthesized on the membrane-bound ribosomes in the rough ER as a preproenzyme that is processed to pro-CD having an apparent molecular mass of 53 kDa by removal of the signal peptide during the passage into the ER. It then undergoes a series of modifications during transport to and through the Golgi. During transport, the pro-CD is proteolytically processed to generate the active forms, mostly by a two-step cleavage. The first cleavage is thought to occur in a prelysosomal compartment to yield a single chain CD with an apparent molecular mass of 44 kDa. The second takes place in lysosomes, yielding a two-chain enzyme (the 31 kDa and 14 kDa chains). Rat CD has a single chain form, whereas CD from human, bovine and porcine consists of two chains. Thus whether CD has a single or a double chain form appears to depend on the species of origin. The two chains are connected with a proteinase-prone peptide loop, which is missing in the precursors of single-chain CD. Available evidence also suggests that targeting of CD to lysosomes is mediated by either mannose-6-phosphate receptor (MPR)-dependent or independent pathways [22–24].

CE is a relatively newly characterized intracellular aspartic proteinase consisting of two identical subunits with a molecular mass of 42 kDa in the active form [25–28]. In the past there has been considerable confusion in the naming of cathepsin E obtained from various sources such as lymphocytes [29], spleen [25], gastric mucosa [26, 30–32], and erythrocytes [27, 33]. The names included cathepsin E-like acid proteinase [25], cathepsin D-type protease [29], cathepsin D-like acid proteinase [26, 31], and slow moving proteinase (SMP) [32]. However, on the basis of detailed biochemical and immunochemical analyses [28, 34–36], these enzymes were found to be identical and have been referred to as cathepsin E [28]. In contrast to CD, CE has a limited distribution in cell types such as lymphoid tissues, gastrointestinal tracts, urinary organs and blood cells [37, 38]. By immunoelectron microscopic studies CE has also been shown to be different from CD in intracellular localization in various mammalian cells [39–41]. Usually CE is not found in the lysosomal compartment of any cell types. Membrane association of CE is found in intracellular canaliculi of gastric parietal cells (human and rat) [40], renal proximal tubule cells (rat) [40], bile canaliculi of hepatic cells (rat) [40], intestinal and tracheobranchial epithelial cells [38, 45], osteoclasts (rat) [41] and erythrocytes (human and rat) [27, 33, 46, 47]. The localization of CE in the endosomal-structures is observed in various cell types such as gastric cells (human and rat) [40], antigen-presenting B cell lymphoblasts (murine) [48] and microglial cells (rat) [49,50]. The enzyme is also found in the cisternae of the ER [40, 51, 52] and the soluble cytosolic compartments of various cell types [39, 40, 51–53], although the possibility that the latter is derived from breakage of vesicles cannot be excluded. Besides its strategic localization, CE is also unique in that its carbohydrate moieties differ in different cell types. CE isolated from erythrocyte membranes (human and rat) [54], microglia (rat) [49] and thymocytes (rat) [49] is *N*-glycosylated with complex type oligosaccharides, whereas the enzyme from spleen (rat) [54] and stomach (human and rat) [26, 55] has high-mannose type oligosaccharides, suggesting that the types of carbohydrate moieties of CE may be cell-specific or may vary with cellular location. On the basis of these observations, it has been postulated that the physiological function of CE is different from that of CD. Therefore, recent studies on CE have been directed towards clarifying its physiological functions.

In this review, I will focus mainly upon studies of CE and discuss some important points in considering its physiological function, although other relevant CD studies will also be discussed. For further CD studies, please refer to other excellent reviews detailing basic aspects of CD and the references cited therein [5, 42–44].

Structural characterization of cathepsin E

CE appears to be unique in existing as a homodimer of two fully catalytically active monomers in its native state, but it is easily converted to a monomeric form retaining catalytic activity under mild reducing conditions [46, 56]. The other aspartic proteinases are primarily present as monomeric enzymes with a molecular mass of about 40 kDa. Complete nucleotide and deduced amino acid sequences for CE from various origins such as human [57], guinea pig [56], rabbit [58], and rat [59] have been reported. Prepro-CE of human, guinea pig, rabbit, and rat consist of 396, 391, 387, and 395 amino acid residues, respectively. The deduced amino acid sequences reveal that prepro-CE consists of three regions; the signal peptide, the propeptide (activation segment) and the mature proteinase domain. The sequences of CE from these species show a high degree of similarity (more than 80% identity) and are apparently different from those of other aspartic proteinases, such as CD (47% identity), pepsinogen (48%), and renin (41%). The unique structural characteristics of pro-CE are conserved in all of the species; the amino-terminal portion of CE contains a Cys residue at position 7 which is responsible for the formation of a disulfide bond between the two identical subunits. The highly conserved tripeptide sequence Asp-Thr-Gly at the active sites of aspartic proteinases is also found in all of these CE species, except that the sequence near the amino-terminal region of rabbit CE is replaced by Asp-Thr-Val. In addition, the putative *N*-linked oligosaccharide attachment site near the amino-terminal region is conserved in all of the species. However, the number of potential *N*-glycosylation sites varies with different species. Human and rabbit CE have a single potential *N*-glycosylation site at position 73 which is common to other species, whereas rat and guinea pig CE possess another at positions 305 and 318, respectively (Fig. 1). CE exists in either the high mannose-type or the complex type and the nature of its oligosaccharide chains appears to be cell specific or to vary with its cellular localization. On the other hand, CD has predominantly the high mannose-type oligosaccharides, which are known to be present on most soluble lysosomal enzymes, and partly the complex type of oligosaccharides [42]. In the course of intracellular transport, the oligosaccharides of CD are phosphorylated to acquire terminal mannose-6-phosphates which serve as markers for lysosome targeting via binding to MPRs [23, 60], although the MPR-independent targeting pathway is also present in some cell types [7, 61, 62]. In contrast, the phosphorylation of the oligosaccharides of CE is not clear.

Biosynthesis, processing and transport of cathepsin E

From biosynthetic studies, CE is known to be synthesized first as a preproenzyme on the membrane-bound ribosomes of the ER and to undergo a variety of post-translational modifications before being sorted to its appropriate cellular destinations [49, 51, 63]. Pro-CE formed after the removal of the signal peptide is modified with *N*-linked oligosaccharides. Its subsequent processing and transport vary with different cell types. At least three different processing systems have been proposed for CE. In the microglia-type processing system, pro-CE is known to be transported to the Golgi complex, where its oligosaccharides are converted to the complex type from the high mannose-type. Then the enzyme is targeted to the endosome-like acidic com-

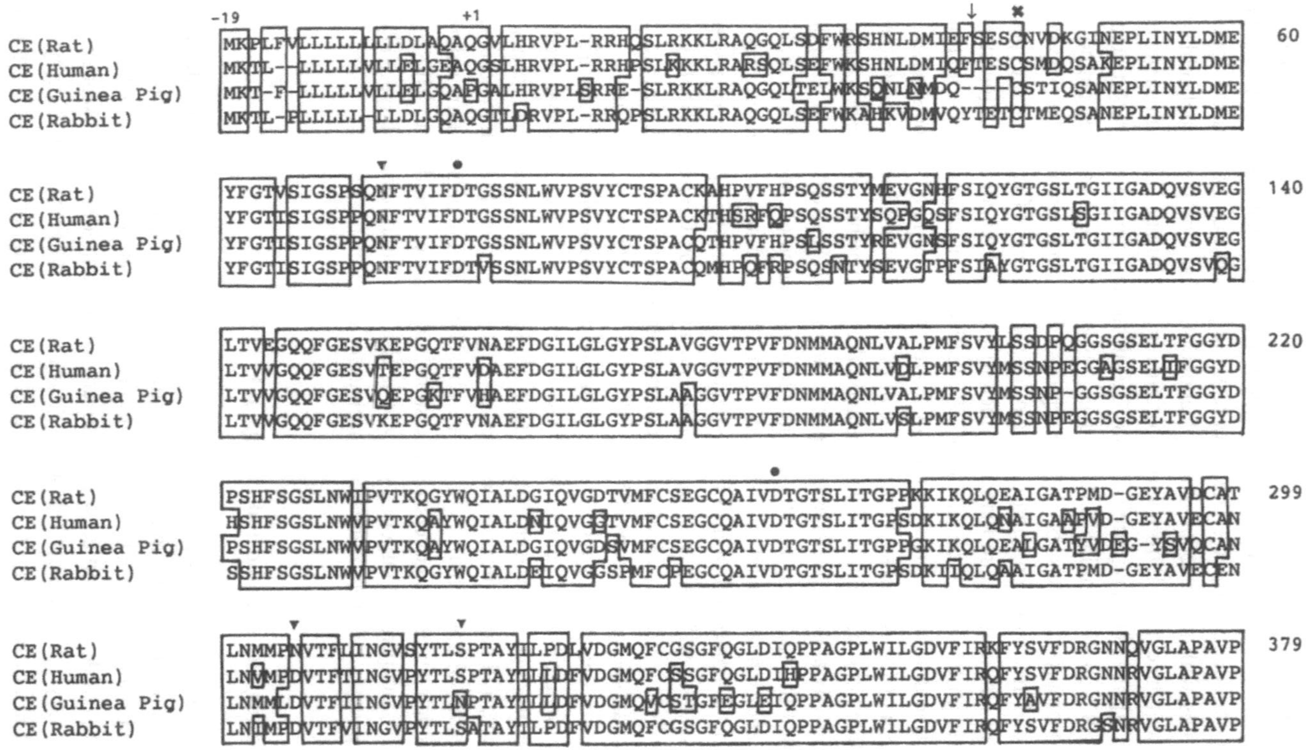

Figure 1. Comparison of the amino acid sequence of rat prepro-CE with those of human, guinea pig, and rabbit prepro-CE species. Identical amino acid residues are boxed. Pro-peptide cleavage site is indicated by an arrow. Cys involved in dimer formation is indicated by a cross. The N-linked oligosaccharide attachment sites are indicated by closed arrowheads. Putative active site Asp residues are indicated by closed circles.

partment, where active CE with an apparent molecular mass of 82 kDa is formed [49, 50]. The processing of CE in rat microglial cells, macrophages and neurons is included in this system. In the thymocyte-type processing system, pro-CE is transported to, and retained in, the *trans*-Golgi (or TGN) where its oligosaccharides are converted to the complex type. During transport the proenzyme is proteolytically processed to the intermediate form of 84 kDa, but not significantly to the mature form. The 84 kDa protein in the *trans*-Golgi region can be mobilized and activated to the mature 82 kDa protein by certain cellular stimuli such as dexamethasone treatment [49, 50, 64], suggesting that the protein in this region serves as a reservoir for response to certain stimuli. The processing of CE in rat thymocytes and murine Friend erythroleukemia cells is included in this system. In the recombinant cell-type processing system, the initially synthesized pro-CE is retained predominantly in the ER and partly in the *cis*- or *medial*-Golgi regions. The proenzyme is not processed to the mature form and its oligosaccharides are not converted to the complex-type from the high-mannose type. However, the protein appears to be mobilized to the *trans*-Golgi region in response to certain cellular stimuli, such as ionomycin treatment, to convert to the 84 kDa intermediate form with the complex type oligosaccharides. The pro-CE in murine A20 lymphoma cells and the recombinant enzyme expressed in CHO, NRK and Cos1 cells are processed by this system.

As compared with CE, studies on the biosynthesis and processing of CD appear to be almost completed. CD is initially synthesized on the membrane-bound ribosomes of the ER as a 51 kDa preproenzyme and provided with two *N*-linked oligosaccharides. After transport to the Golgi complex, pro-CD acquires the mannose-6-phosphate (M6P) recognition signal for targeting to lysosomes by MPRs and is then transported via endosomes to lysosomes [22, 23], where it is ultimately activated to the 44 kDa mature enzyme. In addition to the MRP-dependent targeting system, several lines of evidence indicate that CD can be targeted to lysosomes by the MPR-independent system [7, 22, 24, 65, 66]. It has been postulated that this alternative targeting system acts in the same compartments as the MPR [61] and that the M6P-independent association of pro-CD with cellular membranes is essential for this system [7, 24, 69]. The processing of pro-CD to the active single chain molecule occurs presumably in prelysosomal compartments. Depending on the species, the ultimate proteolytic activation of the single molecule may occur in lysosomes by lysosomal cysteine proteinases to form a two chain form [67].

Enzymatic properties of cathepsin E

Although CE and CD are clearly distinct proteins [28, 70], they have some common biochemical and catalytic features, such as pH optimum, substrate specificity, and susceptibility to various protease inhibitors [25, 27, 71]. However, recent studies have revealed that these two enzymes differ in many biochemical properties. The molecular mass of pro-CE purified from human erythrocyte membranes [54] and guinea pig gastric mucosa [56] has been estimated to be 90 kDa (46 kDa) and 82 kDa (43 kDa) by SDS-PAGE under non-reducing (reducing) conditions, respectively, indicating that pro-CE, like CE, exists as a homodimer. Dimeric pro- and mature CEs are easily converted to the corresponding monomeric forms by treatment with reducing agents, such as 2-mercaptoethanol, dithiothreitol and cysteine [46, 56, 72]. This conversion appears to be reversible, the dimeric form being regenerated after removing the reducing agents [56]. To analyze the molecular basis of the segregation and maturation of CE, the recombinant monomeric [73] and dimeric enzymes [51] have been obtained. These recombinant CEs are catalytically indistinguishable from the naturally occurring dimeric enzymes. The catalytic properties of the recombinant monomeric enzyme are also essentially identical with those of the natural and recombinant dimeric forms [73, 74]. For example, the k_{cat}/K_m values measured for the synthetic substrates Pro-Pro-Thr-Ile-Phe(4-NO_2)-Arg-Leu and Lys-Pro-Ile-Glu-Phe(4-NO_2)-Arg-Leu with both the dimeric and monomeric CE were similar [74]. The specific activities of the monomeric CE against both protein and synthetic substrates were comparable to those of the dimeric CE [51, 73]. The K_i values for various inhibitors, such as pepstatin, Pro-Thr-Glu-Phe(CH_2-NH)Nle-Arg-Leu and Boc-His-Pro-Phe-His-Sta-Leu-Phe-NH_2, with the monomeric CE are comparable to those with the dimeric CE [73]. Further, there is no significant difference in optimal pH for digestion of various proteins and synthetic substrates and in substrate specificity between the monomeric and dimeric forms. In contrast, the physicochemical properties are significantly different between the monomeric and dimeric forms. For example, the monomeric form is more unstable to pH and temperature changes than the dimeric form [56, 73]. From these studies, it has been suggested that the dimerization of CE is not necessarily required to express activity, but that it may be essential to structurally stabilize the molecule *in vivo*.

Proteolytic removal of the propeptide of CE results in the activation of the enzyme. The activation of pro-CE, like other aspartic proteinases, is presumably initiated by a dramatic conformational rearrangement of the propeptide to react with the active site of the enzyme. This propeptide processing has been well studied *in vitro* [54, 56, 75]. Pro-CE is catalytically inactive [54], but is rapidly converted into the active form by brief acid treatment *in vitro*. In the case of human pro-CE, the rate of activation is maximal at pH 3.5–4.0. This activation is accompanied by a reduction in molecular size to 43 kDa. The cleavage sites are identified to be the Met^{36}-Ile^{37} and Phe^{39}-Thr^{40} bonds, which correspond to the amino-termini of the native mature enzyme [51, 54]. This cleavage is rapid and complete within 5 min at pH 4.0 and 37 °C. The activation of pro-CE depends on its protein concentration and is completely inhibited by pepstatin, suggesting that intermolecular reaction is predominantly involved in this activation process. On the other hand, activation of pro-CD also occurs autocatalytically *in vitro* under acidic conditions to yield catalytically active pseudo-CD, which is intermediate in size between pro-CD and the single chain CD [76, 77]. However, there is no evidence for autocatalytic removal of the remainder of the propeptide [78]. Therefore, the inability of pro-CD to generate the mature enzyme by autocatalysis suggests that the autocatalytic activation to pseudo-CD does not occur *in vivo*. [67, 79]. It has rather been suggested that other lysosomal proteinases are required for the activation of pro-CD to the mature enzyme [61, 67, 77]. Available evidence suggests that the propeptide of CD is likely to contribute to the correct folding of the enzyme molecule [80], to membrane association and M-6-P-independent sorting to the lysosome [24, 69].

The functional diversity of cathepsin E

The limited tissue distribution of CE and its cell-specific cellular localization suggests that the CE gene is highly regulated and, consequently, that its expression varies in different tissues and cells in response to local stimuli and demands (Tab. 1). Although the precise function of CE is currently unknown, there is accumulating evidence suggesting its involvement in various impor-

Table 1. Levels of cathepsins E and D in various rat tissues

Sources	Cathepsin E (μg/mg of protein)	Cathepsin D (μg/mg of protein)
Brain tissues		
Cerebral cortex	0.0004*	-
Cerebellum	0.0037*	-
Hippocampus	0.0017*	-
Neostriatum	0.0005*	-
Gastrointestinal tracts		
Stomach	1.05 (1.130*)	0.81
Jejunum	0.03	0.15
Colon	0.10	0.24
Esophagus	n.d.	0.18
Liver	n.d.	0.29
Lymphoid tissues		
Cervical lymph node	0.18	0.22
Thymus	0.19	0.34
Spleen	0.18	0.63
Bone marrow	0.13	0.21
Urinary organs		
Urinary bladder	0.19	0.31
Kidney	0.005	0.22
Secretory tissues		
Submandibular gland	0.006	0.20
Lacrymal gland	n.d.	1.39
Adrenal	n.d.	0.79
Blood cells		
Erythrocytes	0.002	0
Lymphocytes	0.09	0.06
Platelets	0.004	0.05
Peritoneal neutrophils	0.10	0.27
Lung	0.03 (0.04*)	0.39
Heart	n.d. (0.0012*)	0.13
Skeletal muscle	n.d.	0.04
Skin	0.03	0.39

The values indicated by asterisks are from ELISA. n.d. not detectable.

tant biological functions. The highest level of CE is found in the stomach of all the species examined [37, 38]. Immunocytochemical staining revealed that CE in rat and human stomach was concentrated in the surface and foveolar epithelial cells, but that its content was less in the parietal cells of the fundic gland. In contrast, CD was found mainly in the cells at deeper parts of the fundic gland, especially in the deeper parietal cells [32, 38]. Although the physiological significance of the abundance of CE in the foveolar epithelial cells is still not clear, the enzyme may be involved in the adsorption and secretion by the epithelial cells and the turnover of rapidly regenerating epithelial cells. The differential distribution of CE and CD observed in normal human mucosa was reported to be lost in cancerous and precancerous lesions [81, 82]. For example, in carcinoma cells, CE was predominantly localized in the advancing margin of the invasion front with high incidence, and this peculiar localization correlated significantly with the progression of the carcinoma tissue and the incidence of the lymph node metastasis [82]. The results suggest that the expression of the enzyme is involved in the invasive and metastatic activities of gastric carcinoma cells. A similarly peculiar expression of CE was also observed in human pancreatic ductal adenocarcinomas, although the normal duct cells are devoid of CE staining [83]. In addition, CE was found in the pancreatic juice from pancreatic ductal adenocarcinomas but barely detectable in the juice from chronic pancreatitis [83], suggesting that the expression of CE is associated with the pathogenesis of pancreatic ductal adenocarcinoma.

The high levels of CE in lymphoid tissues, such as the lymph node, thymus and spleen, appear to be mainly attributable to the high expression of CE in lymphocytes and macrophage-like cells in these tissues [38, 64, 84, 85]. In these cells, CE appears to be localized mainly to endosomal vesicles and the ER. These observations thus suggest that CE is involved in the basic machinery of the immune response. In this connection, CE has been suggested to play a role in antigen processing for presentation by class II major histocompatibility complex (MHC). For example, in addition to the abundant expression of CE in the antigen presenting cells, such as intestinal and tonsilar M cells [86], Langerhans cells of normal skin and interdigitating reticulum cells of lymph nodes and spleen [84, 85], it has been shown that the proteolytic processing of intact ovalbumin antigen by the murine antigen-presenting B lymphoma cell line A20 for presentation to the T-cell hybridoma is blocked by CE inhibitors, such as Ascaris inhibitor and pepstatin [87]. The possibility that processing of the human invariant chain is mediated by CE has also been suggested [88].

Recently, CE was shown to be able to efficiently hydrolyze various biologically active peptides, such as substance P and related tachykinins, maximally at around pH 5 [58]. The rates of hydrolysis of these peptides by CE were several hundred-fold higher than those by CD, although the cleavage sites by CD were essentially identical with those by CE. Since tachykinins contain the sequence Phe-Xaa-Gly-Leu-Met at the carboxy-terminus that is essential for the biological activity, the specific cleavage of the Phe-Xaa bond by CE appears to cause the inactivation of their biological activities. A similar efficient and specific degradation of fragments of the fibroblast growth factor by CE was also shown [58]. Therefore, CE might be involved in the modulation of signal transduction through degradation of these peptides and ultimately in the regulation of the target cell functions. Further, CE can process precursors to neurotensin and related peptides, which contain the sequence Pro-Xaa-Xaa-hydrophobic amino acid in the carboxy-terminal region [89]. This sequence corresponds to the position P4–P1 that is important for interaction of the precursor molecules with the active sites of CE, and it appears to be responsible for the difference in specificity between CE and CD. The processing of T-kininogen [90] and big endothelin [91] to the respective active peptides by CE has also been suggested.

The recent observations that neuronal lesion and death induced by ageing [92–94] or by transient forebrain ischemia [95] are accompanied by elevation of CE, as well as CD, in both neurons and activated microglial cells in the vulnerable brain regions suggest that CE is involved in the execution of the neuronal death pathway. CE is barely detectable in any brain regions in the embryonic stages and slightly distributed in only a small number of neurons of young rats, such as pyramidal cells of the hippocampus and the cerebral cortex, and Purkinje cells of the cerebellum. However, it is increasingly expressed in degenerating neurons and activated microglia in aged rat brain, attaining maximal levels at 30 months of age [94]. This age-dependent expressed CE was shown to be due predominantly to the mature enzyme and partly to the proenzyme. It is interesting to note that CE in aged neurons colocalized with lipofuscin and the carboxy-terminal fragments of amyloid precursor protein (APP) and further, with CD in lipofuscin-containing lysosomes. From these studies, it has been suggested that the increased expression and lysosomal localization of CE in aged neurons constitute the endosomal/lysosomal proteolytic system, which may be related to lipofucinogenesis and altered intracellular APP metabolism. CE was also shown to be highly expressed in degenerating neurons and activated microglia in the hippocampal CA1 region for up to seven days after transient forebrain ischemia [95]. Taken together, the observation that neuronal lesion and death induced by excessive stimulation of glutamate receptors with excitotoxins is consistent with the induction of the CE gene response followed by the persistent expression of CE in degenerating neurons [96, 97], suggesting that CE is associated with the execution of neuronal death pathways.

Acknowledgements
Contributions by collaborators Drs. Kato Y, Sakai H, Nishishita K, Nakanishi H, Tsukuba T, Okamoto K to the experimental work are gratefully acknowledged. Studies in the laboratory of the author were supported in part by a Grant-in-Aid for Scientific Research from the Ministry of Education, Science and Culture of Japan.

References

1 Blundell TL, Cooper JB, Sali A, Zhu Z (1991) Comparison of the sequences, 3-D structures and mechanisms pepsin-like and retroviral aspartic proteinases. *In*: BM Dunn (ed.): *Structure and Function of the Aspartic Proteinases*. Plenum Press, New York, 443–453

2 Aquilar CF, Dhanaraj V, Guruprasad K, Dealwis C, Badasso M, Cooper JB, Wood SP, Blundell TL (1995) Comparison of the three-dimensional structures, specificities, and glycosylation of renins, yeast proteinase A and cathepsin D. *In*: K Takahashi (ed.): *Aspartic Proteinases: Structure, Function, Biology, and Biomedical Implications*. Plenum Press, New York, 155–166

3 Ringe DC (1994) X-ray structures of retroviral proteases and their inhibitor-bound complexes. *In*: LC Kuo, JA Shafer (eds): *Meth Enzymol* 241: 157–177

4 Bhad TN, Baldwin ET, Liu B, Cheng Y-SE, Erickson JW (1995) X-ray structure of a tethered dimer for HIV-1 protease. *In*: K Takahashi (ed.): *Aspartic Proteinases: Structure, Function, Biology, and Biomedical Implications*. Plenum Press, New York, 439–444

5 Barrett AJ (1977) Cathepsin D and other carboxy proteinases. *In*: AJ Barrett (ed.): *Proteinases in Mammalian Cells and Tissues*. North-Holland, Amsterdam, 209–248

6 Yamamoto K, Ikehara Y, Kawamoto S, Kato K (1980) Characterization of enzymes and glycoproteins in rat liver lysosomal membranes. *J Biochem* 87: 237–248

7 Diment S, Leech MS, Stahl PD (1988) Cathepsin D is membrane-associated in macrophage endosomes. *J Biol Chem* 263: 6901–6907

8 Geuze HJ, Slot JW, Strous GJAM, Hasilik A, von Figura K (1985) Possible pathways for lysosomal enzyme delivery. *J Cell Biol* 101: 2253–2262

9 Neefjes JJ, Ploegh HL (1992) Intracellular transport of MHC class II molecules. *Immunol Today* 13: 179–184

10 Maric MA, Taylor MD, Blum JS (1994) Endosomal aspartic proteinases are required for invariant-chain processing. *Proc Natl Acad Sci USA* 91: 2171–2175

11 van Noort JM, Jacob MJM (1994) Cathepsin D, but not cathepsin B, releases T cell stimulatory fragments from lysozyme that are functional in the context of multiple murine class II MHC molecules. *Eur J Immunol* 24: 2175–2180

12 Pillai S, Zull J (1986) Production of biologically active fragments of parathyroid hormone by isolated Kupffer cells. *J Biol Chem* 261: 14919–14923

13 Tandon AK, Clark GM, Chamness GC, Chirgwin JM, McGuire WL (1990) Cathepsin D and prognosis in breast cancer. *N Engl J Med* 322: 297–302

14 Leto G, Gebbia N, Rausa L, Tumminello FM (1992) Cathepsin D in the malignant progression of neoplastic disease. *Anticancer Res* 12: 235–240

15 Mignatti P, Rifkin DB (1993) Biology and biochemistry of proteinases in tumor invasion. *Physiol Rev* 73: 161–195

16 Sanchez LM, Ferrando AA, Diez-Itza I, Vizoso F, Lopez-Otin C (1993) Cathepsin D in breast secretions from women with breast cancer. *Brit J Cancer* 67: 1076–1081

17 Aaltonen M, Lipponen R, Kosma M, Aaltomaa S, Syrjanen K (1995) Prognostic value of cathepsin D expression in female breast cancer. *Anticancer Res* 15: 1033–1038

18 Fusek V, Vetvicka V (1994) Mitogenic function of human procathepsin D: the role of the propeptide. *Biochem J* 303: 775–780

19 Lah TT, Calaf G, Kalman E, Shinde BG, Russo J, Jarosz D, Zabrecky J, Somers R, Daskal I (1995) Cathepsins D, B and H in breast carcinoma and in transformed human breast epithelial cells (HBEC). *Biol Chem Hoppe-Seyler* 376: 363–367

20 Siman R, Mistretta JT, Durkin MJ, Savage T, Loh S, Trusko S, Scott RW (1993). Processing of the β-amyloid precursor: multiple proteases generate and degrade potentially amyloidgenic fragments. *J Biol Chem* 268: 16602–16609

21 Ladror U, Snyder S, Wang G, Holzman T, Krafft G (1994) Cleavage at the amino acid carboxy termini of Alzheimer's amyloid-β by cathepsin D. *J Biol Chem* 269: 18422–18428

22 Kornfeld S (1986) Trafficking of lysosomal enzymes in normal and disease states. *J Clin Invest* 77: 1–6

23 von Figula K, Hasilik A (1986) Lysosomal enzymes and their receptors. *Ann Rev Biochem* 55: 167–193

24 Rijnboutt S, Aerts HMFG, Geuze HJ, Tager JM, Strous GJ (1991) Mannose-6- phosphate independent membrane association of cathepsin D, glucocerebrosidase, and sphingolipid-activating protein in HepG2 cells. *J Biol Chem* 266: 4862–4868

25 Yamamoto K, Katsuda N, Kato K (1978) Affinity purification and properties of cathepsin E-like acid proteinase from rat spleen. *Eur J Biochem* 92: 499–508

26 Kageyama T, Takahashi K (1980) A cathepsin D-like acid proteinase from human gastric mucosa: purification and characterization. *J Biochem* 87: 725–735

27 Takeda M, Ueno E, Kato Y, Yamamoto K (1986) Isolation, and catalytic and immunochemical properties of cathepsin D-like acid proteinase from rat erythrocytes. *J Biochem* 100: 1269–1277

28 Jupp RA, Richards AD, Kay J, Dunn BM, Wyckoff JB, Samloff IM, Yamamoto K (1988) Identification of the aspartic proteinases from human erythrocyte membranes and gastric mucosa (slow-moving proteinase) as catalytically equivalent to cathepsin E. *Biochem J* 254: 895–898

29 Yago N, Bowers WE (1975) Unique cathepsin D-type proteinase in rat thoracic duct lymphocytes and in rat lymphoid tissues. *J Biol Chem* 250: 4749–4754

30 Roberts NB, Taylor WH (1987) The isolation and properties of a non-pepsin proteinase from human gastric mucosa. *Biochem J* 169: 617–624

31 Muto N, Murayama-Arai K, Tani S (1983) Purification and properties of a cathepsin D-like acid proteinase from gastric mucosa. *Biochim Biophys Acta* 745: 61–69

32 Samloff IM, Taggart RT, Shiraishi T, Branch T, Reid WA, Heath R, Lewis RW, Valler MJ, Kay J (1987) Slow-moving proteinase: isolation, characterization and immunohistochemical localization in gastric mucosa. *Gastroenterology* 93: 77–84

33 Yamamoto K, Marchesi VT (1984) Purification and characterization of acid proteinase from human erythrocyte membranes. *Biochim Biophys Acta* 790: 208–218

34 Tarasova NI, Szecsi PB, Foltman B (1986) An aspartic proteinase from human erythrocytes is immunochemically indistinguishable from a non-pepsin, electrophoretically slow moving proteinase from gastric mucosa. *Biochim Biophys Acta* 880: 96–100

35 Yonezawa S, Tanaka T, Muto N, Tani S (1987) Immunochemical similarity between a gastric mucosa non-pepsin acid proteinase and neutrophil cathepsin E of the rat. *Biochem Biophys Res Commun* 144: 1251–1256

36 Yamamoto K, Ueno E, Uemura H, Kato Y (1987) Biochemical and immunochemical similarity between erythrocyte membrane aspartic proteinase and cathepsin E. *Biochem Biophys Res Commun* 148: 267–272
37 Muto N, Yamamoto M, Tani S, Yonezawa S (1988) Characteristic distribution of cathepsin E which immunologically cross-reacts with the 86-kDa acid proteinase from gastric mucosa. *J Biochem* 103: 629–632
38 Sakai H, Saku T, Kato Y, Yamamoto K (1989) Quantitation and immunohistochemical localization of cathepsins E and D in rat tissues and blood cells. *Biochim Biophys Acta* 991: 367–375
39 Ichimaru E, Sakai H, Saku T, Kunimatsu K, Kato Y, Kato I, Yamamoto K (1990) Characterization of hemoglobin-hydrolyzing acidic proteinases in human and rat neutrophils. *J Biochem* 108: 1009–1015
40 Saku T, Sakai H, Shibata Y, Kato Y, Yamamoto K (1991) An immunocytochemical study on distinct intracellular localization of cathepsin E and cathepsin D in human gastric cells and various rat cells. *J Biochem* 110: 956–964
41 Yoshimine Y, Tsukuba T, Isobe R, Sumi M, Akamine A, Maeda K, Yamamoto K (1995) Specific immunocytochemical localization of cathepsin E at the ruffled border membrane of active osteoclasts. *Cell Tissue Res* 281: 85–91
42 Shewale JG, Takahashi T, Tang J (1985) The primary structure of cathepsin D and the implications for its biological functions. *In*: V Kostka (ed.): *Aspartic Proteinases and Their Inhibitors.* Walter de Gruyter, Berlin. New York, 101–116
43 Fruton JS (1987) Aspartic proteinases. *In*: A Neuberger, K Brocklhurst (eds): *Hydrolytic Enzymes.* Elsevier, Amsterdam, 1–37
44 Koelsch G, Mares M, Metcalf P, Fusek M (1994) Multiple functions of pro-parts of aspartic proteinase zymogens. *FEBS Lett* 343: 6–10
45 Fiocca R, Villani L, Tenti P, Cornaggia M, Finz G, Riva C, Capella C, Bara J, Samloff IM, Solcia E (1990) The foveolar cell component of gastric cancer. *Hum Pathol* 21: 260–270
46 Yamamoto K, Yamamoto H, Takeda M, Kato Y (1988) An aspartic proteinase of erythrocyte membranes: proposed mechanism for activation and further molecular properties. *Biol Chem Hoppe-Seyler* 369: 315–322
47 Yamamoto K, Takeda M, Yamamoto H, Tatsumi M, Kato Y (1985) Human erythrocyte membrane acid proteinase (EMAP): sidedness and relation to cathepsin D. *J Biochem* 97: 821–830
48 Bennett K, Levine T, Ellis JS, Peanasky RJ, Samloff IM, Kay J, Chain BM (1992) Antigen processing for presentation by class II major histocompatibility complex requires cleavage by cathepsin E. *Eur J Immunol* 22: 1519–1524
49 Yamamoto K, Nakanishi H, Tsukuba T, Okamoto K, Sakai H, Nishishita K, Kato Y (1997) Biosynthesis and trafficking of cathepsin E. *In*: Hoppsu-Have VK (ed.): *Proteolysis in Cell Functions.* IOS Press, Amsterdam, 215–222
50 Sastradipura DF, Nakanishi H, Tsukuba T, Nishishita K, Sakai H, Kato Y, Gotow T, Uchiyama Y, Yamamoto K (1998) Identification of cellular compartments involved in processing of cathepsin E in primary cultures of rat microglia. *J Neurochem* 70: 2045–2056
51 Tsukuba T, Hori H, Azuma T, Takahashi T, Taggart RT, Akamine A, Ezaki M, Nakanishi H, Sakai H, Yamamoto K (1993) Isolation and characterization of recombinant human cathepsin E expressed in Chinese hamster ovary cells. *J Biol Chem* 268: 7276–7282
52 Finley EM, Kornfeld S (1994) Subcellular localization and targeting of cathepsin E. *J Biol Chem* 269: 31259–31266
53 Sakai H, Kato Y, Yamamoto K (1992) Synthesis and intracellular distribution of cathepsins E and D in differentiating murine Friend erythroleukemia cells. *Arch Biochem Biophys* 294: 412–417
54 Takeda-Ezaki M, Yamamoto K (1993) Isolation and biochemical characterization of procathepsin E from human erythrocyte membranes. *Arch Biochem Biophys* 304: 352–358
55 Yonezawa S, Takahashi T, Ichinose M, Miki K, Tanaka J, Gasa S (1990) Structural studies of rat cathepsin E: amino-terminal structure and carbohydrate units of mature enzyme. *Biochem Biophys Res Commun* 166: 1032–1038
56 Kageyama T, Ichinose M, Tsukada S, Miki K, Kurokawa K, Koiwai O, Tanji M, Yakabe E, Athauda SBP, Takahashi K (1992) Gastric procathepsin E and progastricsin from guinea pig: purification, molecular cloning of cDNA, and characterization of enzymatic properties with special reference to procathepsin E. *J Biol Chem* 267: 16450–16459
57 Azuma T, Pals G, Mohandas TK, Convreur JM, Taggart RT (1989) Human gastric cathepsin E: predicted sequence, localization to chromosome 1, and sequence homology with other aspartic proteinases. *J Biol Chem* 264: 16748–16753
58 Kageyama T (1993) Rabbit procathepsin E and cathepsin E: nucleotide sequence of cDNA, hydrolytic specificity for biologically active peptides and gene expression during development. *Eur J Biochem* 216: 717–728
59 Okamoto K, Yu H, Misumi Y, Ikehara Y, Yamamoto K (1995) Isolation and sequencing of two cDNA clones encoding rat spleen cathepsin E and analysis of the activation of purified procathepsin E. *Arch Biochem Biophys* 322: 103–111
60 Kornfeld S, Mellman I (1989) The biosynthesis of lysosomes.*Annu Rev. Cell Biol* 5: 483–525
61 Rijnboutt S, Kal AJ, Geuze HJ, Aerts H, Strous GJ (1991) Mannose-6-phosphate- independent targeting of cathepsin D to lysosomes in HepG2 cells. *J Biol Chem* 266: 23586–23592
62 Capony F, Braulke T, Rougeot C, Roux S, Montcourrier P, Rochefort H (1994) Specific mannose-6-phosphate-

independent sorting of pro-cathepsin D in breast cancer cells. *Exp Cell Res* 215: 154–163
63 Yamamoto K, Sakai H, Ezaki M, Kato Y (1994) Biosynthesis, processing and localization of cathepsin E. *In*: N Katunuma, K Suzuki, J Travis, H Fritz (eds): *Biological Functions of Proteases and Inhibitors*. Japan Scientific Society Press, Tokyo, 97–107
64 Nishishita K, Sakai H, Sakai E, Kato Y, Yamamoto K (1996) Age-related and dexamethasone-induced changes in cathepsins E and D in rat thymic and splenic cells. *Arch Biochem Biophys* 333: 349–358
65 Owada M, Neufeld EF (1982) Is there a mechanism for introducing acid hydrolases into liver lysosomes that is independent of mannose-6-phosphate recognition? *Biochem Biophys Res Commun* 105: 814–820
66 Waheed A, Pohlmann R, Hasilik A, von Figura K, van Eisen A, Leroy JG (1982) Deficiency of uridine diphosphate-N-acetylglucosamine 1-phosphotransferase in organs of I-cell patients. *Biochem Biophys Res Commun* 105: 1052–1058
67 Gieselmann V, Hasilik A, von Figura K (1985) Processing of human cathepsin D in lysosomes *in vitro*. *J Biol Chem* 260: 3215–3220
68 Samarel AM, Ferguson AG, Decker RS, Lesch M (1989) Effects of cysteine protease inhibitors on rabbit cathepsin D maturation. *Amer J Physiol* 257: C1069
69 McIntyre GF, Erickson AH (1991) Procathepsins L and D are membrane-bound in acidic microsomal vesicles. *J Biol Chem* 266: 15438–15445
70 Yamamoto K, Kamata O, Katsuda N, Kato K (1980) Immunochemical difference cathepsin D and cathepsin E-like enzyme from rat spleen. *J Biochem* 87: 511–516
71 Yamamoto K, Katsuda N, Himeno M, Kato K (1979) Cathepsin D of rat spleen: affinity purification and properties of two types of cathepsin D. *Eur J Biochem* 95: 459–467
72 Yonezawa S, Fujii K, Maejima Y, Tamoto K, Mori Y, Muto N (1988) Further studies on rat cathepsin E: subcellular localization and existence of the active subunit form. *Arch Biochem Biophys* 267: 176–183
73 Tsukuba T, Sakai H, Yamada M, Maeda H, Hori H, Azuma T, Akamine A, Yamamoto K (1996) Biochemical properties of the monomeric mutant of human cathepsin E expressed in Chinese hamster ovary cells: comparison with dimeric forms of the natural and recombinant cathepsin E. *J Biochem* 119: 126–134
74 Flowler SD, Kay J, Dunn BM, Tatnell PJ (1995) Monomeric human cathepsin E. *FEBS Lett* 366: 72–74
75 Athauda SBP, Takahashi T, Kageyama T, Takahashi K (1991) Autocatalytic processing of procathepsin E to cathepsin E and their structural differences. *Biochem Biophys Res Commun* 175: 152–158
76 Hasilik A, von Figura K, Conzelmann E, Nehrkorn H, Sandhoff K (1982) Lysosomal enzyme precursors in human fibroblasts: activation of cathepsin D precursor *in vitro* and activity of β-hexaminidase A precursor towards ganglioside GM2. *Eur J Biochem* 125: 312–321
77 Conner GE (1989) Isolation of procathepsin D from mature cathepsin D by pepstatin affinity chromatography: autocatalytic proteolysis of the zymogen form of the enzyme. *Biochem J* 263: 601–604
78 Larsen LB, Boisen A, Petersen TE (1993) Procathepsin D cannot autoactivate to cathepsin D at acid pH. *FEBS Lett* 319: 54–58
79 Richo GR, Conner GE (1994) Structural requirement of procathepsin D activation and maturation. *J Biol Chem* 269: 14806–14812
80 Conner GE (1992) The role of the cathepsin D propeptide in sorting to the lysosome. *J Biol Chem* 267: 21737–21745
81 Saku T, Sakai H, Tsuda N, Okabe H, Kato Y, Yamamoto K (1990) Cathepsins D and E in normal, metaplastic, dysplastic, and carcinomatous gastric tissue: an immunohistochemical study. *Gut* 31: 1250–1255
82 Matsuo K, Kobayashi I, Tsukuba T, Kiyoshima T, Ishibashi Y, Miyoshi A, Yamamoto K, Sakai H (1996) Immunohistochemical localization of cathepsins D and E in human gastric cancer: a possible correlation with local invasive and metastatic activities of carcinoma cells. *Hum Pathol* 27: 184–190
83 Azuma T, Hirai M, Ito S, Yamamoto K, Taggart RT, Matsuba T, Yasukawa K, Ueno K, Hayakumo T, Nakajima M (1996) Expression of cathepsin E in pancreas: a possible tumor marker for pancreas, a preliminary report. *Int J Cancer* 67: 492–497
84 Solcia E, Paulli M, Silini E, Fiocca R, Finzi G, Kindl S, Boveri E, Bosi F, Cornaggia M, Capella C (1993) Cathepsin E in antigen-presenting Langerhans and interdigitating reticulum cells: its possible role in antigen processing. *Eur J Histochem* 37: 19–25
85 Paulli M, Feller AC, Boveri E, Kindl S, Berti E, Rosso R, Merz H, Facchetti F et al. (1994) Cathepsin D and E co-expression in sinus histocytosis with massive lymphadenopathy (Rosai-Dorfman disease) and Langerhans' cell histiocytosis: further evidences of a phenotypic overlap between these histiocytic disorders. *Virchows Arch* 424: 601–606
86 Finz G, Cornaggia M, Capella C, Fiocca R, Bosi F, Solcia E, Samloff IM (1993) Cathepsin E in follicle-associated epithelium of intestine and tonsil: localization to M cells and possible role in antigen processing. *Histochemistry* 99: 201–211
87 Bennett K, Levine T, Ellis JS, Peanasky R, Samloff IM, Kay J, Chain BM (1992) Antigen processing for presentation by class II major histocompatibility complex requires cleavage by cathepsin E. *Eur J Immunol* 22: 1519–1524
88 Kageyama T, Yonezawa S, Ichinose M, Miki K, Moriyama A (1996) Potential sites for processing of the human invariant chain by cathepsins E and D. *Biochem Biophys Res Commun* 223: 549–553
89 Kageyama T, Ichinose M, Yonezawa S (1995) Processing of the precursors to neurotensin and other bioactive

peptides by cathepsin E. *J Biol Chem* 270: 19135-19140
90 Sakamoto W, Yoshikawa K, Yokoyama A, Kohri M (1986) T-kinin is released by consecutive cleavage by cathepsin E-like proteinase and 72 kDa proteinase. *Biochim Biophys Acta* 884: 607–609
91 Lees WE, Kalinka S, Meech J, Capper SJ, Cook ND, Kay J (1990) Generation of human endothelin by cathepsin E. *FEBS Lett* 273: 99–102
92 Nakanishi H, Tominaga K, Amano T, Hirotsu I, Inoue T, Yamamoto K (1994) Age- related changes in activities and localizations of cathepsins D, E, B, and L in the rat brain tissues. *Exp Neurol* 126: 119–128
93 Amano T, Nakanishi H, Oka M, Yamamoto K (1995) Increased expression of cathepsins E and D in reactive microglial cells associated with spongiform degeneration in brain stem of senescence-accelerated mouse. *Exp Neurol* 136: 171- 182
94 Nakanishi H, Amano T, Sastradipura DF, Yoshimine Y, Tsukuba T, Tanabe K, Hirotsu I, Ohono T, Yamamoto K (1997) Increased expression of cathepsins E and D in neurons of the aged rat brain and their colocalization with lipofuscin and carboxy- terminal fragments of Alzheimer amyloid precursor protein. *J Neurochem* 68: 739- 749
95 Nakanishi H, Tsukuba T, kondo T, Tanaka T, Yamamoto K (1993) Transient forebrain ischemia induces increased expression and specific localization of cathepsins E and D in rat hippocampus and neostriatum. *Exp Neurol* 121: 215–223
96 Tominaga K, Nakanishi H, yajima M, Yamamoto K (1995) Characterization of cathepsins E and D accumulated at early stages of neuronal damage in hippocampal neurons of rats. *In*: K Takahashi (ed.): *Aspartic Proteinases: Structure, Function, Biology, and Biomedical Implications*. Plenum Press, New York, 341–343
97 Tominaga K, Nakanishi H, Yasuda Y, Yamamoto K (1998) Excitotoxin-induced neuronal death is associated with response of a unique intracellular aspartic proteinase, cathepsin E. *J Neurochem* 71: 2574–2584

Proteases: New Perspectives
V. Turk (ed.)

Cell-associated metalloproteinases

Gary D. Johnson[1] and Judith S. Bond[2]

[1] *Department of Biochemistry, Parke-Davis Pharmaceutical Research, Warner-Lambert Company, Ann Arbor, MI 48105 USA*
[2] *Department of Biochemistry and Molecular Biology, The Pennsylvania State University College of Medicine, Hershey, PA 17033, USA*

Introduction

In the past few years, information about primary and tertiary structures of proteins has revealed the enormous diversity and abundance of metalloproteases, and has led to a new understanding of the number and variety of distinct families that make up the general class of metalloproteases [1–4]. It has become clear that most, if not all, of the metalloproteases contain zinc at their active sites, and employ this metal in catalysis. The majority of the characterized metalloendopeptidases contain the zinc-binding consensus sequence HEXXH in a region of α-helical secondary structure; the two conserved histidines are ligands to the zinc ion via their imidazole side chains. The amino acids designated as X are uncharged and usually hydrophobic. An additional Zn ligand is supplied by a distant Glu side chain in a subgroup of metalloendopeptidases called gluzincins [4]. Examples of gluzincins are thermolysin (EC 3.4.24.27) and neprilysin (EC 3.4.24.11). For the metzincins, another subgroup of metalloendopeptidases, the third Zn ligand is contributed by the last histidine residue in the extended consensus sequence HEXXHXXGFXH. Examples of metzincins are astacin (EC 3.4.24.21) and interstitial collagenase (EC 3.4.24.7). A water molecule involved in catalysis is the fourth zinc ligand in these enzymes. Inverzincins, such as insulysin (EC 3.4.24.56) possess the inverted zinc binding sequence HXXEH, with a downstream Glu residue implicated as a third zinc ligand. Catalysis by zinc metallopeptidases does not involve intermediates covalently bound to protein groups, as in the serine or cysteine proteinases, but rather a zinc-bound water molecule attacks the substrate carbonyl carbon, with hydrolysis assisted by a general base. The general base in the protein, however, has not yet been unambiguously identified for many zinc metallopeptidases. It has been proposed that the conserved Glu residue of the HEXXH consensus sequence performs this function in thermolysin and other related metalloendopeptidases, but recent studies have provided evidence that the catalytic base in thermolysin may actually be a histidine residue, and a tyrosine side chain has been implicated as the general base for serralysins and astacins. A reevaluation of the catalytic mechanisms of the various classes of metalloendopeptidases may be necessary to resolve the controversy and advance our understanding of how they function.

The pKa of the water molecule bound to zinc in metalloendopeptidases is lowered from a value of about 15 to a value in the range of 7–8. Therefore, these enzymes operate in the neutral pH range. Zinc metallopeptidases are non-specifically inhibited by chelators such as 1,10-phenanthroline and EDTA. Activity can be restored by the addition of low concentrations of transition metals, such as Zn^{2+} or Co^{2+}, to metal-depleted enzymes. More specific inhibitors have been generated by the incorporation of zinc-coordinating groups such as thiolate or hydoxamate anions into peptides that bind to metallopeptidase active sites. Many metalloendopeptidases are synthesized as inactive proforms that retain NH_2-terminal prosequences, thereby remaining in an inactive state until their prosequences are removed. Matrix metalloproteinases (also called MMPs or matrixins) are regulated *in vivo* by specific protein inhibitors called tissue inhibitors of matrix metalloproteinases (TIMPs). Physiologically relevant, specific inhibitors of other subgroups of metallopeptidases have not yet been definitively identified.

Zinc metalloendopeptidases are involved in numerous physiological functions, depending on their substrate specificity and the tissues and cellular compartments in which they are expressed. The metalloendopeptidases that were initially characterized are secreted or cell-surface proteins such as collagenases or neprilysin, but it is now recognized that metalloendopeptidases are also expressed intracellularly in the cytosol and in various organelles. In addition, some of these enzymes are present in multiple compartments. For example, insulysin was first purified as a cytosolic enzyme, but was recently characterized as a secreted enzyme, and is also thought to be associated with organelles such as peroxisomes. Thus, metalloendopeptidases are involved not only in the degradation of extracellular peptides and proteins, but also in the intracellular processing of peptides and proteins.

In this article we review the properties of a number of cell-associated eukaryotic, particularly mammalian, zinc metalloendopeptidases. There have been a number of excellent reviews on secreted metalloproteinases, especially the secreted MMPs (e.g. [5–7]); these will not be discussed herein. This article will focus on cell-associated metalloendopeptidases, grouped as intracellular or extracellular depending on where they are thought to be active. The Extracellular cell-associated proteases may have membrane-spanning segments or may associate with plasma membranes through covalent interactions (e.g. disulfide bonds) or non-covalent interactions (e.g. procollagen C-proteinase adheres to membranes). Subcellular localization is an important factor in regulating proteases as this will limit substrates encountered and the functions of the enzymes.

Intracellular metalloendopeptidases

Thimet oligopeptidase, neurolysin, and related peptidases

Thimet oligopeptidase is a neutral metalloendopeptidase of molecular mass 75 kDa that is present in numerous tissues, with the highest expression in the testis, brain, and pituitary gland [8]. It was originally known as Pz-peptidase because of its ability to cleave the Pz-peptide (4-phenylazobenzyloxycarbonyl-Pro-Leu-Gly-Pro-D-Arg), a synthetic substrate for bacterial collagenase. However, its intracellular location precludes any involvement in the *in vivo* turnover of extracellular matrix. Thimet oligopeptidase is stimulated by thiol compounds such as 2-mercaptoethanol at concentrations lower than 0.2 mM, and requires divalent metal ions for activity [9, 10]. Its amino acid sequence, deduced from a cloned cDNA, indicates that it is a zinc metallopeptidase of the gluzincin class [11]. The enzyme has no activity against proteins, reflected in its inability to cleave substrates longer than 18 residues in length [12]. Enzymes with properties similar to thimet oligopeptidase, such as endopeptidase 24.15 and endo-oligopeptidase A, have been described, but it is now accepted that all three of these proteins are identical and should be termed thimet oligopeptidase, or TOP (EC 3.4.24.15). Sensitive fluorogenic assays using internally quenched peptides such as Mcc-Pro-Leu-Gly-Pro-D-Lys(Dnp) are now available for the assay of TOP [13]. As with the Pz-peptide, cleavage occurs at the Leu-Gly bond. N-[1-(RS)-carboxy-3-phenylpropyl]-Ala-Ala-Tyr-*para*-aminobenzoate and related substrate analogs have been found to be potent inhibitors of TOP [14]. No naturally occurring inhibitors of TOP are known.

TOP is primarily a soluble cytosolic enzyme, but 20–25% of TOP activity is associated with membrane fractions of tissue homogenates, possibly with endosomes. An endosomal location would permit TOP to degrade exogenous substrates after their internalization, in addition to the degradation of cytosolic oligopeptides. There have been recent suggestions that TOP in the brain could be involved in the progression of Alzheimer's disease due to its cleavage of the β-amyloid precursor protein to generate amyloidogenic fragments [15, 16].

Neurolysin (oligopeptidase M, EC 3.4.24.16), a metallopeptidase found in the cytosol and synaptosomes, is similar to TOP in many respects. The two enzymes have similar substrate specificities, but can be distinguished from each other by their differing sensitivities to inhibitors [17]. Neurolysin has been reported to be enzymatically indistinguishable from mitochondrial oligopeptidase (MOP, [18]). The structural basis for this has recently been uncovered through sequence analysis of the single gene encoding both neurolysin and MOP [19]. The use of alternative start sites for transcription results in the synthesis of mRNAs that encode proteins differing only at their NH_2-termini. The longer NH_2-terminus of MOP contains a mitochondrial targeting sequence, resulting in its mitochondrial localization. Neurolysin lacks this targeting signal and is therefore expressed in the cytosol. Thus the use of alternative promoters of a single gene allows the production of two highly similar proteins that perform different functions due to the fact that they are expressed in different cellular compartments. Neurolysin is predominantly cytosolic, and its deduced amino acid sequence contains no obvious transmembrane segments [20]. In spite of this, it has been purified from brain synaptic membranes, where it is localized in a glycosyl-phophatidylinositol-independent manner [21, 22]. Its colocalization with neurotensin receptors and ability to cleave neurotensin implicate it in the *in vivo* inactivation of this peptide [23].

Insulysin, insulin-degrading enzyme (EC 3.4.24.56)

The existence of a specific insulin-degrading enzyme (IDE) was first reported forty years ago, but the enzyme was not purified in quantities sufficient to permit its characterization until the 1980's. IDE, or insulysin, has been purified from a number of sources, including the cytosol of skeletal muscle, erythrocytes, and liver [24, 25]. Purified IDE was found to have a subunit molecular mass of 110 kDa and optimal activity at neutral pH. IDE is inhibited by metal chelators such as EDTA and 1,10-phenanthroline; and by sulfhydryl reagents such as N-ethylmaleimide and *para*-chloromercuribenzoate. Insulin is the best-studied *in vitro* substrate for IDE, and several lines of evidence support a role for IDE in the *in vivo* degradation of insulin

[26]. Purified IDE has been shown to degrade other peptides and proteins, including glucagon, atrial natriuretic peptide, transforming growth factor-α, β-endorphin, and insulin-like growth factors, indicating that there may be numerous biological roles for the enzyme [27, 28]. Although insulysin is predominantly cytosolic, biochemical and immunochemical evidence exists for its localization in membranes, endosomes, and peroxisomes [29]. Insulysin has also been purified as an enzyme secreted from thymoma cells [27]. The presence of insulysin both extracellularly and in several intracellular compartments greatly expands the number of potential substrates for this enzyme.

Insulysin has been isolated as a complex with the multicatalytic protease, or proteasome [30]. Furthermore, insulin was found to inhibit the proteasome's degradation of a synthetic substrate, an effect that was not observed when the proteasome and IDE were separated by anion exchange chromatography [31]. Therefore it is possible that insulin binding to IDE is able to downregulate intracellular proteolysis.

Insulysin activity and mRNAs are present in many tissues in addition to liver and kidney, organs known to play a role in insulin degradation [26]. IDE mRNAs have also been detected in heart, skeletal muscle, prostate, intestine, spleen, lung, and thymus, further supporting the idea of multiple physiological roles for insulysin [32]. The highest levels of insulysin mRNA are found in testis, tongue, and brain. Four different insulysin mRNAs, ranging in size from 3.4 to 6.7 kb, have been identified in rat tissues [33].

The amino acid sequence of human insulysin was deduced from its cDNA clone, revealing significant sequence homology with *E. coli* protease III (pitrilysin), but no other known metalloproteinases [34]. Homologous cDNAs have since been cloned from rat and *Drosophila* [32, 35]. Insulysin and related proteins do not contain the canonical zinc-binding sequence HEXXH; instead the inverted sequence HXXEH is present. Site directed mutagenesis studies using human insulysin and bacterial pitrilysin have demonstrated that the conserved residues of the HXXEH consensus sequence are important for zinc binding and catalysis [24, 36]. Metalloproteinases utilizing the inverted HXXEH zinc binding site are termed inverzincins or, alternatively, pitrilysins [4]. Other known inverzincins are the mitochondrial matrix processing peptidase α subunit and N-arginine dibasic convertase.

N-Arginine dibasic convertase (EC 3.4.24.61)

The N-Arginine dibasic convertase (NRD) was originally purified as an activity responsible for the processing of somatostatin precursors. Further characterization of the enzyme has shown that it cleaves a number of somatostatin prohormone fragments on the amino-terminal side of Arg in Arg-Arg and Arg-Lys dibasic sites. Dynorphin and neurotensin precursor fragments were found to be cleaved in the middle of dibasic sequences [37]. Although it is a metalloendopeptidase, it is inhibited by the sulfhydryl reagent N-ethylmaleimide, and the aminopeptidase inhibitors amastatin and bestatin, in addition to metal chelators. NRD has been purified as a soluble enzyme from brain and testis, testis being a much better source of the enzyme. Molecular cloning of the NRD cDNA indicates an open reading frame of 1161 amino acids, consistent with the molecular mass (140 kDa) of the purified enzyme [38]. Other features of the deduced amino acid sequence include an inverzincin zinc-binding motif and a signal sequence, consistent with a putative function of NRD in the secretory pathway.

Mitochondrial processing peptidase (MPP; EC 3.4.24.64) and mitochondrial intermediate peptidase (MIP; EC 3.4.24.59)

The NH_2-terminal targeting signals of nuclear-encoded mitochondrial protein precursors are removed in the mitochondrial matrix by MPP [39]. In mammals and fungi, MPP is a soluble enzyme of the mitochondrial matrix, but is associated with the inner membrane, bound to the cytochrome bc_1 complex, in plants [40]. MPP is a heterodimer of two subunits, α-MPP and β-MPP, both of which are required for precursor processing activity [41]. However, only the β subunit has a conserved inverzincin zinc-binding sequence and is catalytically active [42]. It has been suggested that the α subunit binds the leader peptide and presents it to the β subunit in a structure that promotes its cleavage [43]. Optimal MPP substrates are reported to have a P1' aromatic residue, Arg residues at positions P_2 and P_3, and distal Arg at or near P_{10}, but not all of these features are required for cleavage by MPP [44]. A metallopeptidase related in structure to MPP is involved in precursor processing in chloroplasts [45].

The leader peptides of most imported mitochondrial proteins are removed by MPP in one step. A subset of precursors are first cleaved by MPP, then the remaining octapeptide is processed by mitochondrial intermediate peptidase (MIP, EC 3.4.24.59) to generate mature mitochondrial proteins [46]. MIP is a soluble monomer of 75 kDa present in the mitochondrial matrix. Its thiol-stimulated activity and amino acid sequence indicate a relationship to thimet oligopeptidase [47]. Precursors known to be cleaved *in vivo* by MIP include cytochrome c oxidase subunit IV, ubiquinol-cytochrome c reductase iron-sulfur subunit, malate dehydrogenase, and ribosomal proteins L20 and S28 [46].

ATP-hydrolyzing metalloproteases of mitochondria

Recent work has led to the discovery of three proteases associated with the yeast mitochondrial inner membrane that also hydrolyze ATP; Yme1, Afg3p, and Rca1p [48]. Molecular cloning of these proteins indicates that they possess both ATP-binding and metalloprotease domains. The gene for Yme1 was isolated by complementation of a mutant defective in the degradation of cytochrome c subunits that failed to be incorporated into cytochrome c complexes [49]. Afg3p and Rca1p have been implicated in the assembly of multi-subunit mitochondrial complexes [50]. It has therefore been suggested that these proteins possess an ATP-hydrolyzing chaperone activity used to promote proper folding of mitochondrial proteins and a metalloprotease activity capable of degrading polypeptides that fail to fold properly and are not incorporated into functional complexes. Thus this class of metalloproteases would serve a quality control function in the mitochondrion, preventing the accumulation of misfolded proteins and incompletely assembled complexes in this organelle. Homologs of these enzymes exist in bacteria, but have yet to be identified in higher eukaryotes.

A sterol-regulatory element-binding protein (SREBP) protease – S2P

A gene that encodes a putative metalloprotease, termed S2P, that is required for intramembrane proteolysis of SREBPs, transcription factors important for cholesterol biosynthesis, has recently been cloned [51]. SREBPs (1 and 2) are transcription factors that are ubiquitously expressed. It has been known that sterols regulate the biosynthesis and uptake of cholesterol, but recently the role of proteases in the regulatory process has been elucidated [51, 52]. SREBPs regulate transcription of hydroxy methyl glutaryl (HMG) CoA synthase and other enzymes involved in cholesterol and fatty acid metabolism. The SREBPs are membrane-spanning proteins with an NH_2 terminal domain in the cytosol, a hydrophobic region of approximately 80 amino acids containing two hydrophobic transmembrane segments (the loop between the membrane segments is in the ER), and a COOH-terminal domain in the cytosol. Two proteases involved in the release of SREBPs from the ER membrane have been identified: a sterol-regulated protease that cleaves after a tetrapeptide RXXL in the lumen of the ER, and a membrane-associated protease that cleaves SREBP within the transmembrane segment and releases the NH_2-terminal segment of SREBP which then is transported to the nucleus to stimulate transcription of genes involved in uptake and synthesis of cholesterol and fatty acids. The protease that cleaves the intramembrane segment of SREBP (S2P) is a membrane protein that contains an HEXXH sequence that is essential for activity.

S2P is distinct from the previously described metalloproteases, and defines a new family of metalloproteases. Members of this family have been found in archaea, insects, flatworms, roundworms, hamsters, and humans. No S2P homologue was found in *Saccharomyces cerevisiae*, however, an unrelated hydrophobic metalloprotease, designated Ste24p, has been described [53]. This protease is thought to be located in the ER, and active in the processing of yeast mating pheromone a-factor. Ste24p homologues have been found in bacteria and humans, indicating that this subgroup of enzymes is also conserved throughout evolution.

Extracellular cell-associated metalloendopeptidases

The subunits of most known cell-surface metalloendopeptidases are anchored to the plasma membrane by a single membrane-spanning hydrophobic peptide. The transmembrane domain can be located either near the COOH-terminus or the NH_2-terminus, making the enzymes Type I or Type II integral membrane proteins, respectively. In some instances, the enzymes are anchored to the cell surface by interactions (covalent or non-covalent) with other cell-surface proteins. For the cell-surface metalloproteases that have cytoplasmic domains, this domain is small, and the greater part of their mass, including the catalytic domain, is extracellular. Such proteins are termed ectoenzymes. These metallopeptidases often possess domains in addition to their catalytic domains that modify their function by promoting interactions with other extracellular proteins. Cell-surface metallopeptidases are involved in numerous cellular functions, including the degradation of proteins and peptides in the digestive tract, the inactivation and generation of biologically active peptides at the cell surface, the proteolytic processing of other extracellular proteins, and the release of other ectodomains from the cell membrane.

Meprin A (EC 3.4.24.18) and meprin B (EC 3.4.24.63)

Meprin A was first purified as an azocasein-degrading metalloproteinase from the microvillar membranes of mouse kidney [54]. It is abundant in mammalian kidney and intestine, but has been detected at low levels in other tissues such as thyroid and salivary glands [55]. It is a highly glycosylated oligomeric protein composed of α and/or β subunits that are linked through intersubunit disulfide bonds. The individual subunits range in molecular mass from 82 kDa to 110 kDa, depending on the species and degree of post-translational modification [56]. Secreted and plasma membrane associated forms are known to exist [57]. Meprins are enzymatically active at neutral and alkaline pH. Meprin A hydrolyzes protein substrates such as gelatin and Type IV collagen as well as a number of biologically active peptides [56]. Meprin A also displays arylamidase activity, releasing naphthylamine, amino-methylcoumarin, and *para*-nitroaniline from the COOH-termini of several synthetic peptides [57]. Meprin A has been shown to hydrolyze peptide bonds with virtually any amino acid in the P1 and P1' positions. Meprin B, composed exclusively of β subunits, degrades azocasein but is virtually inactive against most of the small peptide substrates that are efficiently hydrolyzed by Meprin A [58]; an exception is that meprin B degrades gastrin [59]. Like many metalloendopeptidases, meprins are inhibited by thiol compounds, peptide hydroxamates, and metal chelators. The peptide hydroxamate actinonin is a relatively good inhibitor of meprin A (Ki = 0.15 μM), but it is also known to inhibit other metallopeptidases, particularly aminopeptidases [56]. Meprins are not inhibited by thiorphan or phosphoramidon (neprilysin inhibitors), captopril (inhibitor of angiotensin I-converting enzyme) or TIMPs (inhibitors of matrixins).

The abundance of meprins in mammalian kidney proximal tubules and intestine indicates that meprins are involved in the degradation of proteins and peptides of the glomerular filtrate and may participate in the digestion of proteins and peptides in the intestinal tract. Rodent models have been used to study the role of meprins in renal disease. Meprin activities were found to be elevated in streptozocin-induced diabetes [60]. In another study it was observed that meprin biosynthesis is abruptly downregulated in response to experimentally induced hydronephrosis. It was proposed that meprin downregulation could protect the kidney from proteolytic damage in the disease state [61]. The ability of meprin to degrade Type IV collagen implicates meprin in the remodeling of extracellular matrix [62]. Meprin has been detected by immunohistochemical means in salivary glands, thyroid tissue, and the neuroepithelial cells of the inner ear, nasal cochlea, and choroid plexus during embryonic development [55, 63]. The potential roles of meprins in embryonic development have yet to be fully explored. Meprin B has been purified from mouse kidney as a latent proteinase [57]. The presence of significant amounts of inactive meprin B in the kidney could mean that the protein has functions independent of its proteinase activity. Such functions might include serving as a cell surface receptor or promoting the formation of multiprotein complexes at the cell surface through protein-protein interactions. Meprin B has been shown to have enzymatic activity identical to that of the kinase-splitting membranal proteinase, an enzyme that inactivates the catalytic subunit of protein kinase A and cleaves gastrin [59, 64].

cDNAs encoding the meprin α and β subunits have been cloned and sequenced from rat, mouse, and human [57]. As deduced from their cDNAs, the mouse α and β subunits are 42%

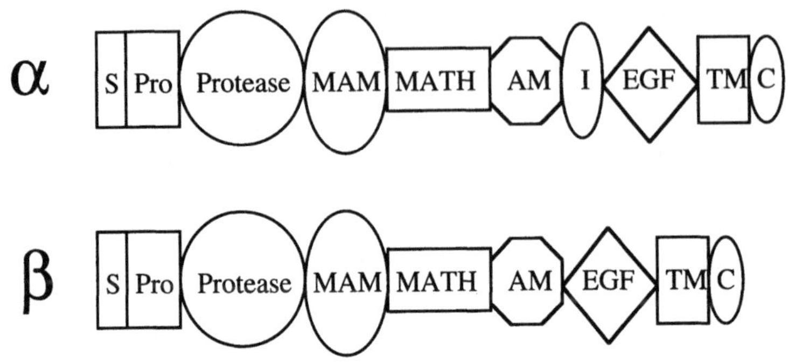

Figure 1. Structural domains of the meprin α and β subunits. The primary translation products (α, 760 amino acids; β, 704 amino acids) of both subunits contain NH_2-terminal signal peptides and prosequences that are not present in the mature, active forms of meprin. The 198 amino acid protease domains, homologous to the crayfish metalloproteinase astacin, have the greatest amount of sequence identity between the meprin subunits. The MAM domains [65] contain cysteine residues that are required for intersubunit disulfide bridge formation [66]. The MATH domain is likely to be an adhesive domain that promotes protein-protein interactions [67]. The AM (AfterMATH) domains have no known functions nor homology to other domains of known function. Both subunits possess domains with homology to epidermal growth factor (EGF), a module present in many extracellular proteins. Both cDNAs encode a potential transmembrane (TM) domain near their COOH-termini, characteristic of Type I integral membrane proteins, and small cytoplasmic domains (C) of unknown function. The principal difference between the two subunits is the presence of a 56 amino acid domain unique to the α subunit (I) inserted between the AM and EGF domains.

identical in amino acid sequence and share a similar arrangement of structural domains. The domains of meprin α and β subunits are compared in Figure 1.

Heterologous expression of meprin subunits has revealed how post-translational modifications affect their function. Expression of α subunits in mammalian cell lines resulted in their secretion, but β subunits behaved as Type I integral membrane proteins. Coexpression of both subunits resulted in the localization of meprin heterodimers to the cell surface, as observed *in vivo* [68]. The secretion of α subunits was found to be a result of proteolytic processing of the α subunit in or near the I domain that occurs in the endoplasmic reticulum, removing its transmembrane domain [69]. When coexpressed with β subunits, α subunits are anchored to the cell surface by intermolecular disulfide bonds with β subunits. Like other enzymes of the astacin family of metalloendopeptidases, meprins retaining propeptides are enzymatically inactive [57]. Incomplete or imprecise removal of the meprin prosequence has been shown to result in a partially active and relatively unstable enzyme, as has been observed for other metzincins such as the matrixins [70].

Procollagen C-proteinases (EC 3.4.24.19): bone morphogenetic protein-1 and tolloid

The fibrillar collagens (Types I, II, and III) are secreted from fibrogenic cells as precursors possessing NH_2- and COOH-terminal propeptides. Proteolytic removal of these propeptides produces insoluble collagen units that associate to form collagen fibrils. Procollagen C-proteinase (pCP) is a secreted zinc metalloproteinase that is responsible for the removal of the COOH-terminal propeptides of procollagen Types I, II, and III. pCP has been purified after its secretion from fibroblasts derived from chick embryo tendons [71]. Its purification from cultured mouse fibroblast medium was aided by affinity chromatography on a column of Type I collagen COOH-terminal propeptide coupled to Sepharose [72]. Depending on its source, the molecular mass of the enzyme has been reported to be between 80- and 120 kDa. pCP is active at neutral pH, is stabilized by Ca^{2+} ions, and removes procollagen COOH-terminal propeptides by a specific cleavage of Ala-Asp or Gly-Asp bonds. There is evidence that pCP also acts as a maturase for prolysyl oxidase, an enzyme required for the crosslinking of collagen fibrils [73]. Thus pCP may play a role in the regulation of several steps of collagen biosynthesis.

NH_2-terminal sequence analysis of pCP revealed that it is identical to Bone Morphogenetic Protein-1 (BMP-1), a putative metalloproteinase cloned as part of a mixture of bone-inducing proteins purified from mammalian bone matrix [74, 75]. cDNAs encoding BMP-1 and a related protein known as tolloid were subsequently cloned from a number of eukaryotes [57]. Human BMP-1 and tolloid mRNAs are transcribed from a single gene by alternative splicing. The mRNAs encode proteins that are identical except that the tolloid mRNA encodes several additional COOH-terminal domains [76]. BMP-1 and tolloid possess NH_2-terminal metalloprotease domains related to astacin and repeats of EGF (epidermal growth factor)-like and CUB (complement components clr/cls, 4egf, BMP-1) domains that may be involved in calcium binding and interactions with other extracellular proteins [77]. Heterologous expression of BMP-1 and tolloid has confirmed that both proteins possess procollagen C-proteinase activity [74, 78]. The domain structures of human BMP-1 and tolloid as deduced from their cDNAs are diagrammed in Figure 2.

Genetic analyses of dorsal-ventral patterning in *Drosophila* indicated that tolloid is involved in the activation of proteins related to transforming growth factor-β [79, 80]. During embryogenesis, the *Drosophila* dorsoventral axis is established by a concentration gradient of decapentaplegic (dpp) activity, a TGFβ-like morphogen. The binding of a protein called short gastrulation (sog) prevents dpp from activating its receptor, but cleavage of sog by tolloid releases active dpp [81]. Analagous pathways have been described in Xenopus and zebrafish, demonstrating a conserved function for tolloid in the embryonic development of eukaryotes [82, 83].

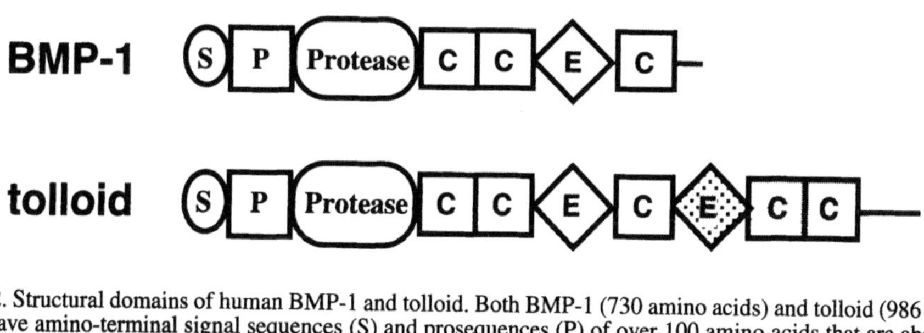

Figure 2. Structural domains of human BMP-1 and tolloid. Both BMP-1 (730 amino acids) and tolloid (986 amino acids) have amino-terminal signal sequences (S) and prosequences (P) of over 100 amino acids that are absent in the mature, enzymatically active proteins. The amino termini of the mature proteins begin with protease domains (about 200 amino acids) homologous to astacin. The COOH-terminal domains are composed of repeats of CUB (C, 110 amino acids) and EGF-like (E, 45 amino acids) domains. The CUB domains were first identified in Complement proteins C1r and C1s, Uegf, and BMP-1 [77]. The first three CUB and first EGF-like domain are present in both BMP-1 and tolloid. Tolloid contains two additional CUB domains and one additional EGF-like domain (stippled) relative to BMP-1.

Metalloprotease-disintegrins

Metalloprotease-disintegrins are Type-I integral membrane proteins with NH_2-terminal metalloprotease, disintegrin, and cysteine-rich domains related to those of the secreted snake venom metalloproteases (SVMPs). Because of this, the proteins are often referred to by the acronyms ADAM (A Disintegrin And Metalloprotease domain) and MDC (Metalloprotease/Disintegrin/Cysteine-rich). In addition to these domains, ADAMs possess a domain related to epidermal growth factor (EGF-like) and a single transmembrane domain near their COOH-termini. A model of an ADAM ectoenzyme anchored to the cell membrane is shown in Figure 3.

ADAMs have a domain organization similar to the larger SVMPs, except that SVMPs lack transmembrane and EGF-like domains. The metalloprotease domains of ADAMs and SVMPs contain about 200 amino acids and are highly related in amino acid sequence, allowing these enzymes to be classified as reprolysins, a subgroup of the metzincins, with the active site consensus sequence HEXXHXXGXXHD [84]. Like SVMPs, nascent ADAMs possess a signal sequence and a prodomain that must be proteolytically removed to generate mature, active enzymes. The prodomains contain a "cysteine switch" peptide which helps to maintain the enzyme in an inactive state [85]. Thus the reprolysins, like the astacins and matrix metalloproteinases, must undergo prosequence removal in order to be fully active. In some cases, ADAMs are processed further to remove the protease domain [86]. Furthermore, some ADAMs have

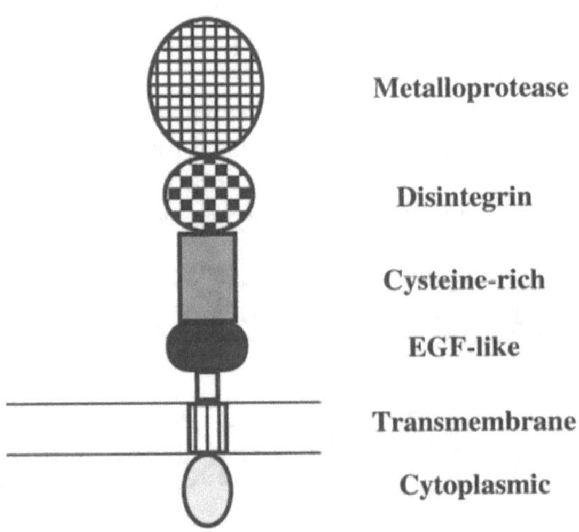

Figure 3. Structural organization of an ADAM at the cell surface. Mature, enzymatically active ADAMs contain NH_2-terminal metalloprotease domains of approximately 200 amino acids, disintegrin domains of about 90 amino acids, cysteine-rich and EGF-like domains of about 150 and 40 residues, respectively. ADAMs are anchored to the cell surface by a single transmembrane domain, and have COOH-terminal cytoplasmic domains of differing sizes. Not shown are the NH_2-terminal signal- and prosequences of nascent ADAMs that are proteolytically processed to generate the mature proteins.

changes in zinc-binding and catalytically important residues that render them inactive, indicating that ADAMs can have functions independent of their catalytic activity and that structural features of their protease domains may have biological functions distinct from their protease activity. Some ADAM functions are likely to be due to the structure of the disintegrin domain. Although related to snake venom disintegrins, most ADAM disintegrin domains lack the Arg-Gly-Asp tripeptide known to bind to the integrin $a_{IIb/b}3$ [87]. It is therefore possible that ADAMs function by binding to a different subset of integrins or to proteins distinct from integrins. Metargidin (ADAM15) is an example of an ADAM possessing the Arg-Gly-Asp peptide in its disintegrin domain [88]. The cysteine-rich domain of ADAMs (and some SVMPs) is not clearly related to domains present in other proteins, but in some cases this domain contains a peptide believed to be involved in the fusion of cell membranes [89]. EGF-like domains are present in many extracellular proteins, but the function of this domain in ADAMs remains unclear at this time. The cytoplasmic domains of some ADAMs possess peptides that are possible phosphorylation sites for protein kinase C or peptides that potentially bind to the *Src*-homology 3, or SH3, domain [90]. In one case an ADAM has been shown to bind a SH3 domain [91]. It is therefore possible that the cytosolic domains of ADAMs could be involved in protein-protein interactions as a part of a signaling pathway or to influence the cellular location of the protein. The cDNAs encoding more than 20 ADAMs have been isolated and sequenced (see [92]), yet the functions of most have not been elucidated. The remainder of this section will focus on several ADAMs for which functions have been recently found.

Tumor necrosis factor-α converting enzyme (TACE)

Domains of many ectoproteins are proteolytically released, or shed, from the cell surface. Among them are proTNF-α, L-selectin, interleukin 6 receptor α subunit, and the β amyloid precursor protein [93]. The precursor of soluble TNF-α is a 26 kDa ectoprotein. Soluble TNF-α is a 17 kDa inflammatory cytokine that is released from the cell surface by proteolysis in a stalk region of proTNF-α located near the plasma membrane. TACE was purified and found to be a membrane-bound metalloprotease of about 85 kDa, and it was cloned independently by two groups to reveal that the enzyme is an ADAM [90, 94]. Native and recombinant TACE are both able to process proTNF-α *in vitro*. It was also demonstrated that mice homozygous for disruptions in the TACE gene are deficient in the metalloprotease activity that releases TNF-α from T cells, supporting a role for TACE in the *in vivo* release of TNF-α from the cell surface [94]. However, TACE may not be the only ADAM capable of processing proTNF-α *in vivo*. It was recently reported that the cell-surface enzyme ADAM10 is capable of releasing TNF-α from cells and displays specificity for the cleavage site found in proTNF-α [95, 96].

It is known that the shedding of ectodomains from cells occurs in response to protein kinase C activators and is sensitive to metalloprotease inhibitors [93]. It is also known that the cleavage responsible for shedding occurs at an accessible stalk region near the cell membrane with no obvious sequence similarity among the cleavage sites of different ectoproteins [97]. This secretase activity could therefore be accomplished by a battery of different enzymes or by a single protease such as TACE with relaxed substrate specificity. Because evidence already exists for the presence of secretases of different mechanistic classes of proteases, it is likely that a number of secretases are present on the cell surface [98]. These secretases may have different substrate specificities or may be regulated in response to different stimuli in order to release different populations of ectodomains from the cell surface.

KUZ/ADAM10

The *kuzbanian* gene of *Drosophila* is required for the partitioning of neural and non-neuronal cells during embryonic development. The product of this gene, KUZ, is an ADAM with a highly conserved mammalian homolog, now known as ADAM10 [99]. Recent studies indicate that KUZ is required for the proteolytic processing of the transmembrane receptor, Notch, in its extracellular domain [100]. This cleavage produces amino- and carboxyl-terminal fragments of Notch that associate to form a functional cell-surface receptor [101]. The processing of Notch by KUZ may occur in the trans-Golgi network, indicating that some ADAMs function in the secretory pathway. Since both Notch and KUZ have conserved mammalian homologs, it is likely that this processing event is conserved in higher eukaryotes to activate a signaling pathway.

Fertilin/PH-30

Fertilin is a heterodimeric protein composed of α and β subunits found on the cell surface of sperm involved in sperm-egg fusion. The precursors of both subunits are full-length ADAMs, but are processed during maturation to remove the protease domains [102]. The available data indicate that their function in sperm-egg fusion is mediated by their disintegrin and cysteine-rich domains. The fertilin β subunit contains sequences in its disintegrin domain that allow binding to components of the egg membrane [87]. Antibodies raised against this peptide were shown to inhibit fertilization, supporting a role for this sequence in egg recognition [103]. Antibodies against egg integrin $\alpha6\beta1$ also inhibit fertilization, making this cell surface molecule a candidate for a receptor to the fertilin β subunit or other sperm ADAMs [104]. The fertilin α subunit contains a peptide related to viral fusion peptides in its cysteine-rich domain which has been shown to interact with membranes [105]. The ADAM meltrin-α has been shown to play a role in myoblast fusion, indicating a general role for ADAMs in cell fusion [89].

Neprilysin (neprilysin, EC 3.4.24.11) and related metallopeptidases

Neprilysin is the best characterized member of a recently recognized class of cell surface metallopeptidases involved in peptide hormone processing and metabolism [106]. These proteins are Type II integral membrane proteins, ectoenzymes anchored to the cell membrane by an uncleaved NH_2-terminal signal sequence. All proteins of this class possess an extracellular domain of about 700 amino acids that retain a core of ten conserved cysteines likely to be involved in intramolecular disulfide bonds, indicating a conserved three-dimensional structure for all. The peptidase domains, the most highly conserved regions, occupy the COOH-terminal 350 residues of the proteins. Amino acid sequence conservation, mutagenesis studies, and inhibition by phosphoramidon indicate that these enzymes are gluzincins, related to thermolysin. There is no requirement for proteolytic processing of a prosequence for activity of these enzymes, distinguishing them from the metzincins. Metallopeptidases related to neprilysin include isoforms of endothelin-converting enzyme (ECE-1 and ECE-2); KELL, an erythrocyte cell surface antigen; and PEX, the product of a gene associated with hypophosphatemic rickets. Figure 4 is a representation of the domain structure of neprilysin and related proteins, showing that the proteins are similar in their extracellular domains, but differ greatly in their small cytoplasmic domains.

Neprilysin is routinely isolated from mammalian kidneys, where it is a major protein of the proximal tubule microvillar membrane [107, 108]. It is a 94 kDa glycosylated ectoenzyme, is most active at neutral pH, and, like thermolysin, cleaves peptides on the amino-terminal side of hydrophobic residues. Neprilysin is a relatively non-specific peptidase able to degrade many biologically active peptides, but has little activity against protein substrates. Mutagenesis and

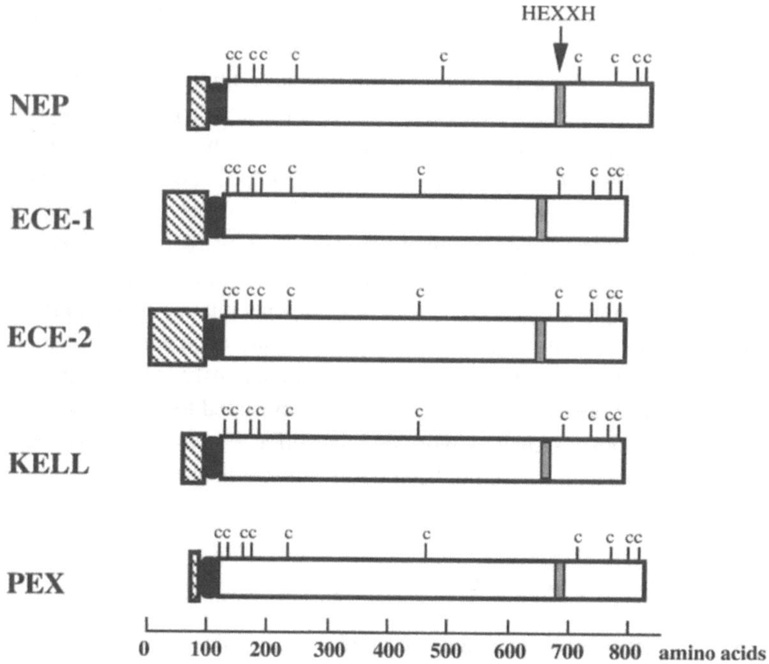

Figure 4. Structural conservation among neprilysin and related proteins. All proteins of this group have NH_2-terminal cytoplasmic domains (cross-hatched rectangles) that differ greatly in size and amino acid sequence, followed by transmembrane domains (black ovals). The extracellular domains (open rectangles) contain ten conserved cysteine residues, denoted by C, that are probably involved in intrasubunit disulfide bonds. The greatest amount of sequence homology is found in the COOH-terminal regions, particularly surrounding the conserved zinc-binding sequence HEXXH, shown as a stippled rectangle.

chemical modification experiments guided by the known structure of thermolysin have identified a number of residues essential for substrate binding and catalysis in neprilysin [109]. His^{583}, His^{587}, and Glu^{646} are zinc ligands. His^{711} and Glu^{584} are residues required for catalysis. Residues believed to be involved in substrate binding include Arg^{102}, Arg^{717}, and Asn^{542}. It is expected that conserved residues in ECE, KELL, and PEX will have similar roles.

Neprilysin is most abundant in the kidney, but is also present at low levels in a number of other tissues, such as the nervous system, lung, testis, placenta, and lymphoid tissues [108]. Three alternatively spliced mRNAs encoding neprilysin have been detected [110]. The mRNAs differ only in their 5' non-coding sequences, and are regulated in a tissue-specific manner as a result of alternative promoter usage [111]. Because it has been rediscovered a number of times in different tissues, neprilysin has also been known by names such as enkephalinase, CALLA (common acute lymphoblastic leukemia antigen), and CD10 [112]. Evidence exists that neprilysin degrades a number of biologically active peptides *in vivo*, thereby inactivating them. Among these are enkephalins, bradykinin, substance P, atrial natriuretic peptide, gastrin releasing peptide, and endothelins [113]. The inactivation of these peptides by neprilysin points to roles for the enzyme in both nociception and the regulation of blood pressure. Recent work indicates that neprilysin activity is involved in the regulation of T cell activation [114].

The clinical relevance of neprilysin has led to a considerable effort in the development of potent, specific inhibitors of the enzyme. Potent peptide inhibitors of neprilysin generally possess a hydrophobic P1' residue and a group such as a thiolate, hydroxamate, or carboxylate positioned to coordinate the active site zinc ion. The design and testing of such compounds has been reviewed in detail elsewhere, so will be summarized here only briefly [109, 113, 115]. The first synthetic, potent neprilysin inhibitors were also found to inhibit angiotensin converting enzyme (ACE), a potentially undesirable side effect. This problem has been overcome by the development of the peptide hydroxamate compounds HACBO-Gly (K_i of 1.7 nM for neprilysin, >10,000 nM for ACE) and RB104 (K_i of 0.03 nM for neprilysin, >10,000 nM for ACE). Both neprilysin and aminopeptidase N are able to inactivate enkephalins by degrading them, and it has been observed that inhibitors of these enzymes have antinociceptive effects. In an attempt to produce analgesic drugs, compounds able to simultaneously inhibit both neprilysin and aminopeptidase N have been developed. Compounds developed to date are not yet as effective at controlling pain as opiates such as morphine, but offer the potential advantages that tolerance and physical dependence have not been observed with their use.

Endothelin-converting enzymes (ECEs)

Endothelins (ETs) are peptides secreted by vascular endothelial cells that are among the most powerful vasoconstrictors known [116]. The endothelins are known to play important roles in the regulation of blood pressure and renal function. Three distinct ET genes have been identified, encoding the precursors of ET-1, ET-2, and ET-3, peptides with similar biological effects. The final step in the generation of ET-1 is the cleavage of its 38 amino acid precursor "big ET-1" at its Trp^{21}-Val^{22} bond to produce active ET-1 (residues 1–21) by an activity termed endothelin-converting enzyme (ECE). ECE-1 (EC 3.4.24.71) was purified from endothelial cells and found to be a membrane-bound metallopeptidase with a subunit molecular mass of 130 kDa [117]. The human ECE-1 cDNA encodes a 758 residue Type II integral membrane protein, an ectoenzyme with 37% identity to neprilysin [118]. As shown in Figure 4, ten Cys residues are conserved between neprilysin and ECE-1, evidence of conserved tertiary structure between the two enzymes. Molecular cloning has revealed that at least three isoforms of ECE-1 exist [119, 120]. The ECE-1 isoforms, probably produced by alternative splicing or alternative promoter usage, differ only in their NH_2-terminal regions; their extracellular domains are identical. The ECE-1α mRNA is the most abundant one in all tissues tested, including lung, liver, kidney, brain, pancreas, and adrenal gland [119]. An additional isoform, ECE-2, with 59% sequence identity to ECE-1, has now been identified [121]. From the available data, it is apparent that the different ECE isoforms are able to process all three big ETs, but that all are most efficient at the conversion of big ET-1.

The ECE isoforms and neprilysin are readily distinguished from each other by structural and enzymatic criteria. ECEs possess two Cys residues not present in neprilysin; at least one of these is responsible for the formation of an intersubunit disulfide bond that makes ECE a disulfide-linked dimer [122]. Neprilysin has a very broad substrate specificity; ECE appears to have a more narrow substrate specificity. For example, neprilysin will cleave endothelins at numerous peptide bonds, but ECE-1 does not digest endothelins further after releasing them from their big ET precursors [123]. However, both ECE-1 and neprilysin are able to hydrolyze bradykinin to release the COOH-terminal dipeptide [124]. ECEs are not significantly inhibited by thiorphan, a potent neprilysin inhibitor (K_i = 4 nM). ECE-2 and neprilysin are highly sensitive to inhibition by phosphoramidon, with Ki values in the range of 2–5 nM, while ECE-1 is much less sensitive to this inhibitor, with a K_i of approximately 1 μM [106]. ECE-1 activity is optimal at neutral pH, while ECE-2 has optimal activity at pH 5.5, being virtually inactive at neutral pH, an unusual property for a metallopeptidase [121]. It has been suggested that ECE-2 functions in an acidified cellular compartment, such as the trans-Golgi network lumen, where it could function in the processing of peptides in the secretory pathway. The highest levels of ECE-2 expression are found in neural tissues, where ECE-1 expression is low, raising the possibility that ECE-2 acts upon peptides distinct from endothelins in neural tissues.

KELL and PEX, putative metallopeptidases

KELL, a 93 kDa glycoprotein expressed on the surface of erythrocytes, is one of the major blood group antigen systems in humans; its expression appears to be restricted to erythroid cells [125]. Incompatibility between KELL antigens can result in hemolytic reactions to blood transfusions and erythroblastosis in newborn children [126]. The KELL cDNA encodes a Type II integral membrane protein with remarkable sequence similarity to neprilysin and ECE, including conserved cysteine residues important for tertiary structure, and conserved residues required for zinc binding and catalysis [127]. This similarity to the functional peptidases ECE and neprilysin indicates that KELL should be an active peptidase, but no substrates have yet been reported for KELL.

The most common genetic cause of rickets in humans is X-linked hypophosphatemia (XLH), which results from impaired renal phosphate reabsorption. A gene believed to be responsible for XLH has been identified through positional cloning. This gene has been designated PEX, for Phosphate-regulating neutral Endopeptidase on the X chromosome [128]. The PEX cDNA encodes an ectoenzyme similar to neprilysin, ECE, and KELL [129]. PEX mRNA is expressed in significant amounts only in bone tissues such as calvaria, teeth, and osteoblasts. Like KELL, PEX retains residues conserved in neprilysin and ECE that are required for zinc binding and catalysis, but has no known substrates either *in vivo* or *in vitro*. The role PEX plays in phosphate absorption and the XLH disease state remains unknown.

Membrane-type matrix metalloproteinases (MT-MMPs)

The matrixins (MMPs) play a role in the turnover of extracellular matrix proteins, and are implicated in normal processes of development and tissue remodeling as well as pathological processes such as arthritis, cancer, and artherosclerosis. There are several reviews about MMPs, and the properties of these enzymes will not be discussed herein [4–7]. Most of the characterized MMPs are secreted proteins, however, there is a subgroup of this family that are membrane bound; the MT-MMPs, membrane-type matrix metalloproteinases, also called MMP-14, -15, -16, -17. The genes for MMP-14, -15, and -16 have been localized to human chromosomes 14q12.2, 16q12.2-q21, and 8q21, respectively [130]. The MT-MMPs are Type I membrane proteins that have the potential to be activated by furin-like enzymes. One of the functions of the MT-MMPs is to activate secreted MMPs. For example, MMP-14 participates in the cell surface activation of progelatinase, MMP-2 (e.g. [131]). The activation requires the formation of a trimolecular complex of MMP-14, proMMP-2, and TIMP-2.

Procollagen N-proteinases

Procollagen N-proteinases (pNPs) catalyze the removal of NH_2-terminal propeptides from procollagen Types I, II, and III. pNP I (EC 3.4.24.14) is specific for procollagen Types I and II, while a different enzyme cleaves the NH_2-terminal propeptides of Type III procollagen[132]. Deficiencies of these enzymes are one cause of Ehlers-Danlos Syndrome, characterized by skin fragility due to abnormally polymerized collagen. The assay and purification of both forms of pNP has been reviewed by Kadler et al. [132]. Both enzymes are zinc metalloendopeptidases that are active at neutral pH and require Ca^{2+} ions for maximal activity. The pNPs cleave procollagen NH_2-terminal propeptides at specific Ala-Gln or Pro-Gln bonds. pNP I has been purified as part of a heterooligomeric complex of 500 kDa, with catalytic activity attributed to subunits of 161 and 135 kDa. Cloning of the pNP I cDNA has shown that the enzyme contains an NH_2-terminal protease domain related to the reprolysins and four repeats of a domain with homology to domains found in properdin and thrombospondins in its COOH-terminal region [133]. The highest levels of pNP I expression are in bone, skin, tendon, and aorta. However, pNP I mRNA and activity can be detected in many other tissues.

Concluding remarks

It is clear that the cell-associated metalloproteases are a remarkably diverse, complex, and highly regulated group of proteases. We are only just beginning to understand the structure and function of these enzymes, and how they participate in multiple physiological and pathological processes.

References

1 Jiang W, Bond JS (1992) Families of metalloendopeptidases and their relationships. *FEBS Lett* 312: 110–114
2 Barrett AJ (ed) (1995) Proteolytic Enzymes: Aspartic and Metallo Peptidases. *Meth Enzymol* 248. Academic Press, San Diego
3 Stöcker W, Grams F, Baumann U, Reinemer P, Gomis-Ruth F-X, McKay DB, Bode W (1995) The metzincins – Topological and sequential relations between the astacins, adamalysins, serralysins, and matrixins (collagenases) define a superfamily of zinc-peptidases. *Protein Sci* 4: 823–840
4 Hooper NM (ed) (1996) *Zinc Metalloproteases in Health and Disease.* Taylor and Francis, London
5 Birkedal-Hansen H (1995) Proteolytic remodeling of extracellular matrix. *Curr Opin Cell Biol* 7: 728–735
6 Stetler-Stevenson WG, Liotta LA, Kleiner DE (1993) Extracellular matrix 6: role of matrix metalloproteinases in tumor invasion and metastasis. *FASEB J* 7: 1434–1441
7 Parks WC, Mecham RP (eds) (1998) *Matrix Metalloproteinases*. Academic Press, San Diego
8 Chu TG, Orlowski M (1985) Soluble metalloendopeptidase from rat brain: action on enkephalin-containing peptides and other bioactive peptides. *Endocrinology* 116: 1418–1425
9 Shrimpton CN, Glucksman MJ, Lew RA, Tullai JW, Margulies EH, Roberts JL, Smith AI (1997) Thiol activation of endopeptidase EC 3.4.24.15. *J Biol Chem* 272: 17395–17399
10 Barrett AJ, Tisljar U (1989) The activities of "Pz-peptidase" and endopeptidase 24.15 are due to a single enzyme. *Biochem J* 261: 1047–1050
11 McKie N, Dando PM, Rawlings ND, Barrett AJ (1993) Thimet oligopeptidase: similarity to "soluble angiotensin II-binding protein". *Biochem J* 295: 57–60
12 Dando PM, Brown MA, Barrett AJ (1993) Human thimet oligopeptidase. *Biochem J* 294: 451–457
13 Tisljar U, Knight CG, Barrett AJ (1990) An alternative quenched fluorescence substrate for Pz-peptidase. *Anal Biochem* 186: 112–115
14 Orlowski M, Michaud C, Molineaux CJ (1988) Substrate-related potent inhibitors of brain metalloendopeptidase. *Biochemistry* 27: 597–602
15 Papastoitsis G, Siman R, Scott R, Abraham CR (1994) Identification of a metalloprotease from Alzheimer's disease brain able to degrade the β-amyloid precursor protein and generate amyloidogenic fragments. *Biochemistry* 33: 192–199
16 Thompson A, Huber G, Malherbe P (1995) Cloning and functional expression of a metalloendopeptidase from human brain with the ability to cleave a β-APP substrate peptide. *Biochem Biophys Res Commun* 213: 66–73
17 Jiracek J, Yiotakis A, Vincent B, Checler F, Dive V (1996) Development of the first potent and selective inhibitor of the zinc endopeptidase neurolysin using a systematic approach based on combinatorial chemistry of phosphinic peptides. *J Biol Chem* 271: 19606–19611
18 Serizawa A, Dando PM, Barrett AJ (1995) Characterization of a mitochondrial metallopeptidase reveals neurolysin as a homologue of thimet oligopeptidase. *J Biol Chem* 270: 2092–2098
19 Kato A, Sugiura N, Saruta Y, Hosoiri T, Yasue H, Hirose S (1997) Targeting of endopeptidase 24.16 to different subcellular compartments by alternative promoter usage. *J Biol Chem* 272: 15313–15322
20 Dauch P, Vincent JP, Checler V (1995) Molecular cloning and expression of rat brain endopeptidase 3.4.24.16. *J Biol Chem* 270: 27266–27271
21 Checler F, Vincent JP, Kitabgi P (1986) Purification and characterization of a novel neurotensin-degrading peptidase from rat brain synaptic membranes. *J Biol Chem* 261: 11274–11281
22 Barrelli H, Vincent JP, Checler F (1993) Rat kidney endopeptidase 24.16. Purification, physico-chemical characteristics and differential specificity towards opiates, tachykinins and neurotensin-related peptides. *Eur J Biochem* 211: 79–90
23 Vincent B, Jiracek J, Noble F, Loog M, Roques B, Dive V, Vincent JP, Checler F (1997) Contribution of endopeptidase 3.4.24.15 to central neurotensin inactivation. *Eur J Pharmacol* 334: 49–53
24 Becker AB, Roth RA (1992) An unusual active site in a family of zinc metalloendopeptidases. *Proc Natl Acad Sci USA* 89: 3835–3839
25 Duckworth WC, Hamel FG (1996) Insulin-degrading enzyme-a new type of metalloprotease. *In*: NM Hooper (ed.): *Zinc Metalloproteases in Health and Disease.* Taylor and Francis, London, 221–240

26 Bondy CA, Zhou J, Chin E, Reinhardt RR, Ding L, Roth RA (1994) Cellular distribution of insulin-degrading enzyme gene expression. Comparison with insulin and insulin-like growth receptors. *J Clin Invest* 93: 966–973
27 Safavi A, Miller BC, Cottam L, Hersh LB (1996) Identification of γ-endorphin-generating enzyme as insulin-degrading enzyme. *Biochemistry* 35: 14318–14325
28 Bennett RG, Hamel FG, Duckworth WC (1997) Characterization of the insulin inhibition of the peptidolytic activities of the insulin-degrading enzyme-proteasome complex. *Diabetes* 46: 197–203
29 Yokono K, Roth R, Baba S (1982) Identification of insulin-degrading enzyme on the surface of cultured human lymphocytes, rat hepatoma cells, and primary cultures of rat hepatocytes. *Endocrinology* 111: 1102–1108
30 Bennett RG, Hamel FG, Duckworth WC (1994) Identification and isolation of a cytosolic proteolytic complex containing insulin degrading enzyme and the multicatalytic proteinase. *Biochem Biophys Res Commun* 202: 1047–1053
31 Duckworth WC, Bennett RG, Hamel FG (1994) A direct inhibitory effect of insulin on a cytosolic proteolytic complex containing insulin-degrading enzyme and multicatalytic proteinase. *J Biol Chem* 269: 24575–24580
32 Baumeister H, Muller D, Rehbein M, Richter D (1993) The rat insulin-degrading enzyme. Molecular cloning and characterization of tissue-specific transcripts. *FEBS Lett* 317: 250–254
33 Kuo WL, Montag AG, Rosner MR (1993) Insulin-degrading enzyme is differentially expressed and developmentally regulated in various rat tissues. *Endocrinology* 132: 604–611
34 Affholter JA, Fried VA, Roth RA (1988) Human insulin-degrading enzyme shares structural and functional homologies with *E. coli* protease III. *Science* 242: 1415–1418
35 Kuo WL, Gehm BD, Rosner MR (1990) Cloning and expression of the cDNA for a *Drosophila* insulin-degrading enzyme. *Mol Endocrinol* 4: 1580–1591
36 Perlman RK, Rosner MR (1994) Identification of zinc ligands of the insulin-degrading enzyme. *J Biol Chem* 269: 33140–33145
37 Chesneau V, Pierotti AR, Barre N, Creminon C, Tougard C, Cohen P (1994) Isolation and characterization of a dibasic selective metalloendopeptidase from rat testes that cleaves at the amino terminus of arginine residues. *J Biol Chem* 269: 2056–2061
38 Pierotti AR, Prat A, Chesneau V, Gaudoux F, Leseney A-M, Foulon T, Cohen P (1994) N-arginine dibasic convertase, a metalloendopeptidase as a prototype of a class of processing enzymes. *Proc Natl Acad Sci USA* 91: 6078–6082
39 Mori M, Miura S, Tatibana M, Cohen PP (1980) Characterization of a protease apparently involved in processing pre-ornithine transcarbamylase of rat liver. *Proc Natl Acad Sci USA* 77: 7044–7048
40 Braun H-P, Emmermann M, Kruft V, Schmitz UK (1992) The general mitochondrial processing peptidase from potato is an integral part of cytochrome c reductase of the respiratory chain. *EMBO J* 11: 3219–3227
41 Geli V (1993) Functional reconstitution in *E. coli* of the yeast mitochondrial matrix peptidase from its two inactive subunits. *Proc Natl Acad Sci USA* 90: 6247–6251
42 Kitada S, Shimokata K, Ogishima T, Ito A (1995) A putative metal binding site in the β subunit of rat mitochondrial processing peptidase is essential for its catalytic activity *J Biochem* 117: 1148–1150
43 Braun H-P, Schmitz UK (1995) Are the "core" proteins of the mitochondrial bc_1 complex evolutionary relics of a processing protease? *Trends Biochem Sci* 20: 171–175
44 Ogishima T, Niidome T, Shimokata K, Kitada S, Ito A (1995) Analysis of elements in the substrate required for processing by mitochondrial processing peptidase. *J Biol Chem* 270: 30322–30326
45 Vander Vere PS, Bennett TM, Oblong JE, Lamppa GK (1995) A chloroplast processing enzyme involved in precursor maturation shares a zinc-binding motif with a recently recognized family of metalloendopeptidases. *Proc Natl Acad Sci USA* 92: 7177–7181
46 Branda SS, Isaya G (1995) Prediction and identification of new natural substrates of the yeast mitochondrial intermediate peptidase. *J Biol Chem* 270: 27366–27373
47 Isaya G, Kalousek F, Rosenberg LE (1992) Sequence analysis of rat mitochondrial intermediate peptidase: similarity to zinc metallopeptidases and to a putative yeast homologue. *Proc Natl Acad Sci USA* 89: 8317–8321
48 Suzuki CK, Rep M, van Dijl JM, Suda K, Grivell LA, Schatz G (1997) ATP-dependent proteases that also chaperone protein biogenesis. *Trends Biochem Sci* 22: 118–123
49 Thorsness PE, White KH, Fox TD (1993) Inactivation of YME1, a member of the ftsH-SEC18-PAS1-CDC48 family of putative ATPase-encoding genes, causes increased escape of DNA from mitochondria in *Saccharomyces cerevisiae*. *Mol Cell Biol* 13: 5418–5426
50 Paul M-F, Tzagoloff A (1995) Mutations in RCA1 and AFG3 inhibit F1-ATPase assembly in *Saccharomyces cerevisiae*. *FEBS Lett* 373: 66–70
51 Rawson RB, Zelenski NG, Nijhawan D, Ye J, Sakai J, Hasan MT, Chang TY, Brown MS, Goldstein JL (1997) Complementation cloning of *S2P*, a gene encoding a putative metalloprotease required for intramembrane cleavage of SREBPs. *Molec Cell* 1: 47–57
52 Brown MS, Goldstein JL (1997) The SREBP pathway: regulation of cholesterol metabolism by proteolysis of a membrane-bound transcription factor. *Cell* 89: 331–340
53 Fujimura-Kanada K, Nouvet FJ, Michaelis S (1997) A novel membrane-associated metalloprotease, Ste24p, is required for the first step of NH_2-terminal processing of the yeast a-factor precursor. *J Cell Biol* 136:

271–285
54 Beynon RJ, Shannon JD, Bond JS (1981) Purification and characterization of a metallo-endopeptidase from mouse kidney. *Biochem J* 199: 591–598
55 Craig SS, Mader C, Bond JS (1991) Immunohistochemical localization of the metalloproteinase meprin in salivary glands of male and female mice. *J Histochem Cytochem* 39: 123–129
56 Wolz RL, Bond JS (1995) Meprins A and B. *Meth Enzymol* 248: 325–345
57 Bond JS, Beynon RJ (1995) The astacin family of metallendopeptidases. *Protein Sci* 4: 1247–1261,
58 Kounnas MZ, Wolz RL, Gorbea CM, Bond JS (1991) Meprin-A and -B Cell surface endopeptidases of the mouse kidney. *J Biol Chem* 266: 17350–17357
59 Chestukhin A, Litovchick L, Muradov K, Batkin M, Shaltiel S (1997) Unveiling the substrate specificity of meprin β on the basis of the site in protein kinase A cleaved by the kinase splitting membranal proteinase. *J Biol Chem* 272: 3153–3160
60 Trachtman H, Greenwald R, Moak S, Tang J, Bond JS (1993) Meprin activity in rats with experimental renal disease. *Life Sci* 52: 1339–1344
61 Ricardo SD, Bond JS, Johnson GD, Kaspar J, Diamond JR (1996) Expression of subunits of the metalloendopeptidase meprin in renal cortex in experimental hydronephrosis. *Amer J Physiol* 270: F669–F676
62 Kaushal GP, Walker PD, Shah SV (1994) An old enzyme with a new function: purification and characterization of a distinct matrix-degrading metalloproteinase in rat kidney cortex and its identification as meprin. *J Cell Biol* 126: 1319–1327
63 Spencer-Dene B, Thorogood P, Nair S, Kenny AJ, Harris M, Henderson B (1994) Distribution of, and a putative role for, the cell-surface neutral metallo-endopeptidases during mammalian craniofacial development. *Development* 120: 3213–3226
64 Chestukhin A, Muradov K, Litovchick L, Shaltiel S (1996) The cleavage of protein kinase A by the kinase-splitting membranal proteinase is reproduced by meprin β. *J Biol Chem* 271: 30272–30280
65 Beckmann G, Bork P (1993) An adhesive domain detected in functionally diverse receptors. *Trends Biochem Sci* 18: 40–41
66 Marchand P, Volkmann M, Bond JS (1996) Cysteine mutations in the MAM domain result in monomeric meprin and alter stability and activity of the proteinase. *J Biol Chem* 271: 24236–24241
67 Uren AG, Vaux DL (1996) TRAF proteins and meprins share a conserved domain. *Trends Biochem Sci* 21: 244–245
68 Johnson GD, Hersh LB (1994) Expression of meprin subunit precursors. Membrane anchoring through the β subunit and mechanism of zymogen activation. *J Biol Chem* 269: 7682–7688
69 Marchand P, Tang J, Johnson GD, Bond JS (1995) COOH-terminal proteolytic processing of secreted and membrane forms of the α subunit of the metalloprotease meprin A. Requirement of the I domain for processing in the endoplasmic reticulum. *J Biol Chem* 270: 5449–5456
70 Johnson GD, Bond JS (1997) Activation mechanism of meprins, members of the astacin metalloendopeptidase family. *J Biol Chem* 272: 28126–28132
71 Hojima Y, van der Rest M, Prockop DJ (1985) Type I procollagen carboxyl-terminal proteinase from chick embryo tendons. *J Biol Chem* 260: 15996–16003
72 Kessler E, Adar R (1989) Type I procollagen C-proteinase from mouse fibroblasts. *Eur J Biochem* 186: 115–121
73 Panchenko M, Stetler-Stevenson WG, Trubetsko OV, Gacheru SN, Kagan HM (1996) Metalloproteinase activity secreted by fibrogenic cells in the processing of prolysyl oxidase. *J Biol Chem* 271: 7113–7119
74 Kessler E, Takahara K, Biniaminov L, Brusel M, Greenspan DS (1996) Bone morphogenetic protein-1: the type I procollagen C-proteinase. *Science* 271: 360–362
75 Wozney JM, Rosen V, Celeste AJ, Mitsock LM, Whitters MJ, Kriz RW, Hewick RM, Wang EA (1988) Novel regulators of bone formation: molecular clones and activities. *Science* 242: 1528–1534
76 Takahara K, Lyons GE, Greenspan DS (1994) Bone Morphogenetic Protein-1 and a mammalian tolloid homologue (mTld) are encoded by alternatively spliced transcripts which are differentially expressed in some tissues. *J Biol Chem* 269: 32572–32578
77 Bork P, Beckmann G (1993) The CUB domain: a widespread module in developmentally regulated proteins. *J Mol Biol* 231: 539–545
78 Li S-W, Sieron AL, Fertala A, Hojima Y, Arnold WV, Prockop DJ (1996) The C-proteinase that processes procollagens to fibrillar collagens is identical to the protein previously identified as bone morphogenetic protein-1. *Proc Natl Acad Sci USA* 93: 5127–5130
79 Childs SR, O'Connor MB (1994) Two domains of the *tolloid* protein contribute to its unusual genetic interaction with *dpp Devel Biol* 162: 209–220
80 Finelli AL, Bossie CA, Xie T, Padgett RW (1994) Mutational analysis of the *Drosophila tolloid* gene, a human BMP-1 homolog. *Development* 120: 861–870
81 Marques G, Musacchio M, Shimell MJ, Wünnenberg-Stapleton K, Cho KWY, O'Connor MB (1997) Production of a DPP activity gradient in the early *Drosophila* embryo through the opposing actions of the sog and tld proteins. *Cell* 91: 417–426
82 Blader P, Rastegar S, Fischer N, Strähle U (1997) Cleavage of the BMP-4 antagonist chordin by zebrafish tolloid. *Science* 278: 1937–1940

83 Piccolo S, Aguis E, Lu B, Goodman S, Dale L, De Robertis EM (1997) Cleavage of chordin by xolloid metalloprotease suggests a role for proteolytic processing in the regulation of Spemann Organizer activity. *Cell* 91: 407–416
84 Fox JW, Bjarnason JB (1996) The reprolysins: a family of metalloproteinases defined by snake venom and mammalian metalloproteinases. *In*: NM Hooper (ed.): *Zinc Metalloproteases in Health and Disease.* Taylor and Francis, London, 47–81
85 Grams F, Huber R, Kress LF, Moroder L, Bode W (1993) Activation of snake venom metalloproteinases by a cysteine switch-like mechanism. *FEBS Lett* 335: 76–80
86 Blobel CP (1997) Metalloprotease-disintegrins: links to cell adhesion and cleavage of TNF-α and Notch. *Cell* 90: 589–592
87 Myles DG, Kimmel LH, Blobel CP, White JM, Primakoff P (1994) Identification of a binding site in the disintegrin domain of fertilin required for sperm-egg fusion. *Proc Natl Acad Sci USA* 91: 4195–4198
88 Krätzschmar J, Lum L, Blobel CP (1996) Metargidin, a membrane-anchored metalloprotease-disintegrin protein with an RGD integrin binding sequence. *J Biol Chem* 271: 4593–4596
89 Yagami-Hiromasa T, Sato T, Kurisaki T, Kamijo K, Nabeshima Y, Fujisawa-Sehara A (1995) A metalloprotease-disintegrin participating in myoblast fusion. *Nature* 377: 652–656
90 Moss ML, Jin S-L-C, Milla ME, Burkhart W, Carter HL, Chen W-J, Clay WC, Didsbury JR, Hassler D, Hoffman CR et al. (1997) Cloning of a disintegrin metalloprotease that processes precursor tumour necrosis factor-α. *Nature* 385: 733–736
91 Weskamp G, Krätzschmar J, Reid MS, Blobel CP (1996) MDC9, a widely expressed cellular disintegrin containing cytoplasmic SH3 ligand domains. *J Cell Biol* 132: 717–726
92 Wolfsberg TG, White JM (1996) ADAMs in fertilization and development. *Dev Biol* 180: 389–401
93 Arribas J, Coodly L, Vollmer P, Kishimoto TK, Rose-John S, Massague J (1996) Diverse cell surface protein ectodomains are shed by a system sensitive to metalloprotease inhibitors. *J Biol Chem* 271: 11376–11382
94 Black RA, Rauch CT, Kozlosky CJ, Peschon JJ, Slack JL, Wolfson MF, Castner BJ, Stocking KL, Reddy P, Srinivasan S et al. (1997) A metalloprotease disintegrin that releases tumour necrosis factor-α from cells. *Nature* 385: 729–733
95 Lunn CA, Fan X, Dalie B, Miller K, Zavodny PJ, Narula SK, Lundell D (1997) Purification of ADAM 10 from bovine spleen as a TNF-α convertase. *FEBS Lett* 400: 333–335
96 Rosendahl MS, Ko SC, Long DL, Brewer MT, Rosenzweig B, Hedl E, Anderson L, Pyle SM, Moreland J, Meyers MA, Kohno T, Lyons D, Lichenstein HS (1997) Identification and characterization of a pro-tumor necrosis factor—processing enzyme from the ADAM family of zinc metalloproteases. *J Biol Chem* 272: 24588–24593
97 Arribas J, Lopez-Casillas F, Massague J (1997) Role of the juxtamembrane domains of the transforming growth factor-α precursor and the β-amyloid precursor protein in regulated ectodomain shedding. *J BiolChem* 272: 17160–17165
98 Hooper NM, Karran EH, Turner AJ (1997) Membrane protein secretases. *Biochem J* 321: 265–279
99 Rooke J, Pan D, Xu T, Rubin GM (1996) KUZ, a conserved metalloprotease-disintegrin protein with two roles in *Drosophila* neurogenesis. *Science* 273: 1227–1231
100 Pan D, Rubin GM (1997) Kuzbanian controls proteolytic processing of Notch and mediates lateral inhibition during drosophila and vertebrate neurogenesis. *Cell* 90: 271–280
101 Blaumueller CM, Qi H, Zagouras P, Artavanis-Tsakonas S (1997) Intracellular cleavage of Notch leads to a heterodimeric receptor on the plasma membrane. *Cell* 90: 281–291
102 Wolfsberg TG, Bazan JF, Blobel CP, Myles DG, Primakoff P, White JM (1993) The precursor region of a protein active in sperm-egg fusion contains a metalloprotease and a disintegrin domain: structural, functional, and evolutionary implications. *Proc Natl Acad Sci USA* 90: 10783–10787
103 Yuan R, Primakoff P, Myles DG (1997) A role for the disintegrin domain of cyritestin, a sperm surface protein belonging to the ADAM family, in mouse sperm-egg plasma membrane adhesion and fusion. *J Cell Biol* 137: 105–112
104 Almeida EAC, Huovila A-PJ, Sutherland AE, Stephens LE, Calarco PG, Shaw LM, Mercurio AM, Sonnenberg A, Primakoff DG, Myles DG et al. (1995) Mouse egg integrin α6β1 functions as a sperm receptor. *Cell* 81: 1095–1104
105 Muga A, Neugebauer W, Hirama T, Surewicz WK (1994) Membrane interactive and conformational properties of the putative fusion peptide of PH-30, a protein active in sperm-egg fusion. *Biochemistry* 33: 4444–4448
106 Turner AJ, Tanzawa K (1997) Mammalian membrane metallopeptidases: neprilysin, ECE, KELL, and PEX. *FASEB J* 11: 355–364
107 Kerr MA, Kenny AJ (1974) The purification and specificity of a neutral endopeptidase from rabbit kidney brush border. *Biochem J* 137: 477–488
108 Li C, Hersh LB (1995) Neprilysin: Assay methods, purification, and characterization. *Meth Enzymol* 248: 253–263
109 Roques BP, Noble F, Daugé V, Fournié-Zaluski M-C, Beaumont A (1993) Neutral endopeptidase 24.11. Structure, inhibition, and experimental and clinical pharmocology. *Pharmcol Rev* 45: 87–146
110 Li C, Booze RM, Hersh LB (1995) Tissue-specifc expression of rat neutral endopeptidase (neprilysin) mRNAs. *J Biol Chem* 270: 5723–5728

111 Ishimaru F, Mari B, Shipp MA (1997) The type 2 CD10/neutral endopeptidase 24.11 promoter: functional characterization and tissue-specific regulation by CBF/NF-Y isoforms. *Blood* 89: 4136–4145
112 Letarte M, Vera S, Tran R, Addis JBL, Onizuka RJ, Quackenbush EJ, Jongeneel CV, McInnes RR (1988) Common acute lymphoblastic leukemia antigen is identical to neutral endopeptidase. *J Exp Med* 168: 1247–1253
113 Beaumont A, Fournié-Zaluski M-C, Roques BP (1996) Neutral endopeptidase 24.11: structure, and design and clinical use of inhibitors. *In*: NM Hooper (ed.): *Zinc Metalloproteases in Health and Disease.* Taylor and Francis, London, 105–130
114 Mari B, Guerin S, Maulon L, Belhacene N, Far DF, Imbert V, Rossi B, Peyron J-F, Auberger P (1997) Endopeptidase 24.11 (CD10/neprilysin) is required for phorbol ester-induced growth arrest in Jurkat T cells. *FASEB J* 11: 869–879
115 Roques BP, Noble F, Crine P, Fournié-Zaluski M-C (1995) Inhibitors of neprilysin: Design, pharmacological and clinical applications. *Meth Enzymol* 248: 263–283
116 Yanagisawa M, Kurihara H, Kimura S, Tomobe Y, Kobayishi Y, Mitsui Y, Yazaki Y, Goto K, Masaki T (1988) A novel, potent, vasoconstrictor peptide produced by vascular endothelial cells. *Nature* 332: 411–415
117 Takahashi M, Matsushita Y, Iijima Y, Tanzawa K (1993) Purification and characterization of endothelin-converting enzyme from rat lung. *J Biol Chem* 268: 21394–21398
118 Shimada K, Matsushita Y, Wakabayashi K, Takahashi M, Matsubara A, Iijima Y, Tanzawa K (1995) Cloning and functional expression of human endothelin-converting enzyme cDNA. *Biochem Biophys Res Commun* 207: 807–812
119 Shimada K, Takahashi M, Ikeda M, Tanzawa K (1995) Identification and characterization of two isoforms of an endothelin-converting enzyme-1. *FEBS Lett* 371: 140–144
120 Schweizer A, Valdenaire O, Nelböck P, Deuschle U, Dumas Milne Edwards JB, Stumpf JG, Löffler BM (1997) Human endothelin-converting enzyme (ECE-1): three isoforms with distinct subcellular localizations. *Biochem J* 328: 871–877
121 Emoto N, Yanagisawa M (1995) Endothelin-converting enzyme-2 is a membrane-bound, phosphoramidon-sensitive metalloprotease with acidic pH optimum. *J Biol Chem* 270: 15262–15268
122 Shimada K, Takahashi M, Turner AJ, Tanzawa (1996) Rat endothelin-converting enzyme-1 forms a dimer through Cys^{412} with a similar catalytic mechanism and a distinct substrate binding mechanism compared with neutral endopeptidase-24.11. *Biochem J* 315: 863–867
123 Vijayaraghavan J, Scicli AG, Carretero OA, Slaughter C, Moomaw C, Hersh LB (1990) The hydrolysis of endothelins by neutral endopeptidase 24.11 (enkephalinase). *J Biol Chem* 265: 14150–14155
124 Hoang MV, Turner AJ (1997) Novel activity of endothelin-converting enzyme: hydrolysis of bradykinin. *Biochem J* 327: 23–26
125 Lee S, Zambas ED, Marsh WL, Redman CM (1993) The human Kell blood group gene maps to chromosome 7q33 and its expression is restricted to erythroid cells. *Blood* 81: 2804–2809
126 Redman CM, Lee S (1995) The Kell blood group system. *Transfusion Clinique et Biologique* 2: 243–249
127 Lee S, Zambas ED, Marsh WL, Redman CM (1991) Molecular cloning and primary structure of Kell blood group protein. *Proc Natl Acad Sci USA* 88: 6353–6357
128 The HYPConsortium (1995) A gene (PEX) with homologies to endopeptidases is mutated in patients with X-linked hypophosphatemic rickets. *Nat Genet* 11: 130–136
129 Du L, Desbarats M, Viel J, Glorieux FH, Cawthorn C, Ecarot B (1996) cDNA cloning of the murine Pex gene implicated in X-linked hypophosphatemia and evidence for expression in bone. *Genomics* 36: 22–28
130 Sato H, Tanaka M, Takino T, Inoue M, Seiki M (1997) Assignment of the human genes for membrane-type-1, -2, -3 matrix metalloproteinase (MMP14, MMP15, MMP16) to 14q12.2, 16q12.2-q21, and 8q21, respectively, by *in situ* hybridization. *Genomics* 39: 412–413
131 Butler GS, Butler MJ, Atkinson SJ, Will H, Tamura T, van Westrum SS, Crabbe T, Clements J, d'Ortho MP, Murphy G (1998) The TIMP2 membrane type1 metalloproteinase "receptor" regulates the concentration and efficient activation of progelatinase A. *J Biol Chem* 273: 871–880
132 Kadler KE, Lightfoot SJ, Watson RB (1995) Procollagen N-Peptidases: Procollagen N-Proteinases. *Meth Enzymol* 248: 756–771
133 Colige A, Li S-W, Sieron AL, Nusgens BV, Prockop DJ, Lapiére CM (1997) cDNA cloning and expression of bovine procollagen I N-proteinase: a new member of the superfamily of zinc-metalloendoproteinases with binding sites for cells and other matrix components. *Proc Natl Acad Sci USA* 94: 2374–2379

Proteases: New Perspectives
V. Turk (ed.)

Proteinases in parasites

Juan José Cazzulo

Instituto de Investigaciones Biotecnológicas, Universidad Nacional de General San Martín, Av. General Paz y Albarellos, INTI, C.C. 30, 1650 San Martín, Prov. Buenos Aires, Argentina.

Introduction

A number of organisms, both protozoa and helminths, are endoparasites pathogenic to man and domestic animals [1]. The study of the proteinases present in these organisms has received considerable interest over the last decade, since they seem to be relevant to essential aspects of the host-parasite relationship, in addition to their obvious participation in the nutrition of the parasite at the expense of the host. As will be noted in the appropriate sections, in some cases these enzymes are involved as major parasite antigens, in the host response, or, conversely, in the protection of the parasite against the host immune system, by specific degradation of immunoglobulins. In some cases where intracellular forms of the parasite are involved, proteinases participate in the penetration into, and/or the release from, the infected mammalian cell. There are examples of direct participation of proteinases in the process of tissue invasion, from the digestive tract or from outside the human body. Last, but not least, the coordinated action of proteinases is involved in protein degradation for nutritional purposes by the parasites, either intracellularly (malarial parasites) or in the digestive tract (helminths, such as schistosomes). The present review will emphasise the recent developments in this field, and will focus on the parasites responsible for prevalent human infections, quoting recent reviews [2–6] whenever possible.

Proteinases in parasitic Protozoa

The most important protozoan parasites include the Trypanosomatids, the Plasmodia, and the 'anaerobic' Protozoa, namely *Entamoeba*, *Trichomonas* and *Giardia*. Trypanosomatids [1] include (1) the Leishmanias, causing several forms of leishmaniasis, which can be fatal (the visceral leishmaniasis or kala azar), grossly disfiguring (the mucocutaneous leishmaniasis or espundia), or relatively mild, localized and in some cases self-healing (some forms of cutaneous leishmaniasis); (2) the African trypanosomes, which cause sleeping sickness in humans (*Trypanosoma brucei gambiense, T. b. rhodesiense)*, and diseases of cattle (nagana, *T. b. brucei, T. congolense, T. vivax)*, horses and camels (surra, *T. evansi*) the latter being of considerable economic importance; (3) *Trypanosoma cruzi*, which is the agent of the American Trypanosomiasis, Chagas disease, endemic in most of Latin America. These parasites have complex life cycles involving a number of differentiation stages, some of which are obligately intracellular, in the haematophagous insect vector and the mammalian host. A role for proteinases in the differentiation steps has been proposed.

The Plasmodia are the causative agents of malaria [1], the most prevalent protozoan infection, responsible for about one million deaths per year. They have complex life cycles, involving asexual and sexual stages, the former in the mammalian host and the latter in the mosquito vector. Proteinases seem to be essential for parasite penetration into, and release from, the infected cell, for hemoglobin digestion and for differentiation. *Entamoeba histolytica* is the agent of amebiasis [1]. This disease affects hundreds of millions of people, a number of whom develop the invasive form of the disease, namely colitis, dysentery and amoebic abscess. Proteinases seem to be essential for invasion, and are considered to be virulence factors directly involved in pathogeny. Trichomonads are agents of sexually transmitted diseases of humans and cattle [1], vaginal trichomoniasis being the most important. These organisms produce and excrete high levels of proteinases. *Giardia lamblia* is an intestinal parasite which represents the most prevalent cause of acute diarrhoea worldwide [1]. In the cases of both Trichomonads and *Giardia* the role played by proteinases in the respective diseases is still doubtful.

Cysteine proteinases

Most of the proteolytic activities reported so far in Trypanosomatids are those of cysteine proteinases (CPs), present in high amounts in amastigotes of *Leishmania mexicana* [4, 6], in epimastigotes of *T. cruzi* [3, 5] and in bloodstream trypomastigotes of *T. brucei* [3, 4, 6]. They are present in moderate amounts in choanomastigotes of *Crithidia fasciculata* [7] and in very low amounts in epimastigotes of *Trypanosoma rangeli* [8] and in promastigotes of *L. mexicana* [2]. The CPs of Trypanosomatids have been classified by Coombs and co-workers into three types, Types I and II, both cathepsin L-like, and Type III, cathepsin B-like [4, 6]. The genes encoding Type I CPs (a) predict a long C-terminal extension, absent in most other CPs described so far, but shared (albeit with very low homology) by some CPs from tomatoes; (b) are multicopy genes, arrayed in tandems, several of which are usually simultaneously expressed, generating a mixture of isoforms and (c) are expressed under developmental regulation. Type I enzymes have been found in all Trypanosomatids studied so far, including the Leishmanias [4, 6], i.e. *Trypanosoma cruzi* (cruzipain [5], also called cruzain [9] or GP57/51 [10]); *Trypanosoma brucei* (trypanopain [2, 11]); *Trypanosoma congolense* (congopain [12]); *Trypanosoma rangeli* [13] and *Crithidia fasciculata* [7]. In the case of cruzipain, a complete tandem set of genes has been sequenced, showing that in the X10.6 strain all the genes are identical, except for that at the 3' end which has a shorter and completely different C-terminal extension [14]. In the case of type I CPs of *L. mexicana* the first two and the last genes in a tandem have truncated C-terminal extensions; the first two are expressed predominantly in metacyclic promatigotes, whereas the last seems to be a pseudogene [15]. Type I CPs are cathepsin L-like endoproteinases with good gelatinolytic activity. The C-terminal extension is in some cases lost by processing, as in *T. brucei* and *L. mexicana*, the mature CPs being similar to other enzymes belonging to the papain family, whereas in other cases, such as the CPs from *T. cruzi* [5], *T. congolense* [12], *T. rangeli* [9] and *C. fasciculata* [8], it is retained in the mature enzyme, which thus has a molecular weight appreciably higher than those of the more usual papain-like CPs. The C-terminal extension of cruzipain can be obtained by self-proteolysis and is found to contain a number of post-translational modifications [5]. Cruzipain [5, 10] and congopain [12] are strongly antigenic and, at least in the case of cruzipain, the C-terminal extension is responsible for this antigenicity [5]. The CPs of epimastigotes of some strains, such as Tul2, of *T. cruzi* [5] and of amastigotes of *L. mexicana* [2] are expressed as complex mixtures of isoforms; in the latter case, a number of CPs, differing in specificity, charge and/or glycosylation, have been identified. This complexity makes the crystallization for structural studies very difficult. A truncated recombinant Type I CP (cruzainΔc), lacking the C-terminal domain and the last nine amino acids of the catalytic domain, has been crystallized and its three-dimensional structure determined by X-ray crystallography [16]. The structure shows considerable similarity with that of some cathepsins, particularly cathepsin L, in good agreement with the sequence homologies.

Type II CPs have so far been detected only in *Leishmania*; their genes are present as single copies, and predict a mature enzyme with a characteristic three-amino acid insertion close to the N-terminus and a very short C-terminal extension. The enzyme lacks gelatinolytic activity [4]. Type III CPs have also been described only in *Leishmania*, are encoded by a single copy gene, and completely lack a C-terminal extension and gelatinolytic activity; this cathepsin B-like enzyme is amphiphilic and possibly associated to membranes [4].

Type I and II CPs are lysosomal enzymes; in addition, there is some evidence for the presence of membrane-associated forms of Type I and Type III CPs [4, 5].

Although the CPs from Trypanosomatids seem to be important for the growth and differentiation of the parasites, as suggested by the effect of irreversible CP inhibitors *in vivo* [2–5], their actual functions are far from recognised. Overexpression of cruzipain has recently been shown to be linked to an increase in the differentiation of epimastigotes to metacyclic trypomastigotes, without an increase in invasion of Vero cells *in vitro*, or in virulence to severe combined immunodeficient (SCID) mice *in vivo* [17]. In the case of *L. mexicana*, Coombs, Mottram and their co-workers [4] have been able to obtain knock-out mutants lacking one or another of the three types of CP. Null mutants for the gene encoding Type II CP did not differ from the wild type in any of the tests performed, thus suggesting that it is not essential for viability. Null mutants for the Type I CP tandems were also able to grow and differentiate quite well, although their infectivity *in vitro* to macrophages was five-fold lower than that of the wild type; re-incorporation of only one of the Type I genes to the mutant restored full infectivity, whereas re-expression individually of two others did not. Double-null mutants for Types I and II CPs had a phenotype similar to that of the Type I null mutants. Mutants null for the Type III CP genes were also considerably less infective in macrophages *in vitro* but only a little less infective than wild type parasites in BALB/c mice. The question of the true importance of the abundant and varied CP array expressed in *L. mexicana* amastigotes is therefore still open, and the validity of using these enzymes as targets for chemotherapy is not clear, despite the fact that, as in the case of *T. cruzi*, irreversible CP inhibitors interfere with both differentiation and intracellular growth of the parasite [4].

In addition to these well-characterized CPs, there are examples probably belonging to other proteinase types, like the ATP-activated 30 kDa CP recently described in *T. cruzi* [18].

Malarial parasites, including the most important protozoan human pathogen, *Plasmodium falciparum*, contain a complex array of proteolytic enzymes, including CPs, serine proteinases (SPs) and aspartic proteinases (APs), some of which seem to be involved in red blood cell (RBC) invasion and rupture, or in digestion of hemoglobin inside the food vacuole [3, 19]. The proposed participation of these enzymes in the processes mentioned is based on results of their blocking by different proteinase inhibitors *in vivo*. A number of CPs have been described in *P. falciparum* and other *Plasmodium* species (summarized in [3]). A 68 kDa CP present in schizonts and merozoites and a 35–40 kDa CP present in mature schizonts have been proposed to participate in RBC invasion and/or rupture [3], together with one or more SPs (see below). A cathepsin L-like 28 kDa CP able to degrade hemoglobin *in vitro* has been identified in trophozoites of *P. falciparum*. Genes presumably encoding this CP have been cloned from *P. falciparum* and *P. vinckei*, and predict substantial homology with cathepsin L [3]. More recently, genes showing at least 45% identity in the sequence of a 167 amino acid stretch corresponding to the mature form of the CP have been amplified by PCR from eight more malarial species [20]; in all cases a characteristic insert close to the C-terminus (amino acid residues 499–527, pre-pro-enzyme numbering, in the case of *P. falciparum*) was present [20]. The *P. falciparum* gene is also characterized by coding for a large pro-enzyme domain, about 280 amino acids long. This pro-domain is shorter (about 210 amino acid residues) in the *P. vinckei* enzyme. Recent studies suggest the presence of as many as five CP genes in the *P. falciparum* genome [21]; the one cloned and sequenced first [22] and described above, however, is likely to correspond to the major CP in the ring and trophozoite stages (those stages actively digesting hemoglobin), named falcipain and considered as a major hemoglobinase activity. Although the lethal effect of CP inhibitors on malarial parasites, both *in vivo* and *in vitro* [3], with the concomitant accumulation of hemoglobin in the digestive vacuole, suggests a major role for a CP in hemoglobin degradation, recent results obtained with recombinant [23] and native [21] falcipain present discrepancies related to the ability of this CP to degrade native hemoglobin *in vivo*. The differences found are probably due to the fact that the "mature" recombinant enzyme [23] retains a substantial portion of the pro-domain [21]. Thus, the actual role of falcipain as a hemoglobinase, as well as the relative roles played by falcipain and the APs, plasmepsins I and II (see below), are still controversial [20, 22, 24].

The pathogenic amoeba *E. histolytica* produces large amounts of CPs, the best characterized ones being those with apparent molecular masses of 22–27 kDa (amoebapain), 26–29 kDa (histolysain), 16 kDa (cathepsin B) and 56 kDa (neutral CP) [25]. So far, at least six genes encoding CPs have been isolated from this parasite [26]. Comparison of actual peptide sequences from purified enzymes with those deduced from DNA sequences indicates that *Eh-CPp1* encodes amoebapain, whereas *Eh-CPp2* encodes histolysain; these mature proteins are 86.5% identical [25]. A third gene, *Eh-CPp3*, originally called ACP1 by McKerrow and co-workers encodes a CP with only 45% identity to the other two. Initial reports suggested that the presence of *Eh-CPp3* might be diagnostic for pathogenic *E. histolytica*, but recently a homolog has been PCR-amplified from several non-pathogenic strains, now classified as *Entamoeba dispar* [27]. Homologs of *Eh-CPp1* and *Eh-CPp2* are found in *E. dispar* strains, as has been reported for most genes identified so far in *E. histolytica*. This fact should not obscure the clear correlation found between high levels of CPs released into the medium, and the pathogenicity of a strain [3, 4, 26]. Pathogenesis of amebiasis requires penetration of the amoebae through the intestinal mucosal cells and the basal membrane. In good agreement with the participation of the amoebal CPs in this process, they are highly active on collagen types I, IV and V, fibronectin and laminin [25]. The specificity of these CPs is cathepsin B-like, an Arg residue in P2 being required for high activity on synthetic peptides [3, 25]. As in most other parasitic protozoa, these enzymes are located in subcellular vesicles similar to lysosomes, and also, to a smaller extent, at the outer cell surface [24].

Trichomonas vaginalis produces, and excretes into the medium, a complex mixture of CPs with apparent molecular mass values ranging from 20 to 96 kDa [2]. One-dimensional gelatin-SDS-PAGE has shown the presence of at least 11 different CP activities in cell lysates, whereas two-dimensional gels indicated the presence of 23 distinct activities [28]. Cloning and sequencing of a number of CP genes from *T. vaginalis*, however, has failed to detect mature CPs predicted to be larger than 24 kDa [28]. Two complete genes (*TvCP1* and *TvCP2*) and two other genes apparently slightly shorter than full length (*TvCP3* and *TvCP4*) were amplified by PCR from cDNAs and sequenced [28]. The same authors found evidence for at least a further two different, not fully sequenced, genes; moreover, the N-terminal sequence of the only CP (23 kDa) purified so far from *T. vaginalis* is different from those predicted from *TvCP1-4* [28]. The sequences obtained had higher homology to mammalian cathepsin L than to cruzipain or the Type I and II CPs from *L. mexicana* [28]. The sequences found for the substrate-binding sites for *TvCP1* and *4* are in good agreement with the finding of some *T. vaginalis* CPs with cathepsin L-like or cathepsin B-like specificity. It is clear, therefore, that the considerable diversity of CPs found in *T. vaginalis* is due, at least partially, to the simultaneous expression of a number of genes. An interesting feature of these CP genes is the absence, or the presence of a very short, signal peptide, despite the fact that all these enzymes seem to be lysosomal and are secreted into the medium. *TvCP1-3* are single-copy genes, whereas *TvCP4* is present as multiple copies [28].

The role of the CP activities present in *T. vaginalis* is still not clear, although at least some membrane-bound forms have been associated with adhesion to epithelial cells [4, 29], and some of the secreted CPs seem to participate in the degradation of immunoglobulins, thus participating in the defensive mechanisms of the parasite against the immune system [29].

The situation in the cattle parasite *Tritrichomonas foetus* seems to be very similar to that in *T. vaginalis*; a number of CPs are produced and excreted into the medium [2], and seven different CP genes have been cloned and sequenced after PCR amplification [30].

The trophozoites of *Giardia intestinalis* also contain a number of proteolytic activities, most of which correspond to CPs [2]. Evidence has been found, however, through the use of inhibitors, for the presence of SPs, APs and an aminopeptidase [31]. Gelatin-SDS-PAGE showed the presence of 18 proteolytic activities, with apparent molecular mass values ranging from 30 to 211 kDa; all of these activities seem to correspond to CPs, since they were enhanced by dithiothreitol and inhibited by CP inhibitors. At least some of them showed distinctive preferences for small fluorogenic substrates [31]. These CP activities seem to be lysosomal. Their functions are still unknown but they might be involved in the degradation of proteins, including host immunoglobulins, or in the transformation of cysts into trophozoites [3].

Serine proteinases

Serine proteinases (SPs) have been reported much less frequently than CPs in parasitic protozoa. A peptidase shown to be indirectly involved in the penetration of the trypomastigotes of *T. cruzi* into the mammalian cell by activating a factor which induces a transient increase of cytosolic Ca^{2+} in the latter cell [32], has recently been cloned and sequenced. The 270 amino acid residues in the C-terminal region show homology to members of the prolyl oligopeptidase family of SPs, leading to its classification as *T. cruzi* oligopeptidase B [33]. The enzyme, present in all the life-cycle stages of *T. cruzi* [34], seems to be the 120 kDa alkaline peptidase initially described by Ashall [34] and later purified to homogeneity [35].

A gene fragment apparently encoding a chymotrypsin-like SP has been amplified by PCR from *T. cruzi* DNA [36].

T. brucei procyclic trypomastigotes contain a putative SP activity two to three-fold higher than that present in bloodstream forms of the same parasite, and show an apparent molecular mass of 80 kDa. This enzyme was inhibited by DFP, leupeptin and TLCK, but not by PMSF, and is devoid of gelatinolytic activity in SDS-PAGE, although it is active on synthetic substrates [37].

A 75 kDa SP seems to be involved, together with the CP previously mentioned, in invasion and/or rupture of RBCs by *P. falciparum* [3]. The enzyme seems to be bound in an inactive state to the membrane of schizonts and merozoites by a glycosylphosphatidylinositol (GPI) anchor, and to be released and activated by a specific phospholipase C [18]. This SP has recently been purified from merozoites and shown to cleave two major proteins of the RBC, glycophorin A and Band 3 protein; this action may facilitate the formation of the parasitophorous vacuole from the RBC plasma membrane [38]. Peptides derived from the Band 3 protein and inhibitory to the 76 kDa SP are efficient inhibitors of parasite penetration into RBCs [38]. In addition, a Ca^{2+}-dependent, membrane-associated SP is responsible for processing the major merozoite surface protein, MSP-1. The polypeptides thus generated seem to be involved in the receptor-ligand interactions which initiate RBC invasion [39].

An SP with plasminogen activator properties has been reported in *Acanthamoeba castellanii*, a free-living amoeba which may produce severe keratitis leading to blindness in soft contact lens wearers; the enzyme may be involved in the pathogenesis of the disease [40].

Metalloproteinases

Metalloproteinases (MPs) have been described in a number of parasitic protozoa, but only one that, present in promastigotes of *Leishmania* spp., has been thoroughly characterized [3, 6, 41]. The enzyme is known as gp63 (63 kDa glycoprotein), PSP (promastigote surface proteinase), or leishmanolysin [41]. gp63 is bound to the promastigote membrane by a GPI anchor, and is the most abundant protein exposed at the promastigote membrane (ca. 5×10^5 molecules per cell). There is, in addition, a soluble, lysosomal, homologous MP with an acidic optimal pH, present at lower levels in amastigotes [6]. The membrane-bound enzyme is a Zn^{2+}-dependent HEXXH MP, with an alkaline pH optimum [8–9], able to degrade components of the extracellular matrix such as fibrinogen. The enzyme specificity has been studied using a number of natural and synthetic peptides, among them a fluorescent peptide substrate, efficiently cleaved between Tyr and Leu residues [41]. Leishmanolysin is N-glycosylated at its three potential sites [42]. Although its three-dimensional structure is not yet available, Raman spectroscopy and circular dichroism indicate a predominantly β structure [41]. The MP is produced as a pro-enzyme, its activation seems to occur by a Cys-switch mechanism, as proposed for matrix MPs [43]. The genes encoding gp63 are multicopy, with a variable number in different species of *Leishmania* and although most of them are placed in a tandem, some are dispersed [6]. The transcripts found in promastigotes contain the predicted GPI anchor addition site (in which Asn 577 is essential for anchoring, 42), whereas those expressed in amastigotes lack this site, as expected from the lysosomal location of leishmanolysin in this parasite form. Moreover, different gp63 proteins are expressed in log phase (63 kDa) or stationary phase (59 kDa) promastigotes of *L. chagasi* [44].

The roles played by leishmanolysin in the host-parasite interactions are still not clear [3, 6]. The enzyme is a ligand for the fucosyl-mannosyl receptor, an acceptor for C3b deposition, and a major surface antigen [3]. Its predominant presence in the promastigote (insect) stage, together with the finding of homologues in monogenetic (i.e. having only one host, an insect) Trypanosomatids such as *C. fasciculata* or *Herpetomonas samuelpessoai*, suggests that leishmanolysin may be important for the interaction with the invertebrate host [6].

Leishmanolysin seems to be not the only MP present in *Leishmania* spp.; recently, two soluble MPs, a carboxy- and an amino-peptidase, have been identified in *L. major* [45].

Among other parasitic protozoa, *E. histolytica* contains a poorly characterized membrane-associated MP with collagenase activity, apparently related to virulence [3].

Aspartyl proteinases

Aspartyl proteinases (APs) have been described and studied in detail, among parasitic protozoa, in malarial parasites. APs with molecular masses ranging from 10–148 kDa have been described in *P. falciparum*, and enzymes belonging to this class are present in other malarial species [3]. Recent studies have focused on two APs, named plasmepsins I and II, which have 73% amino acid identity, but cleave hemoglobin with different specificities [46]. Both of them, however, are able to cleave the bond Phe 33/Leu 34 in the α-chain of native hemoglobin at an α-helical hinge region which seems to unravel the substrate molecule and to allow further proteolysis [46]. Both APs, as well as the CP falcipain are located in the digestive vacuole of *Plasmodium* spp., a lysosome-like organelle responsible for the massive hemoglobin digestion which is characteristic of malarial trophozoites [47]. The genes encoding plasmepsin I [48] and II [49] have been cloned and sequenced. Both enzymes are translated as pro-enzymes, the mature enzymes having a molecular weight of 36.9 kDa. The pro-domain (124–126 amino acid residues) is bigger than those usually found in APs, and has hydrophobic features suggesting that the pro-enzyme may be an integral membrane protein [48]. Both plasmepsins have been expressed as recombinant proteins [46, 49] but only the recombinant plasmepsin II had kinetic properties similar to those of the native enzyme [49]. Recombinant plasmepsin I had considerably different properties [46], perhaps due to the conservation of a substantial part of the pro-domain. The plasmepsins are undoubtedly involved in intravacuolar hemoglobin degradation. Their specific inhibition, as well as the inhibition of falcipain, results in the killing of malarial parasites [3]. There are still, however, discrepancies about the relative roles of the APs and the CP. Although plasmepsins are able to cleave native hemoglobin and start the unravelling of the molecule for further proteolysis (see above), the possibility exists that falcipain is required for the activation of the pro-plasmepsins, as has been suggested for pro-cathepsins B and L in higher organisms [3].

The proteasome

In 1996 three groups reported independently the finding and characterization of proteasomes in *E. histolytica* [50], in *T. brucei* [51] and in *T. cruzi* [52].

E. histolytica trophozoites were shown to contain both a 20S proteasome, made up of a number of subunits with apparent molecular mass values around 30 kDa, and an 11S proteinase, which seems to be a hexamer of identical 60–65 kDa subunits, consisting of two trimers which were still active under weakly denaturing conditions. The 11S proteinase seems to be a new kind of enzyme [50].

The 20S proteasomes of *T. brucei* were purified from both bloodstream and procyclic forms and shown to consist of at least eight different subunits, with apparent molecular mass values ranging from 23–34 kDa and pI values ranging from 4.5–7.0 [51]. Similar proteasomes were purified from epimastigotes of *T. cruzi* [52]. The 670 kDa intact molecules are made up of subunits of 25–35 kDa, with pI values ranging from 4.5–8.5 [52]. In both Trypanosomatids the proteasomes exhibit their characteristic electron microscopic images. In the case of *T. cruzi* the activity of the proteasome was specifically inhibited by lactacysin, which was able to prevent the differentiation of trypomastigotes to amastigotes and the intracellular differentiation of amastigotes to trypomastigotes [52]. This role for the proteasome in protozoan cell remodeling seems not to be restricted to Trypanosomatids, since recent evidence suggests a similar role in *Plasmodium berghei* and in *Entamoeba invadens* [52a].

Proteinases in parasitic helminths

A number of helminths, commonly called worms, are endoparasites of man and domestic animals. Most of them belong to two phyla: the Platyhelminths, including Cestoda (*Taenia, Echinococcus*) and Digenea (*Schistosoma, Fasciola*), and the Nematoda (*Ascaris, Ancylostoma, Trichinella* and the Filariae such as *Onchocerca* and *Dirofilaria*). Some helminths, for instance the schistosomes, are the agents of tropical diseases affecting hundreds of millions of people [1]. These parasites have complex life cycles, with two or even three different hosts, and proteinases seem to be essential for at least some of their different developmental stages.

Proteinases in Schistosoma *spp.*

Schistosomes have a complex life cycle, involving a snail as intermediate host. Proteinases seem to be particularly important for the active penetration of the cercariae liberated by the snail through the human skin, resulting in their transformation into schistosomula, and in the digestion of hemoglobin by the adult worm. The major cercarial proteinase involved in skin penetration is a 28 kDa soluble SP, released by the acetabular glands, which is homologous in sequence, and also similar in specificity, to pancreatic elastases [53]. In addition, there is a similar enzyme, antigenically cross-reactive, anchored by GPI to the tegument of mechanically transformed schistosomulae [54]. A MP activity associated with the surface of schistosomula and adults has also been described [53].

Adult schistosomes feed on RBCs, which are lysed in the digestive tract and hemoglobin is then digested [55]. As in the case of *Plasmodium* spp., discussed previously, there seems to be more than one 'hemoglobinase'; hemoglobin degradation being the result of the concerted action of both CPs and at least one AP [55, 56]. Two CPs, initially identified as immunodominant schistosome antigens, were named Sm31 and Sm32. When the genes encoding them were cloned and sequenced, the former was shown to correspond to a cathepsin B-like CP, whereas the latter, when expressed in insect cells, seemed not to be proteolytically active [57]. Sm32 has recently been shown to be an asparaginyl endoproteinase, with considerable homology to the *Canavalia ensiformis* legumain [4, 55], a CP belonging to a family different from the papain family. It has recently been suggested that Sm32 may be involved in the processing and activation of cathepsins B (Sm31) and L, the AP cathepsin D, and the dipeptidylpeptidase cathepsin C [58]. Although there is still disagreement between the groups of Dalton and Klinkert [59] on which of these proteinases is the major 'hemoglobinase', it seems likely that the 31 kDa cathepsin B, the 24 kDa cathepsin L, and the 36 kDa cathepsin D all participate in the process of hemoglobin digestion.

Proteinases from Fasciola hepatica

This sheep and human pathogen also has a snail as intermediate host, but the cercariae do not actively penetrate the human skin, rather infecting man by ingestion and penetration through the intestinal mucosa [1]. Multiple CP genes have been amplified by PCR and cloned from cDNAs of adult worms [60]. The 28–30 kDa cathepsin L-like CPs [61] present unusual 3-hydroxyproline residues [62]. A 25–26 kDa cathepsin B-like enzyme has also been found in newly excisted juveniles (the infective form of the parasite) and in adults [63]. In addition, a secreted dipeptidylpeptidase, belonging to the SPs, has been identified in juveniles and adults [64]. SP and AP activities are present, but CPs are predominant from a quantitative point of view [65].

Proteinases from Nematodes

Proteinases belonging to all major classes have been described in these organisms [66]. MPs are important in parasites having skin-penetrating larvae such as *Ancylostoma caninum*, which releases 90 and 50 kDa MPs [67]; in addition, this organism presents cathepsin B- and L-like CPs [68]. In the case of the ovine parasite *Haemonchus contortus*, a 46 kDa MP able to digest fibrinogen and fibronectin has been described in L3 and L4 larvae. A CP seems also to participate in the degradation of extracellular matrix proteins by adults [69]. A CP gene family has been described in this parasite [70]. Among the Filariae, the proteinases from *Dirofilaria immitis* have received some attention. An AP with a molecular mass of 42 kDa has been purified to homogeneity from adult worms and shown to be distributed in a number of different tissues [71]. Both MPs and CPs have been identified in secretions and extracts of L3 and L4 larvae and CP inhibitors have been shown to inhibit the third to fourth larval stage moult [72].

Conclusion

Proteinases are present, in many instances at high levels, in human and domestic animal endoparasites, both unicellular (Protozoa) and multicellular (helminths). In some cases there is evidence of their participation in important aspects of the parasite/host relationship. The considerable mass of knowledge obtained over the last 10–15 years opens up the possibility of exploiting the unusual characteristics presented by a number of these enzymes to develop new drugs based on selective proteinase inhibitors [3, 4]. We may hope for important developments in the near future, which would result in direct benefit to mankind if they brought about the cure of important diseases affecting largely the populations of developing countries.

References

1 Cox FEG (ed) (1993) *Modern Parasitology.* 2nd ed. Blackwell Scientific Publications, Oxford
2 Coombs GH, North MJ (eds) (1991) *Biochemical Protozoology.* Taylor and Francis, London
3 McKerrow JH, Sun E, Rosenthal PJ, Bouvier J (1993) The proteases and pathogenicity of parasitic protozoa. *Annu Rev Microbiol* 47: 821–853
4 Robertson CD, Coombs GH, North MJ, Mottram JC (1996) Parasite cysteine proteinases. *Perspect Drug Discov Design* 6: 99–118
5 Cazzulo JJ, Stoka V, Turk V (1997) Cruzipain, the major cysteine proteinase from the protozoan parasite *Trypanosoma cruzi. Biol Chem* 378: 1–10
6 Coombs GH, Mottram JC (1997) Proteinases of Trypanosomes and *Leishmania. In*: G Hide, JC Mottram, GH Coombs, PH Holmes (eds): *Trypanosomiasis and Leishmaniasis.* CAB International, Oxford, 177–197
7 Cazzulo JJ, Labriola C, Parussini F, Duschak V, Martínez J, Franke de Cazzulo BM (1995) Cysteine proteinases in *Trypanosoma cruzi* and other Trypanosomatid parasites. *Acta Chim Slovenica* 42: 409–418
8 Labriola C, Cazzulo JJ (1995) Purification and partial characterization of a cysteine proteinase from *Trypanosoma rangeli. FEMS Microbiol Lett* 129: 143–148
9 Eakin AE, Mills AA, Harth G, McKerrow JH, Craik CS (1992) The sequence, organization and expression of the major cysteine proteinase (cruzain) from *Trypanosoma cruzi. J Biol Chem* 267: 7411–7420
10 Scharfstein J, Schechter M, Senna M, Peralta JM, Mendonça-Previato L, Miles MM (1986) *Trypanosoma cruzi*: characterization and isolation of a 57/51,000 m.w. surface glycoprotein (GP57/51) expressed by epimastigotes and bloodstream trypomastigotes. *J Immunol* 137: 1336–1341
11 Mottram JC, North MJ, Barry JD, Coombs GH (1989) A cysteine proteinase cDNA from *Trypanosoma brucei* predicts an enzyme with an unusual C-terminal extension. *FEBS Lett* 258: 211–215

12 Authié E (1994) Trypanosomiasis and trypanotolerance in cattle: a role for congopain. *Parasitol Today* 10: 360–364
13 Martínez J, Henriksson J, Ridåker M, Cazzulo JJ, Pettersson U (1995) Genes for cysteine proteinases from *Trypanosoma rangeli. FEMS Microbiol Lett* 129: 135–142
14 Tomas AM, Kelly JM (1996) Stage regulated expression of cruzipain, the major cysteine proteinase of *Trypanosoma cruzi*, is independent of the level of RNA. *Mol Biochem Parasitol* 76: 91–103
15 Mottram J, Frame MJ, Brooks DR, Tetley L, Hutchison JE, Souza AE, Coombs GH (1997) The multiple *cpb* cysteine proteinase genes of *Leishmania mexicana* encode isoenzymes that differ in their stage regulation and substrate preferences. *J Biol Chem* 272: 14285–14293
16 McGrath M, Eakin AE, Engel JC, McKerrow JH, Craik CS, Fletterick RJ (1995) The crystal structure of cruzain: a therapeutic target for Chagas' disease. *J Mol Biol* 247: 251–259
17 Tomas AM, Miles MM, Kelly JM (1997) Overexpression of cruzipain, the major cysteine proteinase of *Trypanosoma cruzi*, is associated with enhanced metacyclogenesis. *Eur J Biochem* 244: 596–603
18 Santana JM, Nóbrega OT, Santos Silva MA, Grellier P, Schrével J, Teixeira ARL (1996) Novel acidic and neutral proteinases from *Trypanosoma cruzi*. Are they involved in the mechanism of mammalian host cell invasion? *Mem Inst Oswaldo Cruz* 91: 53–54
19 Braun Breton C, Pereira da Silva LH (1993) Malaria proteases and red blood cell invasion. *Parasitol Today* 9: 92–96
20 Rosenthal PJ (1996) Conservation of key amino acids among the cysteine proteinases of multiple malarial species. *Mol Biochem Parasitol* 75: 255–260
21 Francis SE, Gluzman IY, Oksman A, Banerjee D, Goldberg DE (1996) Characterization of native falcipain, an enzyme involved in *Plasmodium falciparum* hemolobin degradation. *Mol Biochem Parasitol* 83: 189–200
22 Rosenthal PJ, Nelson RG (1992) Isolation and characterization of a cysteine proteinase gene of *Plasmodium falciparum. Mol Biochem Parasitol* 51: 143–152
23 Salas F, Fichmann J, Lee GK, Scott MD, Rosenthal PJ (1995) Functional expression of falcipain, a *Plasmodium falciparum* cysteine proteinase, supports its role as a malarial hemoglobinase. *Infect Immunity* 63: 2120–2125
24 Rosenthal PJ, Meshnick SR (1996) Hemoglobin catabolism and iron utilization by malarial parasites. *Mol Biochem Parasitol* 83: 131–139
25 Scholze H, Tannich E (1994) Cysteine endopeptidases of *Entamoeba histolytica. Meth Enzymol* 244: 512–523
26 Que X, Reed SL (1997) The role of extracellular cysteine proteinases in pathogenesis of *Entamoeba histolytica* invasion. *Parasitol Today* 13: 190–194
27 Mirelman D, Nuchamowitz Y, Böhm-Gloning B, Walderich B (1996) A homologue of the cysteine proteinase gene (ACP or *Eh-CPp3*) of pathogenic *Entamoeba histolytica* is present in non-pathogenic *E. dispar* strains. *Mol Biochem Parasitol* 78: 47–54
28 Mallinson DJ, Lockwood BC, Coombs GH, North MJ (1994) Identification and molecular cloning of four cysteine proteinase genes from the pathogenic protozoan *Trichomonas vaginalis. Microbiology* 140: 2725–2735
29 Provenzano D, Alderete JF (1995) Analysis of human immunoglobulin-degrading cysteine proteinases of *Trichomonas vaginalis. Infec Immun* 63: 3388–3395
30 Mallinson DJ, Livingstone J, Appleton KM, Lees SJ, Coombs GH, North MJ (1995) Multiple cysteine proteinases of the pathogenic protozoan *Tritrichomonas foetus*: identification of seven diverse and differentially expressed genes. *Microbiology* 141: 3077–3085
31 Williams AG, Coombs GH (1995) Multiple protease activities in *Giardia intestinalis* trophozoites. *Int J Parasitol* 25: 771–778
32 Burleigh BA, Andrews NW (1995) A 120-kDa alkaline peptidase from *Trypanosoma cruzi* is involved in the generation of a novel Ca^{2+} signalling factor for mammalian cells. *J Biol Chem* 270: 5172–5180
33 Burleigh BA, Caler EV, Webster P, Andrews NW (1997) A cytosolic serine endopeptidase from *Trypanosoma cruzi* is required for the generation of Ca^{2+} signalling in mammalian cells. *J Cell Biol* 136: 609–620
34 Ashall F (1990) Characterization of an alkaline peptidase of *Trypanosoma cruzi. Mol Biochem Parasitol* 38: 77–88
35 Santana JM, Grellier P, Rodier MH, Schrével J, Teixeira ARL (1992) Purification and characterization of a new 120 kDa alkaline proteinase of *Trypanosoma cruzi. Biochem Biophys Res Commun* 187: 1466–1473
36 Sakanari JA, Staunton CE, Eakin AE, Craik CS, McKerrow JH (1989) New serine proteases from nematode and protozoan parasites: isolation of sequence homologues using generic molecular probes. *Proc Natl Acad Sci USA* 86: 4863–4867
37 Kornblatt MJ, Mpimbaza GWN, Lonsdale-Eccles JD (1992) Characterization of an endopeptidase of *Trypanosoma brucei brucei. Arch Biochem Biophys* 293: 25–31
38 Roggwiller E, Morales Bétoulle ME, Blisnick T, Braun Breton C (1996) A role for erythrocyte band 3 degradation by the parasite gp76 serine protease in the formation of the parasitophorous vacuole during invasion of erythrocytes by *Plasmodium falciparum. Mol Biochem Parasitol* 82: 13–24
39 Blackman MJ, Chappel JA, Shai S, Holder AA (1993) A conserved parasite serine protease processes the *Plasmodium falciparum* merozoite surface protein-1. *Mol Biochem Parasitol* 62: 103–114
40 Mitra MM, Alizadeh H, Gerard RD, Niederkorn JY (1995) Characterization of a plasminogen activator pro-

duced by *Acanthamoeba castellanii*. *Mol Biochem Parasitol* 73: 157–164
41 Bouvier J, Schneider P, Etges R (1995) Leishmanolysin: Surface metalloproteinase of *Leishmania Meth Enzymol* 248: 614–633
42 McGwire BS, Chang K-P (1996) Posttranslational regulation of a *Leishmania* HEXXH metalloprotease (gp63). *J Biol Chem* 271: 7903–7909
43 Macdonald MH, Morrison CJ, McMaster WR (1995) Analysis of the active site and activation mechanism of the *Leishmania* surface metalloproteinase GP63. *Biochim Biophys Acta* 1253: 199–207
44 Roberts SC, Wilson ME, Donelson JE (1995) Developmentally regulated expression of a novel 59-kDa product of the major surface protease (Msp or gp63) gene family of *Leishmania chagasi*. *J Biol Chem* 270: 8884–8892
45 Schneider P, Glaser TA (1993) Characterization of two soluble metalloexopeptidases in the protozoan parasite *Leishmania major*. *Mol Biochem Parasitol* 62: 223–232
46 Luker KE, Francis SE, Gluzman IY, Goldberg DE (1996) Kinetic analysis of plasmepsins I and II, aspartic proteases of the *Plasmodium falciparum* digestive vacuole. *Mol Biochem Parasitol* 79: 71–78
47 Olliaro P, Goldberg DE (1995) The *Plasmodium* digestive vacuole: metabolic headquarters and choice drug target. *Parasitol Today* 11: 294–297
48 Francis SE, Gluzman IY, Oskman A, Knickerbocker A, Mueller R, Bryant ML, Sherman DR, Russell DG, Goldberg DE (1994) Molecular characterization and inhibition of a *Plasmodium falciparum* aspartic hemoglobinase. *EMBO J* 13: 306–317
49 Dame JB, Reddy GR, Yowell CA, Dunn BM, Kay J, Berry C (1994) Sequence, expression and modeled structure of an aspartic proteinase from the human malaria parasite *Plasmodium falciparum*. *Mol Biochem Parasitol* 64: 177–190
50 Scholze H, Frey S, Cejka Z, Bakker-Grunwald T (1996) Evidence for the existence of both proteasomes and a novel high molecular weight peptidase in *Entamoeba histolytica*. *J Biol Chem* 271: 6212–6216
51 Hua S-B, To W-Y, Nguyen TT, Wong M-L, Wang CC (1996) Purification and characterization of proteasomes from *Trypanosoma brucei*. *Mol Biochem Parasitol* 78: 33–46
52 González J, Ramalho-Pinto FJ, Frevert U, Ghiso J, Tomlinson S, Scharfstein J, Corey EJ, Nussenzweig V (1996) Proteasome activity is required for the stage-specific transformation of a protozoan parasite. *J Exp Med* 184: 1909–1918
52a Gantt SM, Myung JM, Briones MR, Li WD, Corey EJ, Omura S, Nussenzweig V, Sinnis P (1998) Proteasome inhibitors block development of *Plasmodium* spp. *Antimicrob Agents Chemother* 42: 2731–2738
53 McKerrow JH, Doenhoff MJ (1988) Schistosome proteases. *Parasitol Today* 4: 334–340
54 Ghendler Y, Parizade M, Arnon R, McKerrow JH, Fishelson Z (1996) *Schistosoma mansoni*: evidence for a 28-kDa membrane-anchored protease on *Schistosomula*. *Exp Parasitol* 83: 73–82
55 Dalton JP, Smith AM, Clough KA, Bridley PJ (1995) Digestion of haemoglobin by Schistosomes: 35 years on. *Parasitol Today* 11: 299–303
56 Becker MM, Harrop SA, Dalton JP, Kalinna BH, McManus DP, Brindley PJ (1995) Cloning and characterization of the *Schistosoma japonicum* aspartic proteinase involved in hemoglobin degradation. *J Biol Chem* 270: 24496–24501
57 Gotz B, Klinkert M-Q (1993) Expression and partial characterization of a cathepsin B-like enzyme (Sm31) and a proposed "haemoglobinase" (Sm32) from *Schistosoma mansoni*. *Biochem J* 290: 801–806
58 Dalton JP, Brindley PJ (1996) Schistosome asparaginyl endopeptidase Sm32 in hemoglobin digestion. *Parasitol Today* 12: 125
59 Klinkert M, Kunz W (1996) Digestion of hemoglobin by Schistosomes. *Parasitol Today* 12: 165
60 Heussler VT, Dobbelaere DAE (1994) Cloning of a protease gene family of *Fasciola hepatica* by the polymerase chain reaction. *Mol Biochem Parasitol* 64: 11–23
61 Dowd AJ, Smith AM, McGonigle S, Dalton JP (1994) Purification and characterization of a second cathepsin L proteinase secreted by the parasitic trematode *Fasciola hepatica*. *Eur J Biochem* 223: 91–98
62 Wijffels GL, Panaccio M, Salvatore L, Wilson L, Walker ID, Spithill TW (1994) The secreted cathepsin L-like proteinases of the trematode, *Fasciola hepatica*, contain 3-hydroxyproline residues. *Biochem J* 299: 781–790
63 McGinty A, Moore M, Halton DW, Walker B (1993) Characterization of the cysteine proteinases of the common liver fluke *Fasciola hepatica* using novel active-site directed affinity labels. *Parasitology* 106: 483–493
64 Carmona C, McGonigle S, Dowd E, Smith AM, Coughlan S, McGowran E, Dalton JP (1994) A dipeptidyl peptidase secreted by *Fasciola hepatica*. *Parasitology* 109: 113–118
65 Wijffels GL, Salvatore L, Dosen M, Waddington J, Wilson L, Thompson C, Campbell N, Sexton J, Wicker J, Bowen F, Friedel T, Spithill TW (1994) Vaccination of sheep with purified cysteine proteinases of *Fasciola hepatica* decreases worm fecundity. *Exp Parasitol* 78: 132–148
66 Sakanari JA (1990) *Anisakis* – from the platter to the microfuge. *Parasitol Today* 6: 323–327
67 Hawdon JM, Jones BF, Perregaux MA, Hotez PJ (1995) *Ancylostoma caninum:* metalloprotease release coincides with activation of infective larvae *in vitro*. *Exp Parasitol* 80: 205–211
68 Harrop SA, Sawangjaroen N, Prociv P, Brindley PJ (1995) Characterization and localization of Cathepsin B proteinases expressed by adult *Ancylostoma caninum* hookworms. *Mol Biochem Parasitol* 71: 163–171
69 Rhoads ML, Fetterer RH (1996) Extracellular matrix degradation by *Haemonchus contortus*. *J Parasitol* 82: 379–383

70 Pratt D, Armes LG, Hageman R, Reynolds V, Boisvenue RJ, Cox GN (1992) Cloning and sequence comparison of four distinct cysteine proteases expressed by *Haemonchus contortus* adult worms. *Mol Biochem Parasitol* 51: 209–218
71 Sato K, Yamaguchi H, Waki S, Suzuki M, Nagai Y (1995) *Dirofilaria immitis*: Immunohistochemical localization of acid proteinase in the adult worm. *Exp Parasitol* 81: 63–71
72 Richer JK, Hunt WG, Sakanari JA, Grieve RB (1993) *Dirofilaria immitis*: Effect of fluoromethyl ketone cysteine protease inhibitors on the third- and fourth-stage molt. *Exp Parasitol* 76: 221–231

Proteases: New Perspectives
V. Turk (ed.)

Cell-surface proteases in cancer

Lisa L. Demchik[1] and Bonnie F. Sloane[1,2]

[1]*Cancer Biology Program and* [2]*Department of Pharmacology, School of Medicine, Wayne State University, 540 E. Canfield, Detroit, MI 48201*

Summary. Proteases of the serine, cysteine, aspartic, and metallo classes are capable of degrading extracellular matrix and basement membrane components thus contributing to the invasive and metastatic properties of tumor cells. The localization of proteases to tumor cell surfaces may be important for their ability to invade and may also be a way to regulate protease expression and activity. Proteases found to be associated with the tumor cell surface and which are also correlated with tumor invasion and metastasis include: urokinase plasminogen activator, cathepsin B, and matrix metalloproteinase-2. Cell surface receptors have been identified for urokinase plasminogen activator and matrix metalloproteinase-2. Although regulation of protease expression and activity can occur at several levels, the interaction of proteases with cell surface receptors may be most crucial to protease function. Increased expression of proteases, cell surface receptors, and endogenous inhibitors have been observed in many cancers.

Introduction

Malignant tumors have the ability to invade through basement membranes and the stromal extracellular matrix, form new blood vessels, and metastasize to distant sites [1]. Tumor cells are attached to basement membrane (BM) and extracellular matrix (ECM) surfaces via tumor cell surface receptors of the integrin and non-integrin type [2]. These receptors recognize membrane components such as laminin, fibronectin, and type IV collagen. A localized area of degradation can be produced at the site of tumor cell attachment to the BM/ECM. Invasive tumor cells then migrate through the degraded basement membrane into the surrounding normal stromal tissue [3]. Some normal physiological processes also involve invasion and migration through the BM/ECM. These processes include trophoblast invasion during implantation of the placenta, angiogenesis, pathological invasion by parasites and bacteria, embryo morphogenesis, and wound healing.

There are five major classes of proteases: serine, threonine, cysteine, aspartic and metallo. Of these, primarily serine, cysteine, and metallo proteases have been implicated in tumor invasion and metastasis. Because of the number of components that make up the BM/ECM and the substrate specificity of each protease, proteolysis of the matrix may involve the simultaneous action of numerous proteases and/or the sequential action of a protease cascade. Due to space limitations, this chapter will focus on specific surface-associated serine, cysteine, and metallo proteases. Regulation of the proteases involved in tumor invasion and metastasis is complex and occurs at multiple levels. In the majority of solid tumors and tumor cell lines, proteases are overexpressed in comparison to normal tissue. In some tumors, the increased expression occurs in the fibroblasts of the surrounding stroma and invading macrophages rather than the tumor cells themselves. A number of agents such as growth factors, cytokines, oncogenes, and tumor promoter genes can induce the expression of proteases. Protease expression can be regulated at the level of transcription and translation and, once translated, can occur at the level of processing and trafficking. For example, in tumor cells, cathepsin B has been found to be redistributed to different vesicular compartments, secreted from the cell, and associated with the plasma membrane [4]. The proteases are synthesized as inactive enzymes that require cleavage into the active form. Regulation of this process depends on several conditions including the amount of enzyme activator present, the amount of proenzyme present, and the location of the proenzyme in relation to its activator [2]. Overproduction of the proenzyme is necessary but not sufficient to produce an invasive phenotype.

Location plays an important role in the activation of proenzymes. The recent identification and functional analyses of serine protease receptors, metalloproteinase receptors, and cell surface-associated cathepsins suggests that activation of many proenzymes might be associated with the cell surface. Plasma membrane protrusions from cancer cells, i.e. invadopodia, can degrade BM/ECM [5]. Secreted proteases and integral membrane proteases have been localized to invadopodia suggesting that proteases are present at discrete areas on the tumor cell surface where their degradative actions may be most effective. Tumor cells tightly regulate proteolytic degradation of the BM/ECM to allow cell attachment and migration. In many tumors, increased expression of inhibitors is observed and presumed to be necessary to balance increases in protease expression. Protease inhibitors may also act to focus proteolytic activity at the invasive edge of the tumor by inhibiting protease activities elsewhere. Therefore, localization of proteases and their inhibitors to the cell surface in areas of cell-matrix interaction could well play a key role in tumor invasion.

Serine proteases: plasminogen activators and plasmin

The plasmin-plasminogen system participates in a number of cellular invasion processes including tumor invasion. The components of the plasmin-plasminogen system include urokinase-type plasminogen activator (uPA), tissue-type plasminogen activator (tPA), and plasminogen, which is activated to plasmin by either uPA or tPA [6, 7, 8]. Plasmin is a broad-spectrum serine protease and can be involved in tumor invasion directly through cleavage of the ECM components fibronectin, laminin, and collagen IV and/or indirectly by activation of latent proteases such as procollagenases and prostromelysin [7]. Thus, a small amount of plasmin could produce a cascade of proteolytic events contributing to malignancy.

A variety of cells produce uPA, including neoplastic cells. uPA, like plasmin, is a serine protease and is secreted as a single-chain proenzyme form termed prourokinase (pro-uPA). Pro-uPA is converted into the two chain active form by proteolysis. The known activators of pro-uPA *in vivo* are plasmin and kallikrein, but other proteases such as cathepsin B [9] and cathepsin L [10] have been shown to activate prourokinase *in vitro*. Cathepsins B and L cleave pro-uPA at the same site as plasmin and kallikrein, therefore it is possible that activation by the cathepsins could also occur *in vivo*. Besides activating plasminogen, uPA can also activate other proteases such as matrix metalloproteinase-2, thereby contributing to the cascade effect [11].

uPA and plasmin bind to cell surface receptors

Both uPA and pro-uPA bind to the cell surface urokinase plasminogen activator receptor (uPAR) with high affinity [12]. uPAR binds uPA and pro-uPA via a sequence located within the growth factor domain in the amino terminus of the enzyme [13]. Receptor bound pro-uPA can be activated by plasmin [14]. Estreicher et al. [15] discovered that binding of uPA is species-specific, i.e. human uPA binds only to the human uPAR and mouse uPA binds only to mouse uPAR. Receptor bound uPA is retained on the cell surface for long periods of time without being degraded or internalized [16]. The receptor is anchored in the plasma membrane by a glycosyl-phosphatidylinositol (GPI) moiety [17]. Endocytosis of these glycolipid-anchored proteins does not normally occur, which could explain the slow turnover rate of the uPA-uPAR complex. Recently, Higazi et al. [18] observed that the association of uPA with its receptor causes a conformational change in the complex allowing interactions with other cell surface components and preventing the clearance of uPA that normally occurs via binding of uPA to the α_2-macroglobulin receptor/LDL (low density lipoprotein) receptor-related protein. For further information about uPAR and its chemistry the following reviews can be consulted [12, 19, 20]. Several studies have documented the important role that uPAR plays in proteolytic degradation of the ECM by plasmin. Some of the functions of uPAR are: (1) to bind uPA at the cell surface where the active enzyme remains for long periods of time, (2) to bind pro-uPA making it readily accessible to activation by plasmin, (3) to increase the efficiency of plasminogen activation, (4) to reduce interactions with uPA inhibitors (see below), and (5) to localize the enzyme to focal contacts and invadopodia (for review see [7]).

Regulation of uPA occurs on several levels, including activation of pro-uPA, binding to uPAR, and inhibition by endogenous inhibitors. The two major inhibitors of uPA are plasminogen activator inhibitors 1 and 2 (PAI-1 and -2). After these inhibitors bind to the uPA-uPAR complex, the entire complex is internalized and degraded by the cell. uPA is not the only component of this system for which receptor binding has been reported. Plasminogen also binds to the cell surface, but a specific receptor has not yet been characterized [21–23]. Like the uPA-uPAR complex, the plasminogen binding protein binds both the proenzyme and the active form and retains the active enzyme at the cell surface. Bound plasmin is resistant to its inhibitor α_2-antiplasmin [21, 24, 25]. Ellis et al. [25] demonstrated that for efficient activation by uPA, it was important that plasminogen be cell-surface associated. These investigators concluded that the cell surface is the primary site for both the activation to plasmin and its proteolytic activity (Fig. 1). The activation of plasminogen by uPA and the activation of pro-uPA by plasmin can form an amplification loop in which activation of proenzymes proceeds faster as more active enzyme is produced.

uPA and uPAR expression in cells in vitro

uPARs were first identified on migratory monocytic cells [26, 27]. Stoppelli et al. [28] showed that epidermoid carcinoma cells express both uPA and uPAR, thus binding of uPA to its receptor can occur in an autocrine fashion. Reiter et al. [29] observed that cultured human colon car-

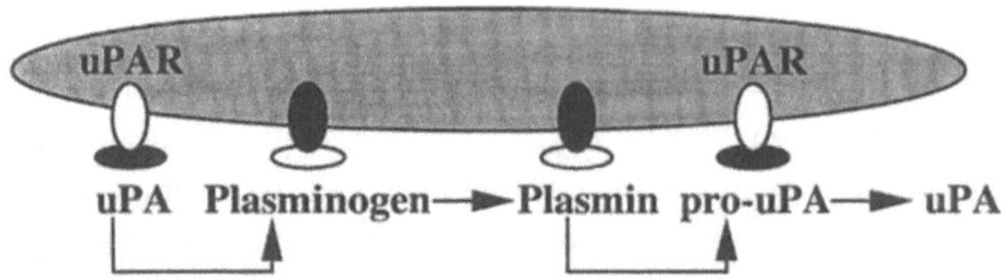

Figure 1. The interaction of plasmin and urokinase-type plasminogen activator (uPA) at the cell surface. Both pro-urokinase-type plasminogen activator (pro-uPA) and uPA bind to urokinase-type plasminogen activator receptors (uPARs) on the cell surface. Activation of plasminogen occurs at these sites. Plasmin can then activate pro-uPA or degrade extracellular matrix.

cinoma cells express uPAR, but not uPA. In these cells, uPARs bind exogenous uPA in a paracrine fashion. In general, white blood cells and cultured tumor cells have been shown to express uPAR, whereas most stromal fibroblasts associated with tumors, but only some tumor cells, express uPA [30]. Normal cells (that are normally not invasive) do not express uPAR [30].

uPA was initially localized to the extracellular membrane of HT1080 cells at focal contact sites (cell-substratum) and cell-cell contact sites [31] and was subsequently shown to be bound to uPAR [26]. In accordance with this evidence, we can hypothesize that plasminogen activation occurs at specific regions on the cell surface where there are uPARs. Myohanen et al. [32] confirmed the localization of receptor bound uPA at focal adhesions and cell-cell contact sites in human fibroblasts and rhabdomyosarcoma cells. In the same studies, unoccupied uPAR sites were distributed diffusely on the cell surface. Therefore, only at the focal contact sites are all the elements necessary for the degradation of the matrix by plasmin present and is plasmin activity focused to the degradation front by the uPAR-uPA complex.

Increased levels of uPA, plasmin, and uPARs correlate with tumor malignancy but further evidence is required to determine the direct contribution of these proteases to the invasive behavior of tumors. HT29 cells unable to express uPA were transfected with a human uPA cDNA and the degradative ability of the transfected cells analyzed using radiolabeled ECM [29]. The parental cell line does not degrade ECM components, whereas the transfected lines do. Inhibitors of uPA as well as antibodies that block binding to uPAR result in significant inhibition of degradation. Thus, the binding of uPA to its receptor plays an important and possibly essential role in ECM degradation. uPA inhibitors or antibodies which block uPA binding to its receptor have been shown to inhibit the invasion of human breast carcinoma cells [33], prostate cells [34], melanoma cells [35], promyeloid leukemia cells [36], and glioblastoma cells [37] through the reconstituted basement membrane Matrigel. Reduced invasion indicates that activation of plasmin is inhibited, since the proteolytic action of plasmin is required for ECM degradation. The above studies suggest that binding of uPA to its receptor is important for invasion and that plasmin is directly involved in the degradation of the ECM.

uPA, uPAR, and plasmin expression in vivo

Needham et al. [38] first observed uPAR in cell membranes of 13 out of 29 human breast carcinomas. Since then, uPAR expression has been described in many human carcinomas including lung [39], ovarian [40], prostate [34], and colorectal carcinomas [41]. As discovered *in vitro*, not all tumor cells express uPA, but they all express uPAR. Pyke et al. [41] observed that human colon carcinomas do not express uPA, yet stromal fibroblasts associated with the tumor, especially those in areas of degradation, express uPA. uPAR is primarily expressed in malignant cells at the leading edge of the invading tumor. Immunohistochemical staining of breast carcinomas revealed the distribution of uPA and uPAR to be similar to that in colon carcinomas [42]. Stromal fibroblasts express uPA whereas the tumor cells express low amounts of uPA and high amounts of uPAR. Ossowski [43] used the *ex vivo* modified chick embryo chorioallantoic membrane (CAM) assay to show that uPAR is necessary for human tumor cell invasion. The overexpression of uPA in the absence of binding to uPAR does not result in enhanced invasion. This group also described how paracrine interactions between uPA and its receptor enhance invasion [44]. Both paracrine and autocrine interactions between uPA and uPAR are observed *in vivo* and uPA binding to uPAR appears to be important for tumor cell invasion.

Detection of the levels of expression of uPA and uPAR may be important as a diagnostic tool. In human gliomas, the amounts of uPA and uPAR are significantly higher in anaplastic astrocytomas and glioblastomas than in low-grade gliomas and benign brain tissue [45]. Studies of squamous cell lung cancer tissue show an association between high levels of uPAR and a shorter overall patient survival [39]. In ovarian cancer, increased amounts of uPA and uPAR are seen in tumor cells and tumor-associated stromal cells, with a greater amount of uPA and uPAR at metastatic sites [40]. The intensity of uPAR staining in gastric carcinoma cells is greater in metastatic tumor cells than in primary tumor cells [46]. These studies suggest that uPA and uPAR may be useful as markers of tumor invasion and metastasis. In support of the above observations, Crowley et al. [34] observed that inhibition of uPAR and uPA in human prostate cells prevents metastasis formation in mice. Recently, soluble uPAR devoid of uPA was found in the ascites fluid of patients with ovarian cancer [47]. This uPAR is identical to cell-associated uPAR. The authors suggest that its presence in the ascites fluid is due to receptor shedding, and that soluble uPAR may be a useful diagnostic marker for ovarian carcinoma.

Cysteine proteases: cathepsin B

Increased expression of the cysteine protease cathepsin B (CB) in human carcinomas correlates with tumor progression. CB exhibits both endopeptidase and exopeptidase activity [48, 49] and has the ability to degrade multiple components of the ECM including laminin [50], fibronectin, and type IV collagen [48]. *In vitro* studies have shown that CB may be involved in the activation of proteases such as uPA [9]. CB is synthesized as an inactive preproenzyme that is first sorted to the endoplasmic reticulum where the pre-sequence is cleaved and then sorted to the Golgi complex for modification of oligosaccharides and binding to mannose 6-phosphate receptors. Activation occurs in an endosomal compartment [51]. Two active forms of CB have been identified: a 31 kDa single chain form and a double chain form consisting of a 25/26 kDa heavy chain and a 5 kDa light chain [52, 53]. In normal cells, CB is localized in lysosomes in the perinuclear region or is secreted at low levels [53, 54]. In carcinoma cells and other cells involved in degradation, lysosomes containing CB are localized both perinuclearly and at the cell periphery [55, 56] and increased secretion of CB (pro and active) is observed. The ability of CB to cleave numerous extracellular substrates and the change in CB localization in cells involved in degradation suggests that CB may play a role in tumor invasion.

Localization of CB in vitro

Treatment of malignant cells with 12-(S)-hydroxyeicosatetraenoic acid (a product of arachidonic acid metabolism) triggers a release of active CB from the cells [57]. Along with the release of CB, vesicles staining for CB are redistributed to the cell periphery. Exposing cells to pH 6.5 also induces movement of CB to the cell periphery. After exposure to pH 6.5, highly malignant cells exhibit less staining for CB at the periphery and increased secretion of active CB [58]. The *in vitro* change in pH mimics the *in vivo* situation where a more acidic environment is found in tumor regions, thus a small change in pH may induce secretion of active CB *in vivo*. Increased secretion of both latent proCB and active CB from human and animal tumors such as colon carcinoma [59], breast carcinoma [60], glioma [61], and murine melanoma [62] has been observed. The inactive proform found in the extracellular milieu may be autocatalytically activated or it may be activated by other proteases including uPA and cathepsin D [63, 64]. Active CB localized at the cell periphery or present as a soluble extracellular enzyme may play a role in ECM degradation.

In model systems such as MCF-10A human breast epithelial cells, the movement of CB to the cell periphery can be induced by the *ras* oncogene. [65]. The parental MCF-10A cell line is non-tumorigenic. The MCF-10AneoT variant, which has been transfected with the c-H-*ras* oncogene, is preneoplastic, hyperplastic, locally invasive, and progresses to tumors in approximately thirty percent of mice. Immunocytochemical staining of MCF-10A cells reveals that CB is located in the perinuclear region [56, 66]. In contrast, CB staining in MCF-10Aneo T cells is at the cell periphery as well as in the perinuclear region. The MCF-10AneoT cells also exhibit staining for CB on the cell surface, specifically in patches on the basal surface. No surface staining is observed on MCF-10A cells. Similar patterns of cell surface staining for CB are observed in breast carcinoma cell lines including MCF7 and BT20, but the extent of staining is much greater [56, 67]. Arkona and Wiederanders [68] observed a direct association of CB with the plasma membrane in metastatic bone tissue. They suggest that binding of CB to the plasma membrane occurs through association with the general protease inhibitor α_2-macroglobulin, which then binds to the α_2-macroglobulin receptor/LDL-receptor related protein on the cell surface. The above observations suggest there is altered localization of CB within the cell coincident with the tumor cell's ability to degrade the ECM. We hypothesize that CB in vesicles at the cell periphery is secreted to provide a pool of active CB that binds to the cell surface (Fig. 2).

One way to evaluate the role of CB in tumor invasion is by the use of specific inhibitors, thus eliminating the possibility that other proteases are contributing to the observed degrada-

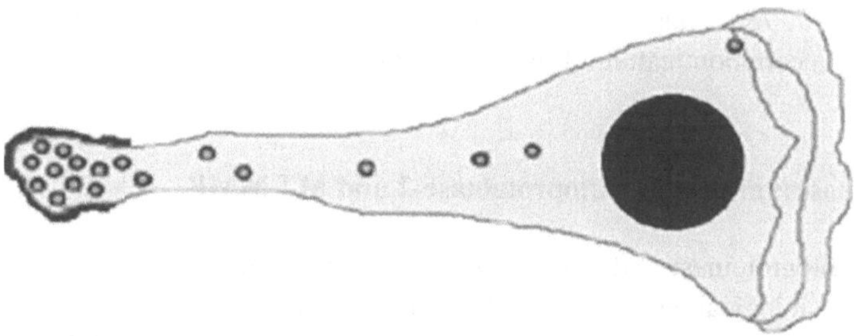

Figure 2. Vesicles containing cathepsin B can be translocated to the cell surface in tumor cells and to the tail of motile cells as indicated in this cartoon. Extracellular staining for cathepsin B occurs on the cell surface adjacent to the translocated vesicles.

tion. We recently observed CB surface staining localized to the basal surface of breast carcinoma cells and glioma cells in areas of laminin degradation [69]. A highly selective CB inhibitor, CAO74 [70], could inhibit the laminin degradation to some extent. Panchal et al. [71] recently presented a new approach to kill tumor cells by taking advantage of the overexpression of CB. Pro-hemolysins were designed by a combinatorial approach to be activated by tumor cell surface-associated CB. These prohemolysins do kill tumor cells. These results confirm our findings that CB is associated with the tumor cell surface in a proteolytically active form.

CB localization in vivo

Using subcellular fractionation, Sloane et al. [72] observed the association of CB with the plasma membrane fraction of a metastatic murine melanoma, whereas CB is not found in the plasma membrane fraction of normal tissue. Further examination revealed that membrane-associated CB has reduced inhibitor binding, appears to be more stable at extracellular pH than unbound CB, and that only treatment with detergent or butanol releases CB from the membrane [73]. This study did not identify a specific CB binding receptor, but it did rule out binding to the membrane by mannose 6-phosphate receptors. A role for CB in the invasive process has been demonstrated in human carcinomas. Poole [60] observed increased secretion of CB at the invading edges of human breast carcinomas. Activity assays on microdissected specimens of human colon carcinoma revealed increased CB activity at the invasive edge [74]. Campo et al. [75] observed CB staining of colon carcinomas to be polarized to the basal region of tumor cells whereas in normal colonic mucosa CB staining is apical. Increased CB staining in colon carcinoma cells and stromal cells associated with the tumor correlates with increased malignancy and reduced patient survival. A correlation between increased CB expression and increased malignancy is also observed in gliomas [76]. The leading edge of the tumor stains more intensely for CB placing the protease at the site of local infiltration. Similar correlations between CB staining/location in tumor tissues and malignancy have been observed in prostate and in bladder carcinomas [77, 78]. Thus, increased CB expression and localization of active CB to the cell periphery is a common feature of the malignant phenotype.

Metalloproteinases: matrix metalloproteinase-2 and MT-MMP

The matrix metalloproteinase (MMP) family contains many members that are considered to be central in the physiological process of ECM degradation. MMPs have several common features including a zinc-containing catalytic site, production as a proenzyme, the ability to cleave ECM components, and inhibition by tissue inhibitors of metalloproteinases (TIMPs) (for review see [79–81]). MMP-2 (also known as the 72 kDa type IV collagenase or gelatinase A) and MMP-9 (also know as the 92 kDa type IV collagenase or gelatinase B) contain an additional domain with sequence similarity to the gelatin-binding domain of fibronectin. Latent MMP-2 and MMP-9 can form a complex with the endogenous inhibitors TIMP-2 [82] and TIMP-1 [83], respectively, which may help to prevent autocatalysis. Activated MMP-2 and MMP-9 have the ability to cleave numerous components of the ECM including gelatins, collagen, elastin, fibronectin, and others [84].

Interaction between MMP-2 and MT-MMP

Many proteases have the potential to activate the MMPs, thus initiating a cascade of proteolytic events. Until recently, no specific enzyme had been shown to activate proMMP-2 *in vivo*. Considerable evidence had accumulated suggesting that proMMP-2 is associated with the cell surface [85–87]. Treatment of cells with concanavalin A or phorbol esters can induce activation of proMMP-2 [85–88]. Activation involves the interaction of the carboxyl-terminal domain of MMP-2 with a membrane-associated component and is inhibited by complex formation with TIMP-2. This membrane-associated component has now been identified as a novel MMP and named membrane-type matrix metalloproteinase (MT1-MMP) [89]. Three other MT-MMPs were later identified: MT2-MMP [90], MT3-MMP [91, 92], and MT4-MMP [92]. All four MT-MMPs have the ability to activate proMMP-2, however, MT1-MMP is the primary activator. Activation of the MT-MMPs themselves has not yet been characterized, but may involve a furin-dependent mechanism due to the presence of an RX–R sequence in MT-MMPs [93].

The interactions among MMP-2, MT1-MMP, and TIMP-2 are complex. Cao et al. [94] determined that the transmembrane domain functions to link MT1-MMP to the cell surface and is essential for the activation of MMP-2. Soluble MT1-MMP lacking the transmembrane domain does not activate MMP-2. TIMP-2 alone and bound to proMMP-2 was also observed to bind to the cell surface, possibly via a TIMP-2 receptor [95]. Interestingly, two different groups reported that MT1-MMP complexed with TIMP-2 is secreted from a human breast carcinoma cell line [96, 97]. Although the mechanism is not completely understood, the MT1-MMP:TIMP-2 complex is able to bind proMMP-2, making the proMMP-2 accessible for activation by other MT1-MMPs (Fig. 3). Many questions remain about the mechanism of proMMP-2 activation and cell surface interaction: Can MT1-MMP activate the

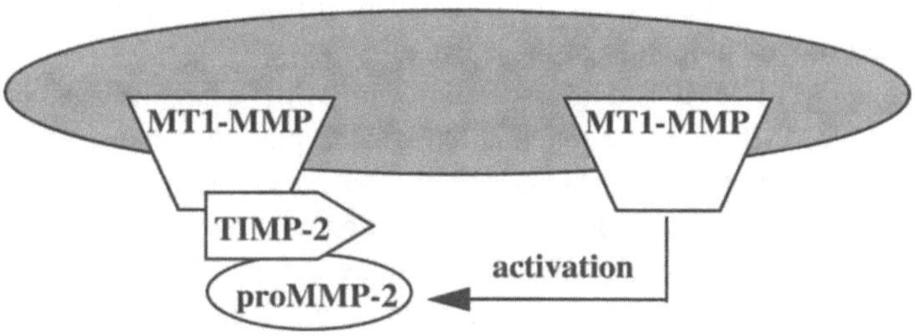

Figure 3. Activation of pro-MMP-2 occurs on the cell surface. TIMP-2 bound to MT1-MMP may block the active site of proMMP-2 via binding to its carboxyl terminus. Meanwhile an adjacent MT1-MMP is in position to activate proMMP-2.

TIMP-2:proMMP-2 complex? Why do some cells secrete a TIMP-2:MT1-MMP complex? Can the secreted complex activate proMMP-2? Is the binding and activation of proMMP-2 mediated only by MT1-MMP or is there a second binding site for either TIMP-2 or MMP-2? Does the fact that TIMP-2 associates with either proMMP-2 or MT1-MMP suggest that it may function both as an inhibitor and as a factor in the activation process? The presence of activated MMP-2 on the cell surface may help to promote localized proteolysis of the ECM in areas of cell-matrix contact.

Expression of MMP-2 and MT-MMP in vitro

Many studies have shown a positive correlation between MMP-2 expression and malignancy [81, 99, 100]. Active, cell surface-bound MMP-2 is detected on invadopodia of transformed chicken embryo fibroblasts [101]. Invadopodia may act to localize MMP-2 degradation by concentrating the enzyme to areas of tumor invasion. An answer to one of the questions posed above concerning a possible receptor for MMP-2 may be found in a study by Brooks et al. [102]. Both *in vitro* and *in vivo* studies localized active MMP-2 to the cell surface of melanoma cells via interaction with the integrin $\alpha_v\beta_3$. Some studies have suggested that a cell surface receptor for MMP-2 besides the MT-MMPs may exist. Perhaps integrin $\alpha_v\beta_3$ is the receptor that localizes MMP-2 to cell-matrix contact sites.

Expression of MT1-MMP mRNA and protein has been observed in a number of human carcinomas. The activation of proMMP-2 by MT1-MMP may play an important role in tumor cell invasion and metastasis. For example, transfection of a mouse tumor cell line with MT1-MMP results in increased subcutaneous tumor growth and metastasis formation in the lungs [92]. Emonard et al. [103] first suggested a tumor cell membrane-associated receptor for MMP-2 when they discovered that two breast carcinoma cell lines bind exogenous MMP-2. Induction of MT1-MMP expression by concanavalin A was also found to correlate with MMP-2 activation and invasive behavior of cervical cancer cells [104]. Young et al. [105] discovered that in addition to its role in activation of the latent enzyme, cell surface binding of MMP-2 by MT1-MMP increases the catalytic properties of the active enzyme.

Characterization of MT1-MMP and its association with MMP-2 have changed our hypotheses on activation of proMMP-2 and the role that MT1-MMP plays in tumors. Yamamoto et al. [106] observed a correlation between MT1-MMP expression and activation of proMMP-2 in glioma cell lines. Their results also suggested that activation required a cell surface receptor for proMMP-2 alone or complexed with TIMP-2. The activity of both MT1-MMP and MMP-2 in glioma cells is tightly regulated. Small amounts of TIMP-2 cause increased binding of proMMP-2, whereas excessive amounts result in inhibition of binding [106]. Recently, Lohi et al. [107] concluded that activation of proMMP-2 requires factors other than MT1-MMP. These investigators could induce the expression of a 60 kDa form of MT1-MMP with concanavalin A or phorbol ester in fibrosarcoma cells and embryonic lung fibroblasts. However, the active form of MMP-2 was observed only in fibrosarcoma cells expressing a 43 kDa form of MT1-MMP. In order to characterize MT1-MMP, Pei and Weiss [108] constructed proMT1-MMP mutants lacking the transmembrane domain. These mutants could be activated and will degrade ECM components as well as proMMP-2. Such observations begin to answer some of the questions posed above, but they also reveal that more information is needed on the processes that regulate MT-MMP activation and the mechanism(s) by which proMMP-2 is activated.

Expression of MMP-2 and MT-MMP in vivo

Studies of human breast, colon, and basal cell carcinomas found that MMP-2 mRNA is localized primarily to stromal fibroblasts adjacent to the invasive tumor edge [109, 110], whereas MMP-2 protein expression is localized to tumor cells [81, 100]. Polette et al. [111] did not detect MMP-2 mRNA in breast carcinoma cells but did in stromal fibroblasts located near non-invasive and well-differentiated invasive tumor cells. The MMP-2 protein is overexpressed and membrane-associated in breast carcinomas; TIMP-2 protein expression is detected in stromal cells surrounding the tumor and correlates with tumor recurrence [112]. Similar results have been seen in colon, gastric, and breast carcinomas: TIMP-2 protein in stromal cells, MMP-2 protein in tumor cells, and mRNA for both MMP-2 and TIMP-2 primarily in stromal cells [111, 113]. The MMP-2 transcribed by the stromal cells could associate with the MT-MMPs found on tumor cells via a paracrine mechanism. In contrast to observations that localize MT-MMPs to tumor cells [114], Okada et al. [115] observed MT1-MMP transcripts to be localized to tumor-associated stromal cells of human colon, breast, and head and neck carcinomas. These authors suggest that MT1-MMP is possibly cleaved from the surface of the stromal cells and the soluble protein then binds to a membrane receptor on the cancer cells. Although these observations of protease localization appear to be contradictory, they suggest that binding to MT-MMP by MMP-2 occurs through either a paracrine or an autocrine mechanism.

The importance of MMP-2 binding and activation by MT1-MMP in tumor cell invasion is only now being studied. MT1-MMP expression in gliomas correlates with proMMP-2 activation and expression during malignant progression [106]. In lung carcinomas, activation of MMP-2 correlates with the expression of MT1-MMP and with lymph node metastasis, suggesting the importance of MT1-MMP in the metastatic process [116]. Nomura et al. [114] observed that MT1-MMP mRNA and protein is expressed exclusively in human gastric carcinoma tissue. The expression and activity of MMP-2 in gastric carcinoma coincides with the expression of MT1-MMP in tumor cells. MT1-MMP was observed to be highly expressed in invasive cervical cancer and lymph node metastases [104]. The many observations of MT1-MMP expression associated with metastases suggest that MT1-MMP may prove to be important as a diagnostic indicator. The binding of proMMP-2 to the membrane bound protease MT1-MMP could serve multiple purposes: (1) cleavage of proMMP-2 into an active enzyme; (2) localization of MMP-2 on the tumor cell surface; and (3) degradation of adjacent BM/ECM by active, cell-surface bound MMP-2. Thus, MT1-MMP expression and activation of MMP-2 appear to contribute to the local proteolysis of ECM components by tumor cells.

Conclusions

Clearly, no single protease or class of proteases is responsible for the invasion of normal tissue by tumors. A number of different proteases are expressed by invasive tumors, both simultaneously and through a cascade effect, contributing to the overall degradation of matrix. Each tumor may express a different variety of proteases depending on tumor heterogeneity, the presence of inflammatory cells, stage-specific increases, the complexity of the ECM components encountered or all of these. Substrate specificity is broad, allowing substitution of proteases, yet each protease degrades specific types of ECM components, so protease actions can also be complementary. The expression of proteases can vary by tumor type. For example, the aspartic protease cathepsin D is expressed primarily by breast carcinomas. Other proteases have a wider distribution of expression, such as CB, uPA, plasmin, and MMP-2. Their wide distribution and their cell surface location suggest that these four proteases play a significant role in invasion and metastasis. The clustering of the proteases, receptors, and inhibitors suggests that the tumor can regulate and focus expression of proteolytic activity. Recent studies from a number of laboratories suggest that proteases may also be involved in earlier steps of tumor cell progression such as angiogenesis or cell proliferation. Whether proteases associated with the cell surface function primarily in invasive processes or contribute to other proteolytic events required for tumor progression will require further study.

Acknowledgments
The authors would like to thank Dr. Edith Elliott, Dr. Rafael Fridman and Jennifer Koblinski for their critical reading of this manuscript. This work was supported by U.S. Public Health Grants CA 36481 and 56586.

References

1 Mignatti P, Rifkin DB (1993) Biology and biochemistry of proteinases in tumor invasion. *Physiol Rev* 73: 161–195
2 Stetler-Stevenson WG, Aznavoorian S, Liotta LA (1993) Tumor cell interactions with the extracellular matrix during invasion and metastasis. *Annu Rev Cell Biol* 9: 541–573

3 Liotta LA, Steeg PS, Stetler-Stevenson WG (1991) Cancer metastasis and angiogenesis: an imbalance of positive and negative regulation. *Cell* 64: 327–336
4 Berquin IM, Sloane BF (1995) Cysteine proteases and tumor progression. *Perspect Drug Discov Design* 2: 371–388
5 Chen W-T, Lee C-C, Goldstein L, Bernier S, Liu CHL, Lin C-Y, Yeh Y, Monsky WL, Kelly T, Dai M et al. (1994) Membrane proteases as potential diagnostic an therapeutic targets for breast malignancy. *Breast Cancer Res Treat* 31: 217–226
6 Polllanen J, Stephens RW, Vaheri A (1991) Directed plasminogen activation at the surface of normal and malignant cells. *Adv Cancer Res* 57: 273–328
7 Kwaan HC (1992) The plasminogen-plasmin system in malignancy. *Cancer Metastasis Rev* 11: 291–311
8 Dano K, Andreasen PA, Grondahl-Hansen J, Kristensen P, Nielsen LS, Skriver L (1985) Plasminogen activators, tissue degradation and cancer. *Adv Cancer Res* 44: 140–239
9 Kobayashi H, Schmitt M, Goretzki L, Chucholowski N, Calvete J, Kramer M, Gunzler WA, Janicke F, Graeff H (1991) Cathepsin B efficiently activates the soluble and the tumor cell receptor-bound form of the proenzyme urokinase-type plasminogen activator (Pro-uPA). *J Biol Chem* 266: 5147–5152
10 Goretzki L, Schmitt M, Mann K, Calvete J, Chucholowski N, Kramer M, Gunzler WA, Janicke F, Graeff H (1992) Effective activation of the proenzyme form of the urokinase-type plasminogen activator (pro-uPA) by the cysteine protease cathepsin L. *FEBS Lett* 297: 112–118
11 Keski-Oja J, Lohi J, Tuuttila A, Tryggvason K, Vartio T (1992) Proteolytic processing of the 72,000-Da type IV collagenase by urokinase plasminogen activator. *Exper Cell Res* 202: 471–476
12 Ellis V, Pyke C, Eriksen J, Solberg H, Dano K (1992) The urokinase receptor: involvement in cell surface proteolysis and cancer invasion. *Ann N Y Acad Sci* 667: 13–31
13 Appella E, Robinson EA, Ullrich SJ, Stoppelli MP, Corti A, Cassani G, Blasi F (1987) The receptor-binding sequence of urokinase: a biological function for the growth-factor module of proteases. *J Biol Chem* 262: 4437–4440
14 Cubellis MV, Nolli ML, Cassani G, Blasi F (1986) Binding of singe-chain prourokinase to the urokinase receptor of human U937 cells. *J Biol Chem* 261: 15819–15822
15 Estreicher A, Wohlwend A, Belin D, Schleuning W-D, Vassalli J-D (1989) Characterization of the cellular binding site for the urokinase-type plasminogen activator. *J Biol Chem* 264: 1180–1189
16 Stoppelli MP, Corti A, Solfientini A, Cassani G, Blasi F, Assoian RK (1985) Differentiation-enhanced binding of the amino-terminal fragment of human prourokinase plasminogen activator to a specific receptor on U937 monocytes. *Proc Natl Acad Sci USA* 82: 4939–4943
17 Ploug M, Ronne E, Behrendt N, Jensen AL, Blasi F, Dano K (1991) Cellular receptor for urokinase plasminogen activator: carboxyl-terminal processing and membrane anchoring by glycosyl-phosphatidylinositol. *J Biol Chem* 266: 1926–1933
18 Higazi AA-R, Upson RH, Cohen RL, Manuppello J, Bognacki J, Henkin J, McCrae KR, Kounnas MZ, Strickland DK, Preissner KT et al. (1996) Interaction of single-chain urokinase with its receptor induces the appearance and disappearance of binding epitopes within the resultant complex for other cell surface proteins. *Blood* 88:542–551
19 Behrendt N, Ronne E, Dano K (1995) The structure and function of the urokinase receptor, a membrane protein governing plasminogen activation on the cell surface. *Biol Chem Hoppe-Seyler* 376: 269–279
20 Ploug M, Behrendt N, Lober D, Dano K (1991) Protein structure and membrane anchorage of the cellular receptor for urokinase-type plasminogen activator. *Sem Thrombosis Hemostasis* 17: 183–193
21 Plow EF, Freaney DE, Plescia J, Miles LA (1986) The plasminogen system and cell surfaces: evidence for plasminogen and urokinase receptors on the same cell type. *J Cell Biol* 103: 2411–2420
22 Burtin P, Fondanecke M-C (1988) Receptor for plasmin on human carcinoma cells. *J Nat Cancer Inst* 80: 762–765
23 Miles LA, Plow EF (1988) Plasminogen receptors: ubiquitous sites for regulation of fibrinolysis. *Fibrinolysis* 2: 61–71
24 Stephens RW, Pollanen J, Tapiovaara H, Leung KC, Sim PS, Salonen EM, Ronne EBehrendt N, Dano KVaheri A (1989) Activation of prourokinase and plasminogen on human sarcoma cells: a proteolytic system with surface-bound reactants. *J Cell Biol* 108: 1987–1995
25 Ellis V, Behrendt N, Dano K (1991) Plasminogen activation by receptor-bound urokinase: a kinetic study with both cell-associated and isolated receptor. *J Biol Chem* 266: 12752–12758
26 Blasi F (1993) Urokinase and urokinase receptor: a paracrine/autocrine system regulating cell migration and invasiveness. *BioEssays* 15: 105–111
27 Vassalli JD, Baccino D, Belin D (1985) A cellular binding site for the Mr 55,000 form of the human plasminogen activator, urokinase. *J Cell Biol* 100: 86–92
28 Stoppelli MP, Tacchetti C, Cubellis MV, Corti A, Hearing VJ, Cassani G, Appella E, Blasi F (1986) Autocrine saturation of pro-urokinase receptors on human A431 cells. *Cell* 45: 675–684
29 Reiter LS, Kruithof EKO, Cajot J-F, Sordat B (1993) The role of the urokinase receptor in extracellular matrix degradation by HT29 human colon carcinoma cells. *Int J Cancer* 53: 444–450
30 Buo L, Bjornland K, Karlsrud TS, Kvale D, Kjonniksen I, Fodstad O, Brandtzaeg P, Johansen HT, Aasen AO (1994) Expression and release of plasminogen activators, their inhibitors and receptor by human tumor cell

lines. *Anticancer Res* 14: 2445–2452
31 Pollanen J, Hedman K, Nielsen LS, Dano K, Vaheri A (1988) Ultrastructural localization of plasma membrane-associated urokinase-type plasminogen activator at focal contacts. *J Cell Biol* 106: 87–95
32 Myohanen HT, Stephens RW, Hedman K, Tapiovaara H, Ronne E, Hoyer-Hansen G, Dano K, Vaheri A (1993) Distribution and lateral mobility of the urokinase-receptor complex at the cell surface. *J Histochem Cytochem* 41: 1291–1301
33 Luparello C, Del Rosso M (1996) *In vitro* anti-proliferative and anti-invasive role of amino-terminal fragment of urokinase-type plasminogen activator on 8701-BC breast cancer cells. *Eur J Cancer* 32A: 702–707
34 Crowley CW, Cohen RL, Lucas BK, Liu G, Shuman MA, Levinson AD (1993) Prevention of metastasis by inhibition of the urokinase receptor. *Proc Natl Acad Sci USA* 90: 5021–5025
35 Stahl A, Mueller BM (1994) Binding of urokinase to its receptor promotes migration and invasion of human melanoma cells *in vitro*. *Cancer Res* 54: 3066–3071
36 Kobayashi H, Gotoh J, Hirashima Y, Fujie M, Sugino D, Terao T (1995) Inhibitory effects of a conjugate between human urokinase and urinary trypsin inhibitor on tumor cell invasion *in vitro*. *J Biol Chem* 270: 8361–8366
37 Mohanam S, Sawaya R, McCutcheon I, Ali-Osman F, Boyd D, Rao JS (1993) Modulation of *in vitro* invasion of human glioblastoma cells by urokinase-type plasminogen activator receptor antibody. *Cancer Res* 53: 4143–4147
38 Needham GK, Sherbet GV, Farndon JR, Harris AL (1987) Binding of urokinase to specific receptor sites on human breast cancer membranes. *Brit J Cancer* 55: 13–16
39 Pedersen H, Brunner Francis D, Osterlind K, Ronne E, Hansen HH, Dano K, Grondahl-Hansen J (1994) Prognostic impact of urokinase, urokinase receptor, and type 1 plasminogen activator inhibitor in squamous and large cell lung cancer tissue. *Cancer Res* 54: 4671–4675
40 Schmalfeldt B, Kuhn W, Reuning U, Pache L, Dettmar P, Schmitt M, Janicke F, Hofler H, Graeff H (1995) Primary tumor and metastasis in ovarian cancer differ in their content of urokinase-type plasminogen activator, its receptor, and inhibitors types 1 and 2. *Cancer Res* 55: 3958–3963
41 Pyke C, Kristensen P, Ralfkiaer E, Grondahl-Hansen J, Eriksen J, Blasi F, Dano K (1991) Urokinase-type plasminogen activator is expressed in stromal cells and its receptor in cancer cells at invasive foci in human colon adenocarcinomas. *Amer J Pathol* 138: 1059–1067
42 Carriero MV, Franco P, Del Vecchio S, Massa O, Botti G, D'Aiuto G, Stoppelli MP, Salvatore M (1994) Tissue distribution of soluble and receptor-bound urokinase in human breast cancer using a panel of monoclonal antibodies. *Cancer Res* 54: 5445–5454
43 Ossowski L (1988) *In vivo* invasion of modified chorioallantoic membrane by tumor cells: the role of cell surface-bound urokinase. *J Cell Biol* 107: 2437–2445
44 Ossowski L, Clunie G, Masucci M-T, Blasi F (1991) *In vivo* paracrine interaction between urokinase and its receptor: effect on tumor cell invasion. *J Cell Biol* 115: 1107–1112
45 Yamamoto M, Sawaya R, Mohanam S, Rao VH, Bruner JM, Nicolson GL, Rao JS (1994) Expression and localization of urokinase-type plasminogen activator receptor in human gliomas. *Cancer Res* 54: 5016–5020
46 Hong SI, Park IC, Son YS, Lee SH, Kim BG, Lee JI, Lee TW, Kook YH, Min YI, Hong WS (1996) Expression of urokinase-type plasminogen activator, its receptor, and its inhibitor in gastric adenocarcinoma tissues. *J Korean Med Sci* 11: 33–37
47 Pedersen N, Schmitt M, Ronne E, Nicoletti MI, Hoyer-Hansen G, Conese M, Giavazzi R, Dano K, Kuhn W, Janicke F, Blasi F (1993) A ligand-free, soluble urokinase receptor is present in the ascitic fluid from patients with ovarian cancer. *J Clin Invest* 92: 2160–2167
48 Buck MR, Karustis DG, Day NA, Honn KV, Sloane BF (1992) Degradation of extracellular-matrix proteins by human cathepsin B from normal and tumour tissues. *Biochem J* 282: 273–278
49 Aronson NN, Barrett AJ (1978) The specificity of cathepsin B. *Biochem J* 171: 759–765
50 Lah TT, Buck MR, Honn KV, Crissman JD, Rao NC, Liotta LA, Sloane BF (1989) Degradation of laminin by human tumor cathepsin B. *Clin Exp Metastasis* 7:461–468
51 Von Figura K, Hasilik A (1986) Lysosomal enzymes and their receptors. *Annu Rev Biochem* 55: 167–193
52 Moin K, Day NA, Sameni M, Hasnain S, Hirama T, Sloane BF (1992) Human tumour cathepsin B: Comparison with normal liver cathepsin B. *Biochem J* 285: 427–434
53 Hanewinkel H, Glossl J, Kresse H (1987) Biosynthesis of cathepsin B in cultured normal and I-cell fibroblasts. *J Biol Chem* 262: 12351–12355
54 Rozhin J, Robinson D, Stevens MA, Lah T, Honn KV, Ryan RE, Sloane BF (1987) Properties of a plasma membrane-associated cathepsin B-like cysteine proteinase in metastatic B16 melanoma variants. *Cancer Res* 47: 6620–6628
55 Page AE, Warburton MJ, Chambers TJ, Pringle JAS, Hayman AR (1992) Human osteoclastomas contain multiple forms of cathepsin B. *Biochim Biophys Acta* 1116: 57–66
56 Sameni M, Elliott E, Ziegler G, Fortgens PH, Dennison C, Sloane BF (1995) Cathepsins B and D are localized at the surface of the human breast cancer cells. *Pathol Oncol Res* 1: 43–53
57 Honn KV, Timar J, Rozhin J, Bazaz R, Sameni M, Ziegler G, Sloane BF (1994) A lipoxygenase metabolite, 12 (*S*)-HETE, stimulates protein kinase C-mediated release of cathepsin B from malignant cells. *Exp Cell Res* 214: 120–130

58 Rozhin J, Sameni M, Ziegler G, Sloane BF (1994) Pericellular pH affects distribution and secretion of cathepsin B in malignant cells. *Cancer Res* 54: 6517–6525
59 Maciewicz RA, Wardale RJ, Etherington DJ, Paraskeva C (1989) Immunodetection of cathepsins B and L present in and secreted from human pre-malignant and malignant colorectal tumour cell lines. *Int J Cancer* 43: 478–486
60 Poole AR, Tiltman KJ, Recklies AD, Stoker TAM (1978) Differences in secretion of the proteinase cathepsin B at the edges of human breast carcinomas and fibroadenomas. *Nature* 273: 545–547
61 McCormick D (1993) Secretion of cathepsin B by human gliomas *in vitro*. *Neuropathol Appl Neurobiol* 19: 146–151
62 Qian F, Bajkowski AS, Steiner DF, Chan SJ, Frankfater A (1989) Expression of five cathepsins in murine melanomas of varying metastatic potential and normal tissues. *Cancer Res* 49: 4870–4875
63 Dalet-Fumeron V, Guinec N, Pagano M (1993) *In vitro* activation of pro-cathepsin B by three serine proteases: leucocyte elastase, cathepsin G, and the urokinase-type plasminogen activator. *FEBS Lett* 332: 251–254
64 Van Der Stappen JWJ, Williams AC, Maciewicz RA, Paraskeva C (1996) Activation of cathepsin B, secreted by a colorectal cancer cell line requires low pH and is mediated by cathepsin D. *Int J Cancer* 67: 547–554
65 Miller RF, Soule HD, Tait L, Pauley RJ, Wolman SR, Dawson PJ, Heppner GH (1993) Xenograft model of progressive human proliferative breast disease. *J Nat Cancer Inst* 85: 1725–1732
66 Sloane BF, Moin K, Sameni M, Tait LR, Rozhin J, Ziegler G (1994) Membrane association of cathepsin B can be induced by transfection of human breast epithelial cells with c-Ha-*ras* oncogene. *J Cell Sci* 107: 373–384
67 Krepela E, Bartek J, Skalkov D (1987) Cytochemical and biochemical evidence of cathepsin B in malignant transformed and normal breast epithelial cells. *J Cell Sci* 87: 145–154
68 Arkona C, Wiederanders B (1996) Expression, subcellular distribution and plasma membrane binding of cathepsin B and gelatinases in bone metastatic tissue. *Biol Chem* 377: 695–702
69 Sloane BF (1996) Suicidal tumor proteases. *Nature Biotech* 14: 826–827
70 Murata M, Miyashita S, Yokoo C, Tamai M, Hanada K, Hatayama K, Towatari T, Nikawa T, Katunuma N (1991) Novel expoxysuccinyl peptides: selective inhibitors of cathepsin B *in vitro*. *FEBS Lett* 280: 307–310
71 Panchal RG, Cusack E, Cheley S, Bayley H (1996) Tumor protease-activated, pore-forming toxins from a combinatorial library. *Nature Biotech* 14: 852–856
72 Sloane BF, Rozhin J, Johnson K, Taylor H, Crissman JD, Honn KV (1986) Cathepsin B: association with plasma membrane in metastatic tumors. *Proc Natl Acad Sci USA* 83: 2483–2487
73 Rozhin J, Robinson D, Stevens MA, Lah TT, Honn KV, Ryan RE, Sloane BF (1987) Properties of a plasma membrane-associated cathepsin B-like cysteine proteinase in metastatic B16 melanoma variants. *Cancer Res* 47: 6620–6628
74 Emmert-Buck MR, Roth MJ, Zhuang Z, Campo E, Rozhin J, Sloane BF, Liotta LA, Stetler-Stevenson WG (1994) Increased gelatinase A (MMP-2) and cathepsin B activity in invasive tumor regions of human colon cancer samples. *Amer J Pathol* 145: 1285–1290
75 Campo E, Munoz J, Miquel R, Palacin A, Cardesa A, Sloane BF, Emmert-Buck MR (1994) Cathepsin B expression in colorectal carcinomas correlates with tumor progression and shortened patient survival. *Amer J Pathol* 145: 301–309
76 Rempel SA, Rosenblum ML, Mikkelsen T, Yan P-S, Ellis KD, Golembieski WA, Sameni M, Rozhin J, Ziegler G, Sloane BF (1994) Cathepsin B expression and localization in glioma progression and invasion. *Cancer Res* 54: 6027–6031
77 Sinha AA, Wilson MJ, Gleason DF, Reddy PK, Sameni M, Sloane BF (1995) Immunohistochemical localization of cathepsin B in neoplastic human prostate. *Prostate* 26: 171–178
78 Visscher DW, Sloane BF, Sameni M, Babiarz JW, Jacobson J, Crissman JD (1994) Clinicopathologic significance of cathepsin B immunostaining in transitional neoplasia. *Modern Pathol* 7: 76–81
79 Matrisian LM (1992) The matrix-degrading metalloproteinases. *BioEssays* 14: 455–463
80. Himelstein BP, Canete-Soler R, Bernhard EJ, Dilks DW, Muschel RJ (1994–95) Metalloproteinases in tumor progression: the contribution of MMP-9. *Invas Metast* 14: 246–258
81. Stetler-Stevenson WG (1994–95) Progelatinase A activation during tumor cell invasion. *Invas Metast* 14: 259–268
82 Goldberg GI, Marmer BL, Grant GA, Eisen AZ, Wilhelm SM, He C (1989) Human 72-kilodalton type IV collagenase forms a complex with tissue inhibitor of metalloproteinase designated TIMP-2. *Proc Natl Acad Sci USA* 86: 8207–8211
83 Wilhelm SM, Collier IE, Marmer BL, Eisen AZ, Grant GA, Goldberg GI (1989) SV40-transformed human lung fibroblasts secrete a 92 kDa type IV collagenase which is identical to that secreted by normal human macrophages. *J Biol Chem* 264: 17213–17221
84 Corcorna ML, Hewitt RE, Kleiner DE, Stetler-Stevenson WG (1996) MMP-2: expression, activation, and inhibition. *Enzyme Protein* 49: 7–19
85 Strongin AY, Marmer BL, Grant GA, Goldberg GI (1993) Plasma membrane-dependent activation of the 72-kDa type IV collagenase is prevented by complex formation with TIMP-2. *J Biol Chem* 268: 14033–14039
86 Brown PD, Levy AT, Margulies IMK, Liotta LA, Stetler-Stevenson WG (1990) Independent expression and cellular processing of Mr 72,000 type IV collagenase and interstitial collagenase in human tumorigenic cell lines. *Cancer Res* 50: 6184–6191

87 Ward RV, Atkinson SJ, Slocombe PM, Docherty AJP, Reynolds JJ, Murphy G (1991) Tissue inhibitor of metalloproteinases-2 inhibits the activation of 72 kDa progelatinase by fibroblast membranes. *Biochim Biophys Acta* 1079: 242–246
88 Yu M, Sato H, Seiki M, Thompson EW (1995) Complex regulation of membrane-type matrix metalloproteinase expression and matrix metalloproteinase-2 activation by concanavalin A in MDA-MB-231 human breast cancer cells. *Cancer Res* 55: 3272–3277
89 Sato H, Takino T, Okada Y, Cao J, Shinagawa A, Yamamoto E, Seiki M (1994) A matrix metalloproteinase expressed on the surface of invasive tumour cells. *Nature* 370: 61–65
90 Will H, Hinzmann B (1995) cDNA sequence and mRNA tissue distribution of a novel human matrix metalloproteinase with a potential transmembrane segment. *Eur J Biochem* 231: 602–608
91 Takino T, Sato H, Shinagawa A, Seiki M (1995) Identification of the second membrane-type matrix metalloproteinase (MT-MMP-2) gene from a human placenta cDNA library. *J Biol Chem* 270: 23013–23020
92 Seiki M, Sato H, Okada Y (1997) Membrane-type matrix metalloproteinase one (MT1-MMP) in cancer. *In*: N Katunuma, H Fritz, H Kido, J Travis (eds): *Medical aspects of proteases and protease inhibitors*. IOS Press, Amsterdam, 195–204
93 Sato H, Takino T, Okada Y, Cao J, Shinagawa A, Yamamoto E, Seiki M (1994) A matrix metalloproteinase expressed on the surface of invasive tumour cells. *Nature* 370: 61–65
94 Cao J, Sato H, Takino T, Seiki M (1995) The C-terminal region of membrane type matrix metalloproteinase is a functional transmembrane domain required for pro-gelatinase A activation. *J Biol Chem* 270: 801–805
95 Emmert-Buck MR, Emonard HP, Corcoran ML, Krutzsch HC, Foidart J-M, Stetler-Stevenson WG (1995) Cell surface binding of TIMP-2 and proMMP-2/TIMP-2 complex. *FEBS Lett* 364: 28–32
96 Imai K, Ohuchi E, Aoki T, Nomura H, Fujii Y, Sato H, Seiki M, Okada Y (1996) Membrane-type matrix metalloproteinase 1 is a gelatinolytic enzyme and is secreted in a complex with tissue inhibitor of metalloproteinases 2. *Cancer Res* 56: 2707–2710
97 Strongin AY, Collier I, Bannikov G, Marmer BL, Grant GA, Goldberg GI (1995) Mechanism of cell surface activation of 72-kDA type IV collagenase. *J Biol Chem* 270: 5331–5338
98 Atkinson S, Crabbe T, Cowell S, Ward RV, Butler MJ, Sato H, Seiki M, Reynolds JJ, Murphy G (1995) Intermolecular autolytic cleavage can contribute to the activation of progelatinase A by cell membranes. *J Biol Chem* 270: 30479–30485
99 Liotta LA, Stetler-Stevenson WG (1990) Metalloproteinases and cancer invasion. *Sem Cancer Biol* 1: 99–106
100 Thompson EW, Yu M, Bueno J, Jin L, Maiti SN, Palao-Marco FL, Pulyaeva H, Tamborlane JW, Tirgari R, Wapnir I, Azzam H (1994) Collagen induced MMP-2 activation in human breast cancer. *Breast Cancer Res Treat* 31: 357–370
101 Monsky WL, Kelly T, Lin C-Y, Yeh Y, Stetler-Stevenson WG, Mueller SC, Chen W-T (1993) Binding and localization of Mr 72,000 matrix metalloproteinase at cell surface invadopodia. *Cancer Res* 53: 3159–3164
102 Brooks PC, Stromblad S, Sanders LC, Von Schalscha TL, Aimes RT, Stetler-Stevenson WG, Quigley JP, Cheresh DA (1996) Localization of matrix metalloproteinase MMP-2 to the surface of invasive cells by interaction with integrin $\alpha v\beta 3$. *Cell* 85: 683–693
103 Emonard HP, Remacle AG, Noel AC, Grimaud J-A, Stetler-Stevenson WG, Foidart J-M (1992) Tumor cell surface-associated binding site of the Mr 72,000 type IV collagenase. *Cancer Res* 52: 5845–5848
104 Gilles C, Polette M, Piette J, Munaut C, Thompson EW, Birembaut P, Foidart J-M (1996) High level of MT-MMP expression is associated with invasiveness of cervical cancer cells. *Int J Cancer* 65: 209–213
105 Young TN, Pizzo SV, Stack MS (1995) A plasma membrane-associated component of ovarian adenocarcinoma cells enhances the catalytic efficiency of matrix metalloproteinase-2. *J Biol Chem* 270: 999–1002
106 Yamamoto M, Mohanam S, Sawaya R, Fuller GN, Seiki M, Sato H, Gokaslan ZL, Liotta LA, Nicolson GL, Rao JS (1996) Differential expression of membrane-type matrix metalloproteinase and its correlation with gelatinase A activation in human malignant brain tumors *in vivo* and *in vitro*. *Cancer Res* 56: 384–392
107 Lohi J, Lehti K, Westermarck J, Kahari V-M, Keski-Oja J (1996) Regulation of membrane-type matrix metalloproteinase-1 expression by growth factors and phorbol 12-myristate 13-acetate. *Eur J Biochem* 239: 239–247
108 Pei D, Weiss SJ (1996) Transmembrane-deletion mutants of the membrane-type matrix metalloproteinase-1 process progelatinase A and express intrinsic matrix-degrading activity. *J Biol Chem* 271: 9135–9140
109 Poulsom R, Hanby AM, Pignatelli M, Jeffrey RE, Longcroft JM, Rogers L, Stamp GW (1993) Expression of gelatinase A and TIMP-2 mRNAs in desmoplastic fibroblasts in both mammalian carcinoma and basal cell carcinoma of the skin. *J Clin Pathol* 46: 429–436
110 Poulsom R, Pignatelli M, Stetler-Stevenson WG, Liotta LA, Wright PA, Jeffery RE, Longcroft JM, Rogers L, Stamp GWH (1992) Stromal expression of 72 kDa type IV collagenase (MMP-2) and TIMP-2 mRNAs in colorectal neoplasia. *Amer J Pathol* 141: 389–396
111 Polette M, Clavel C, Cockett M, de Bentzmann SG, Murphy G, Birembaut P (1993) Detection and localization of mRNAs encoding matrix metalloproteinases and their tissue inhibitor in human breast pathology. *Invas Metast* 13: 31–37
112 Visscher DW, Hoyhtya M, Ottosen SK, Liang C-M, Sarkar FH, Crissman JD, Fridman R (1994) Enhanced expression of the tissue inhibitor of metalloproteinase-2 (TIMP-2) in the stroma of breast carcinomas correlates with tumor recurrence. *Int J Cancer* 59: 339–344

113 Hoyhtya M, Fridman R, Komarek D, Porter-Jordan K, Stetler-Stevenson WG, Liotta LA, Liang C-M (1994) Immunohistochemical localization of matrix metalloproteinase 2 and its specific inhibitor TIMP-2 in neoplastic tissues with monoclonal antibodies. *Int J Cancer* 56: 500–505
114 Nomura H, Sato H, Seiki M, Mai M, Okada Y (1995) Expression of membrane-type matrix metalloproteinase in human gastric carcinomas. *Cancer Res* 5: 3262–3266
115 Okada A, Bellocq J-P, Rouyer N, Chenard M-P, Rio M-C, Chambon P, Basset P (1995) Membrane-type matrix metalloproteinase (MT-MMP) gene is expressed in stromal cells of human colon, breast, and head and neck carcinomas. *Proc Natl Acad Sci USA* 92: 2730–2734
116 Tokuraku M, Sato H, Murkami S, Okada Y, Watanabe Y, Seiki M (1995) Activation of the precursor of gelatinase A/72 kDa type IV collagenase/MMP-2 in lung carcinomas correlates with the expression of membrane-type matrix metalloproteinase (MT-MMP) and with lymph node metastasis. *Int J Cancer* 64: 355–359

Proteases: New Perspectives
V. Turk (ed.)

Insect proteinases

Gerald Reeck[1], Brenda Oppert[2], Michael Denton[1], Michael Kanost[1], James Baker[2] and Karl Kramer[2]

[1]*The Department of Biochemistry, Kansas State University, Manhattan, KS 66506, USA*
[2]*The Grain Marketing and Production Research Center, Agricultural Research Service, U.S. Department of Agriculture, Manhattan, KS 66502, USA*

Overview

In this chapter we will examine the following question: to what extent are proteinases in insects similar biochemically and physiologically to proteinases in more thoroughly studied organisms, particularly mammals? Of course, we can offer only a partial answer, but only in recent years has it seemed reasonable to even ask this question since the amount of detailed information on insect proteinases has been quite limited.

The importance of gaining a detailed understanding of insect proteinases is clear. From a practical standpoint, insects have enormous impact in agriculture and medicine, both human and animal. Proteinases are obvious targets for controlling insects, for instance by incorporation of inhibitor transgenes into plants [1]. At a more basic level, insects occupy an interesting position in evolution, as rather advanced invertebrates. Comparison of their proteins, including proteinases, to genetically related proteins from other invertebrates and from vertebrates should provide a great deal of insight into molecular evolution.

Even within the class Insecta, there is fertile ground for evolutionary studies because of the profusion of species. This class has 750,000 known species and may contain at least 10 million more species yet to be identified [2]. In contrast, the class Mammalia contains 4,000 known species and a large number of unknown species is unlikely. This extraordinary density in a small region of phylogenetic space (a single class, Insecta) means that insects provide a tremendous resource for evolutionary studies. Molecular analysis will no doubt prove extremely valuable as a tool in studying insect evolution at the population level, either in natural populations or in populations under selective pressure in the laboratory. Because of insects' short generation time, changes in gene frequency can occur rapidly in insect populations. For instance, in a period of about 40 years, eleven recognizably different types of greenbugs have arisen in the field [3] as a result of selection pressure established by resistant sorghum and wheat cultivars. The interaction of insects with food sources has been a predominant factor in insect evolution and in the proliferation of insect species [2]. Thus, detailed studies of digestive enzymes, including digestive proteinases, are likely to be crucial to understanding the adaptation of insects to their food sources and to developing new insect control measures.

The small size of insects coupled with a genetic (and therefore biochemical) complexity comparable to that of vertebrates has limited progress in insect biochemistry. Development of more increasingly sensitive analytical tools in biochemistry (particularly in protein structure) has allowed many studies of insect proteins in recent years that would not have been feasible a decade or two ago and, perhaps even more importantly, the development of recombinant DNA methods has allowed problems to be tackled that would otherwise have simply remained intractable.

Insects' small size has also had a distinct advantage in that they are excellent for genetic studies. This has been recognized for decades in the fruit fly, *Drosophila melanogaster*, a dipteran. More recently, a coleopteran species, the red flour beetle, *Tribolium castaneum*, has been developed as a genetic system [4], and the silkworm, *Bombyx mori*, a lepidopteran, is another species in which many genes have been identified and mapped with classical and molecular techniques [5]. Thus, three excellent genetic systems exist in three separate insect orders and offer investigators the powerful combination of classical genetics, molecular genetics, and biochemistry.

In this chapter we will examine insect proteinases first in their physiological contexts, starting with protein digestion in the gut and then turning to processes in the hemolymph. Next we will review studies on purified proteinases. Then we will examine molecular genetic data – nucleotide sequences that encode (or are presumed to encode) insect proteinases. Finally, we will close with a summary and a look to the future.

Protein digestion in the insect gut

In recent years there has been a dramatic increase in research conducted on digestive systems in insects. This surge has been stimulated in part by an increased realization of the importance of the gut tract as the main interface between pest insects and plants, including genetically transformed plants, and has been allowed in part by newly-developed techniques in biochemistry and molecular biology. The result of this research has been a significant increase in our understanding of the variety and complexity of digestive mechanisms in insects, particularly those related to compartmentalization, specificity, regulation, and secretion of gut digestive enzymes. Most of this research has focused on gut proteinases. Several earlier reviews on protein digestion and digestive processes in insects are available [6–14].

Compartmentalization

The digestive process in insects is spatially and temporally compartmentalized. The alimentary tract of insects is divided into three regions, the foregut, midgut, and hindgut. The foregut and hindgut are ectodermally derived, while the midgut is endodermal in origin. The foregut includes the mouth, pharynx, esophagus, and often a crop region. The crop often serves as a storage site for ingested food, but initial stages of digestion can also occur there through the action of digestive fluid that moves anteriorly from the midgut. However, in most insects, proteinase synthesis, secretion, and hydrolytic action on ingested dietary protein occur primarily within the midgut region. There is little evidence for protein digestion in the hindgut region, the main site of water resorption in insects. In addition to protein digestion in the intestinal tract, there are some species in which proteinases are present in the salivary gland secretion and other species, primarily parasitic or predaceous insects, in which proteinases are present in preoral digestive fluids regurgitated from the crop onto the prey [6, 15, 16].

Peritrophic matrix

In most insects, the ventriculus, or midgut, is compartmentalized into an ectoperitrophic space and endoperitrophic space by the peritrophic matrix (also called the peritrophic membrane), an acellular material secreted by certain epithelial cells along the length of the ventriculus or by a specialized group of cells in the anterior region of the ventriculus [12]. Generally, ingested food is enclosed within the peritrophic matrix, although the degree to which this occurs is highly variable amongst insects of different phylogenetic origins. The semipermeable nature of the peritrophic matrix allows for an important function during digestion, the partitioning of digestive proteinases between the ecto- and endoperitrophic spaces within the ventriculus. This same property of the peritrophic matrix is also hypothesized to be the basis for a countercurrent recirculation of digestive enzymes from the posterior region to the anterior region of the ventriculus, a process thought to be responsible for the conservation of midgut hydrolases (see [9]). The peritrophic matrix also functions as a solid phase matrix for immobilization of digestive enzymes including proteinases. It is not known whether the enzymes are actually bound to the peritrophic matrix or are only entrapped within the matrix itself [17].

In addition to the ventricular compartmentalization provided by the peritrophic matrix, digestive proteinases are differentially secreted and are differentially active along the length of the ventriculus from the anterior to the posterior regions. For example, Peterson et al. [18] found that chymotrypsin mRNA was located in the middle and anterior regions of the ventriculus of the tobacco hornworm, *Manduca sexta*. They also found that whereas trypsin mRNA was located in the middle midgut cells, the secreted enzyme was predominantly in the anterior midgut. A cysteine proteinase of the maize weevil, *Sitophilus zeamais*, occurs in the gastric caeca but not the midgut [19]. In *Drosophila*, an embryonic trypsin's mRNA was localized to the posterior midgut [20].

Another form of localization may be the trapping of many digestive enzymes within the glycocalyx [9], an electron dense material consisting of glycosaminoglycans, which coats the microvilli on the luminal side of the ventricular epithelium [21].

Diversity of digestive proteinases

Protein digestion in insects occurs in three generalized phases: proteins in ingested food are subjected to an initial digestion in the endoperitrophic luminal space of the ventriculus through the action of enzymes such as the trypsin-like serine proteinases; the oligomers and small peptides produced by these proteinases are further hydrolyzed in the ectoperitrophic space and on microvillar membranes by the action of exopeptidases, such as aminopeptidase; and final digestion of di- and tripeptides produced by exopeptidases occurs through the action of dipeptidases that are restricted to midgut cells or the glycocalyx. Details of this generalized scheme vary considerably depending on species [9].

Insects have evolved to fill many ecological niches and in the process have developed an extensive diversity with respect to gut enzymology. Wolfson and Murdock [22] studied 23 species and found that the pH optimum for proteinase activity in these species ranged from pH 3 to pH 12. In addition to dramatic differences between widely divergent species, complexes of endoproteinases can be present within a single species [23–27]. These complexes of proteinases provide an efficient and effective hydrolytic mechanism for turnover or digestion of dietary protein in a manner similar to that present in vertebrates.

The four major mechanistic classes of proteinases have been documented in insects: serine proteinases (EC 3.4.21), cysteine proteinases (EC 3.4.22), aspartyl proteinases (EC 3.4.23), and metalloproteinases (EC 3.4.24). Among serine proteinases, enzymes with trypsin-like and chymotrypsin-like specificities are common in insects, particularly among those species with alkaline midguts, and many of their properties with regard to substrate specificities and sensitivity to naturally occurring inhibitors are similar, although not identical, to those of the vertebrate enzymes. Enzymes with elastase-like specificities may be less common in insects, but evidence for their presence has recently been presented [25, 28, 29]. Cysteine proteinases are predominant in many heteropterans [30] and in many coleopterans with slightly acidic midguts [31]. Less is known about aspartyl proteinases in insects, but there is evidence for aspartyl proteinases in several species, including bruchid beetles [32] and the housefly, *Musca domestica [*33], and the red flour beetle, *T. castaneum* [34].

Determination of mechanistic class

Insect proteinases are most commonly assigned to a mechanistic class on the basis of optimal pH and, more importantly, sensitivity to inhibitors that have been believed to be class-specific. Optimal pH, however, and as well as susceptibility to inhibitors can be significantly different for insect proteinases and their presumed homologs from vertebrates [35, 36].

Among the inhibitors that have been taken for years to be most diagnostic of proteinases' mechanistic classes are phenylmethanesulfonylfluoride (PMSF) and E-64 (L-*trans*-epoxy-succinyl-L-leucylamido(4-guanidino) butane). Recently it has become clear that these compounds are not selective for serine proteinases and cysteine proteinases, respectively, as had been believed, and one has to be cautious in assigning a mechanistic class based solely on results obtained using E-64 or PMSF. For example, although E-64 is certainly capable of inhibiting cysteine proteinases, Sreedharan et al. [37] have demonstrated that it can also inhibit bovine trypsin, and others have shown that it can inhibit trypsin-like activity in insects [38, 39]. Novillo et al. [40] found differences as low as one order of magnitude between the concentration of E-64 needed to inhibit cysteine proteinases from the Colorado potato beetle, *Leptinotarsa decimlineata*, and the concentration needed to inhibit serine proteinases from guts of a lepidopteran, the European corn borer, *Ostrinia nubilalis*, or from bovine pancreas.

Michaud et al. [41] have suggested that serine and cysteine proteinases can best be distinguished using gelatin-containing polyacrylamide gels and E-64, PMSF, and other inhibitors. The efficacy of the electrophoretic technique is based on the irreversibility of the inhibition of E-64 with cysteine proteinases and of PMSF with serine proteinases. Sreedharan et al. [37] and Novillo et al. [40] have shown that E-64 inhibits trypsin and trypsin-like proteinases by a reversible competitive mechanism which is unstable during gelatin-polyacrylamide gel electrophoresis. It is nevertheless possible that serine proteinases from some insects may be still more susceptible to E-64 and may even react covalently with it. Therefore, the characterization of cysteine proteinases based only on inhibition by E-64 or of serine proteinases based only on inhibition by PMSF may not be valid in all instances.

Ideally, a mechanistic class would be assigned based on detailed kinetic analyses with purified enzymes. The reality is that insect proteinases have proven quite difficult to purify and investigators have been forced to attempt to determine mechanistic classes with partially purified enzymes or even enzymes in crude extracts. Such assignments should be regarded as provisional.

For example, we have conducted inhibitor sensitivity profiles for proteinase activities in midgut extracts from coleopteran and lepidopteran species [42–44]. Quantitative inhibition was obtained only by using mixtures of serine proteinase and cysteine proteinase inhibitors, which indicated the presence of variable amounts of both classes of proteinases depending on the species examined. Mixtures of these classes of inhibitors suppressed insect growth when fed to larvae in semi-artificial diets [45].

Zymogens

In vertebrates, trypsin, chymotrypsin, and elastase are secreted as zymogens or proenzymes by the pancreas and enter into the duodenum. Enteropeptidase (enterokinase) cleaves an amino-terminal hexapeptide from trypsinogen to form active trypsin, which then activates trypsinogen, chymotrypsinogen, and proelastase, as well as procarboxypeptidases. Initial attempts to document the presence of zymogens of digestive enzymes in insects through biochemical procedures were unsuccessful [6], but very strong evidence for zymogens has been obtained over the last several years from nucleotide sequences of cloned cDNAs or genes that encode insect proteinases (see the molecular genetics section below). Putative zymogens have been found for numerous digestive serine proteinases (of the chymotrypsin/trypsin family) and for cysteine proteinases [19, 46, 47].

Mechanisms for control of enzyme levels

Ingestion of dietary protein by both phytophagous and hematophagous insects increases the levels of proteinases in their midguts [48, 49]. Postulated regulatory mechanisms responsible for this increase in proteinase activity have been recently reviewed by Lehane et al. [10] and include nervous, hormonal, paracrine, or prandial mechanisms. The latter term, which is intended to denote a direct interaction of an element of a meal with the cells involved in digestive enzyme synthesis or regulatory control, is a suggested replacement for the term secretagogue [50].

While most evidence indicates that innervation of the midgut is limited to gut musculature, it is experimentally difficult to differentiate between primary and secondary events initiated by paracrine or prandial mechanisms. The insect midgut contains numerous putative endocrine cells. As reviewed by Lehane et al. [50], these cells are similar in morphology to known vertebrate endocrine cells and epitopes from these cells cross react with antibodies from vertebrate neuropeptides. Nevertheless, there is currently no clear evidence that paracrine secretion from these cells is involved in regulatory control of enzyme activity in the insect midgut.

With the recent application of immunochemical and molecular techniques, significant progress has been made in our understanding of insect digestive enzyme control, particularly with respect to regulatory mechanisms of secretion of serine proteinase in hematophagous insects. In mosquitoes, *early* and *late* trypsins are involved in a regulated, time-dependent digestion of the blood meal. In the mosquito *Aedes aegypti*, three trypsin genes have been isolated, one of which is constitutively expressed (*early*) and two that are induced (*late*). Transcription of the early gene is apparently part of the normal post-emergence maturation process and is controlled by juvenile hormone titer [51]. Production of small peptides by the *early* trypsin in *A. aegypti* is thought to initiate transcription of the *late* trypsins [52, 53] and there is evidence that DNA regions upstream from the *late* trypsin genes function as promoter elements [52]. Both the quality and quantity of dietary protein are involved in transcriptional regulation of the *late* gene in this species [54].

In the mosquito *Anopheles gambiae*, trypsin genes are arranged as a clustered gene family of seven coding sequences [55]. Two trypsin genes are induced by blood feeding and five genes are constitutively expressed. In *A. gambiae*, conserved, non-coding sequences are found upstream from all trypsin genes. These regions may function as binding sites for factors that either activate or suppress transcription.

In contrast to these species of mosquitoes, where transcriptional control of proteinase synthesis apparently occurs, the increase in trypsin activity after blood feeding in the blood sucking stable fly, *Stomoxys calcitrans*, is apparently under translational control. Trypsin synthesis following a blood meal was reduced in the presence of cycloheximide, an inhibitor of translation, but was not affected by cordycepin or 5-fluorouracil, two inhibitors of transcription [56]. In *S. calcitrans*, trypsin is stored in the opaque cells of the midgut as an inactive form but is secreted as the enzyme. This secretion is stimulated in a time and dose-dependent manner by soluble protein in the diet but not by amino acids or poly-L-amino acids [57].

Secretion of digestive proteinases

In insects, midgut cells are primarily responsible for both synthesis and secretion of digestive proteinases. Secretory mechanisms for digestive enzymes in insects have been extensively reviewed by Terra and Ferreira [9], who outlined evidence for three main secretory routes in midgut cells: exocytosis, in which secretory products are contained in vesicles that fuse with the plasma membrane, thus releasing their product; microapocrine secretion, where digestive enzymes are initially an integral part of the membrane of small vesicles that migrate to the cell microvilli and the enzymes are processed to a soluble form within the vesicles which eventually bud laterally from the microvilli; and apocrine secretion, in which vesicles containing enzymes migrate to the tips of the microvilli and eventually pinch off, releasing some of their contents into the lumen. Portions of the vesicles that contain bound enzymes may subsequently be incorporated into the peritrophic matrix.

Exocytotic secretion and microapocrine secretion have been described in several insect orders [9]. Vesicles that bud from the side of microvilli in the lepidopteran *Erinnyis ello* contain soluble forms of α-amylase and trypsin [58]. In *M. domestica*, there is evidence that trypsin is secreted by a special type of exocytosis [59]. Trypsin bound to vesicle membranes is solubilized after exocytosis, apparently because the neutral pH of the luminal contents results in a conformational change in the peptide anchor. In *M. domestica*, amylase is also secreted by exocytosis following processing from a bound form to a soluble form within the vesicles themselves [60]. Digestive carbohydrases in the yellow mealworm, *Tenebrio molitor* (Coleoptera), are secreted exocytotically from anterior midgut cells, while carboxypeptidase and trypsin are secreted similarly in posterior midgut cells [61].

Jordão et al. [62] used antibodies against *M. domestica* trypsin to study trypsin secretion in *S. calcitrans.* Trypsin was primarily immunolocalized in the ectoperitrophic space. About 10% of the trypsin was associated with the peritrophic matrix. Intact secretory granules as well as cell debris were present in the ectoperitrophic space, providing evidence for apocrine secretion from the opaque zone midgut cells in this species.

Interactions with protoxins and proteinase inhibitors

Larval digestive proteinases have been characterized to better understand the mode of action of the insecticidal toxins of *Bacillus thuringiensis* (Bt). Serine proteinases solubilize and activate Bt protoxins early in the physiological process of toxicity [63]. One mode of resistance of insects to the effects of Bt toxins is alternation in the proteinases that activate the protoxins [64, 65]. We have studied a resistant strain of the Indianmeal moth, *Plodia interpunctella*, which lacks a major gut proteinase that activates Bt protoxins [39, 66]. The lack of this proteinase is linked genetically to larval survival on diets supplemented with Bt protoxin [67].

The interaction of digestive enzymes with proteinaceous inhibitors lies outside the scope of this chapter. We have recently reviewed this topic elsewhere [1].

Physiological contexts other than digestion in the gut

Proteinases that digest egg-yolk proteins

Several egg-specific proteins, including vitellin, are stored in yolk granules in insect eggs as a source of amino acids for embryonic development [68]. Proteinases in the egg hydrolyze these proteins to provide amino acids that are used by the embryo to synthesize new proteins [69].

In a sense, these are special digestive enzymes that hydrolyze maternally-provided nutrient proteins within the egg. Vitellin-degrading proteinases that have been purified or molecularly cloned from insects are either serine or cysteine proteinases.

A vitellin-degrading proteinase was purified from *B. mori* eggs [70] and its cDNA isolated [71] The sequence of the vitellin-degrading proteinase is similar to members of the chymotrypsin/trypsin family. A different serine proteinase with specificity for cleavage at Arg or Lys residues, which was purified from eggs of *B. mori* [72], has specificity for degradation of the egg-specific protein, a protein distinct from vitellin. Both of these serine proteinases from *B. mori* are synthesized by the embryo. A cDNA has also been cloned for a serine carboxypeptidase that is synthesized in the fat body of the mosquito, *A. aegypti,* and then transported to oocytes [73]. This proteinase exists in the eggs as a zymogen until embryogenesis begins, when it is proteolytically activated.

Cysteine proteinases that digest egg-yolk proteins have been purified from eggs of *B. mori* [74], *M. domestica* [46], *D. melanogaster* [75], and the German cockroach, *Blatella germanica* [76]. These enzymes have acidic pH optima and have proteolytic properties similar to mammalian cysteine-type cathepsins. The sequence of a cDNA clone for the silkworm enzyme indicates that it is a member of the papain family [77]. The cysteine proteinases in eggs are apparently maternal gene products that are imported into the eggs as zymogens, and then activated through proteolytic processing to remove a propeptide [77–79].

Proteolytic processes in insect hemolymph

The coagulation and complement systems in vertebrate plasma comprise a complex network of serine proteinases, in which sequential proteolytic activation of a cascade of proteinase zymogens amplifies a response to wounding or infection. The final result of these pathways is formation of a clot to stop further bleeding (coagulation) or production of protein complexes with antimicrobial activities (complement). Because excessive activation of these protective mechanisms can harm the host organism, the coagulation and complement pathways are regulated by high plasma concentrations of specific serine proteinase inhibitors, which are predominantly members of the serpin superfamily [80]. Serine proteinase cascades regulated by serpins have also evolved as protective reactions in the hemolymph of arthropods, presumably including insects. The hemolymph coagulation system has been best studied in horseshoe crabs, and we will briefly review this work to provide a context in which to discuss what is known in insects.

When horseshoe crab hemocytes are exposed to bacterial lipopolysaccharide or β-(1-3) glucan (from fungal cell walls), they release five clotting factors stored in membrane-bound granules. Four of these factors are serine proteinase zymogens, and the other is coagulogen, a clotting protein precursor analogous to fibrinogen [81]. Exposure to lipopolysaccharide causes conversion of factor C to its active form (factor $\bar{C}$) and exposure to β-(1-3) glucan converts factor G to active factor $\bar{G}$. These proteinase zymogens act as sensors for the presence of microorganisms in the hemolymph. The pathway has two branches, which converge at the activation of a proteinase called clotting enzyme. In the glucan-stimulated branch, factor activates a proclotting enzyme to produce clotting enzyme, which is functionally analogous to thrombin.

Clotting enzyme cleaves coagulogen to produce coagulin, which forms an insoluble gel-like clot. In the lipopolysaccharide-activated branch of the clotting pathway, factor $\bar{C}$ activates factor B, and factor $\bar{B}$ cleaves the proclotting enzyme. All of these proteins have been purified, their cDNAs cloned, and the activation reactions biochemically characterized in a monumental body of work from Iwanaga's laboratory (reviewed in [81]).

Factors B, C, G and proclotting enzyme each contain a carboxyl-terminal serine proteinase domain that is a member of the chymotrypsin/trypsin family. Factor C (123 kDa) contains a series of other domains that are believed to be involved in recognition of lipopolysaccharide and subsequent autoactivation [82]. Factor G contains two subunits, an α-subunit with a series of carbohydrate-binding domains for recognition of glucans, and a β-subunit that contains the serine proteinase domain [83]. Factor B and proclotting enzyme contain a "clip" domain with similar amino acid sequences and three conserved disulfide bonds [84, 85]. The complexity of these enzymes, with their numerous non-proteinase regulatory domains, is reminiscent of enzymes from the vertebrate coagulation and complement systems. In addition, these proteinases are regulated by serpins produced by horseshoe crab hemocytes [81, 86].

Another hemolymph system for defense against infection in arthropods involves a cascade of serine proteinases that results in activation of a phenoloxidase zymogen (pro-phenoloxidase) [87]. Phenoloxidase catalyzes the hydroxylation of monophenols to produce diphenols, and the oxidation of diphenols to quinones. The quinones may be directly toxic to pathogens and parasites, and they can also polymerize to form melanin capsules that surround parasites. As in the horseshoe crab clotting system, prophenoloxidase activation can be stimulated by lipopolysaccharide or β-(1-3) glucans. Recognition of these microbial polysaccharides results in activation of a serine proteinase (prophenoloxidase activating enzyme: PPAE) that cleaves prophenoloxidase at a specific peptide bond near its amino terminus, which activates this enzyme.

Understanding of the molecules involved in prophenoloxidase activation is at a rudimentary stage. It is not known with certainty how many steps may occur between recognition of polysaccharide by a hemolymph protein and activation by PPAE, although two different serine proteinase activities have been detected after treating silkworm plasma with bacterial cell walls [88]. Enzymes with PPAE activity have been purified from hemolymph of a crayfish [89], *B. mori* cuticle [90], and a homogenate of *D. melanogaster* pupae [91]. They are all~30 kDa serine proteinases that cleave very specifically at an Arg-Phe bond approximately 50 residues from the amino terminus of prophenoloxidase, activating the prophenoloxidase zymogen. Like blood clotting, phenoloxidase activation is normally regulated *in vivo* as a local reaction of brief duration. Also comparable to blood clotting, the regulation may be due in part to serine proteinase inhibitors in plasma [86]. A serpin from hemolymph of *M. sexta* is able to inhibit the activity of a serine proteinase associated with the prophenoloxidase activation process [92].

It is likely that processes exist in insect hemolymph that are not yet identified but involve proteinases. For example, several (putative) serine proteinases have been characterized by protein purification or by molecular cloning but have not yet been assigned a role in a known physiological process. A 39 kDa serine proteinase purified from plasma of *B. mori* exists as a zymogen that is proteolytically activated upon exposure to β-(1-3) glucan [93] and then is rapidly inactivated by a plasma serpin [94]. However, this enzyme (which cleaves specifically after Arg residues) does not appear to be part of the prophenoloxidase activation pathway. Thus, exposure to microbial polysaccharides may stimulate more than one process involving activation of serine proteinase zymogens. Several cloned DNAs with sequences encoding apparently non-digestive serine proteinases have recently been obtained from *D. melanogaster* [95] and from the mosquito *A. gambiae* [96, 97], but their physiological roles are not yet known. In addition, two different cDNAs that encode an amino-terminal clip domain followed by a serine proteinase domain, similar to the structure of horseshoe crab proclotting enzyme and Factor B, have been isolated from hemocytes of *M. sexta* (Jiang H, Kanost M R, unpublished results). This group of proteinases with functions yet to be discovered hints at the existence of potentially complex systems of proteinases regulating a variety of processes in insect hemolymph.

A 29 kDa cysteine proteinase with sequence similarity to mammalian cathepsin B has been purified from hemocytes of the flesh fly, *S. peregrina*, and its cDNA has been cloned [98, 99]. This enzyme is secreted from pupal hemocytes and participates in dissociation of fat body tissue during metamorphosis. It is probable that the extensive destruction and remodeling of tissues that occur during insect metamorphosis involves the action of this and other proteinases, with their expression or activity regulated by developmental cues.

Developmental proteinases

Mutations that disrupt embryonic development in *D. melanogaster* have led to the discovery of several genes that encode proteins with carboxyl-terminal domains that are members of the chymotrypsin family of serine proteinases. These include *masquerade*, which appears to be involved in development of the nervous system and in muscle attachment [100] and *nudel, snake,* and *easter,* which are part of a cascade that determines the dorsal-ventral axis of the embryo [101]. The products of *snake* and *easter* have a structure resembling the horseshoe crab clotting factor B and proclotting enzyme, with an amino-terminal clip domain followed by a serine proteinase domain [102]. Proteinase activities have not yet been demonstrated biochemically for any of these proteins and it is likely that *masquerade* does not have proteolytic activity because the active-site serine has been replaced with glycine. However, the functions of the putative proteinases in the dorsal pathway are consistent with their action as proteinases of high specificity in analogy to the horseshoe crab enzymes.

Studies on purified insect proteinases

Overview

In this section we will highlight selected studies on insect proteinases that draw attention to either notable similarities or differences with mammalian proteinases. We have attempted to be all inclusive in a table that we have placed on the World Wide Web. In it, we have tried to treat all insect proteinases that have been purified to apparent homogeneity. This table can be found at the following URL: http://bru.usgmrl.ksu.edu/proteinases. Comments about the table and suggested additions to it are welcomed at bso@ksu.edu.

In what follows we will describe studies on purified and, in some cases, partially purified insect proteinases with a view to comparing their properties with their mammalian counterparts and thus call attention to the distinctive features of the insect enzymes. We start by examining a particularly important body of early work on cocoonase.

Cocoonase

In addition to its physiological importance, this enzyme has a historical significance in the field of insect proteinases, since it was the first insect proteinase to be purified and well characterized. The foundational studies, which were carried out in the late 1960 s and early 1970 s, clearly demonstrated that this insect proteinase was a recognizable homolog of well studied mammalian proteinases and, further, that the insect proteinase was synthesized in the form of a zymogen that was activated by limited proteolysis.

Cocoonase is a serine proteinase used by adult silkmoths to soften cocoons and permit their escape [103, 104]. Cocoonase digests only one of two cocoon proteins, the glue-like protein sericin, which cements the silk thread protein, fibroin, into an impenetrable layer. Prococoonase is produced by giant zymogen cells in a specialized tissue derived from modified mouthparts, the galea [105]. The zymogen is secreted as a single major product to the surface of the galea, after which it is efficiently activated by cleavage of a single peptide bond catalyzed by enzymes from the molting fluid [106].

The amino-terminal sequence of the *Antheraea polyphemus* proenzyme has a 13-amino acid residue zymogen activation peptide, KKTPNRTNDDGGK [107]. In contrast, bovine trypsinogen has a 6-residue activation peptide with four Asp residues immediately before the cleavage site, a feature that is characteristic of mammalian trypsins. The active-site peptide of cocoonase, containing the nucleophilic Ser residue, is rather similar to the corresponding region of bovine trypsin [108], with eleven identities over a 16-residue stretch. The discovery of this strong similarity in the active site region underscored the preservation of the basic catalytic mechanism of trypsin during the large span of evolution of organisms that includes insects and mammals.

Other serine proteinases

The proteinase isolated most frequently from insects has been trypsin, by which we mean a serine proteinase, demonstrably or presumably of the chymotrypsin/trypsin family, and with a strong preference for Lys or Arg at position P1 of substrates or inhibitors. There have also been numerous reports of insect chymotrypsins.

Enzyme purifications have clearly established that individual insect species have families of digestive serine proteinases. From the locust *Locusta migratoria* [109, 110] both a trypsin and chymotrypsin have been isolated. Similarly, from the honeybee *Apis mellifica*, gut proteinases with trypsin-like and chymotrypsin-like specificities have been purified [111]. Interestingly, some enzymes were present only in adult workers and drones, possibly reflecting the composition of the diets of different castes of the honeybee. Larvae of the dermestid black carpet beetle, *Attagenus megatoma*, had several trypsin and chymotrypsin proteinases [24]. These were divided into anionic or cationic enzymes. The anionic trypsin differed from the two cationic trypsins in its pH activity curve, substrate kinetics, and response to soybean trypsin inhibitor. However, the anionic and cationic forms of chymotrypsin were similar to each other and to mammalian chymotrypsin in their enzyme kinetics and inhibition pattern. Digestive serine proteinases have been best characterized from *M. sexta*. A trypsin and a chymotrypsin have been purified from midgut tissue and the cDNAs that encode these enzymes have been cloned and their nucleotide sequences determined [18, 36].

Insect serine proteinases can differ markedly from mammalian trypsins in their reactivities towards substrates and inhibitors. A purified proteinase from the midgut of the silkworm, *B. mori*, was inhibited by diisopropylfluorophosphate, N-tosyl-l-phenylalanine chloromethyl ketone, and chymostatin and was therefore classified as chymotrypsin-like [35]. However, this enzyme did not hydrolyze classical low-molecular-weight synthetic substrates for mammalian chymotrypsins. Chymotrypsin from the cockroach, *Periplaneta americana*, was also unaffected by N-tosyl-l-phenylalanine chloromethyl ketone, a classic irreversible inhibitor of mammalian chymotrypsin. Benzoaxinones, substrates for vertebrate chymotrypsin, were found to be potent inhibitors of the cockroach chymotrypsin [112]. Serine proteinases are important in solubilizing and activating the protoxins produced by *B. thuringiensis*. Trypsin from the regurgitated gut fluid of the spruce budworm, *Choristoneura fumiferana*, was identified as a major Bt protoxin-hydrolyzing enzyme [113]. Bovine trypsin and *C. fumiferana* trypsin yielded different proteolysis products when Bt protoxin was used as the substrate, but it was unclear whether these toxin products have different toxicities.

Studies of purified enzymes have also demonstrated differences between insect and mammalian enzymes in such properties as pH optimum, stability and effect of pH on activation. Enzyme stability clearly varies among serine proteinases in insects. A trypsin-like enzyme from the European corn borer, *Ostrinia nubilalis*, was stable at acidic pH and exhibited unusual stability in the presence of chaotropic compounds and high temperatures [114]. Three trypsin-like enzymes were purified from the gut of larvae of the army worm, *Spodoptera litura* [115], and each has a unique temperature at which it exhibits maximal activity. One had maximum activity at 60 °C, which was similar to some other insect trypsins. The other two *S. litura* trypsins had optima close to those for porcine (55 °C) or bovine (50 °C) trypsins. Hypodermin A, a major gut proteinase with trypsin-like specificity purified from larvae of the parasitic fly *Hypoderma lineatum*, is relatively stable at elevated temperatures [116]. This may account for its accumulation in insect larvae that remain in their host for approximately eight months.

Trypsins from the hornets *Vespa orientalis* and *V. crabro* differed from bovine trypsin in that they were rapidly and irreversibly inactivated at low pH, resistant to autolysis at pH 8, and not stabilized or activated by calcium [117]. With respect to stability to autolysis and lack of a calcium effect, the hornet trypsins resemble trypsin from the African lungfish [118]. Lepidopteran trypsins and chymotrypsins have pH optima of 10–11. This may be why, among its positively charged side chains, *M. sexta* trypsin contains only arginine [36], whereas bovine trypsin contains 14 lysine and only two arginine residues. It was proposed that the arginine residues, with their very high pK, help to stabilize *M. sexta* trypsin by remaining charged in the high alkaline conditions of the midgut of this lepidopteran.

A putative activation peptide for *M. sexta* chymotrypsin is almost three times longer than those found in mammalian chymotrypsins [18]. In addition, the *M. sexta* activation peptide has an acidic isoelectric point, whereas the mature enzyme has a basic pI. The authors compared this situation to mammalian pepsinogen A, which has a basic activation peptide and an acidic mature enzyme. Perhaps secretion of *M. sexta* chymotrypsinogen into an alkaline midgut disrupts salt bridges between the acidic activation peptide and the basic protein, exposing the enzyme's active site.

Interesting differences in disulfides or cysteine residues have been reported between purified insect proteinases and their mammalian homologs. The locust enzymes were reported to have an absence of half-cysteines [109, 110]. A midgut trypsin-like proteinase was isolated from *Tenebrio molitor* and found to not cross-react with antibodies against bovine trypsin [119]. It contained two disulfides compared to the six disulfides of bovine trypsin. Trypsin from the gut fluid of *B. mori* [120] has been sequenced at the protein level. It contains two cysteinyl residues at novel positions, located near elements of the catalytic triad. The two cysteinyl residues are apparently not in close proximity to each other. Inhibition of the enzyme by mercuric chloride strongly suggests that the cysteinyl residues are not linked as a disulfide.

Proteinases are involved in cuticle degradation [121]. Samuels et al. [122] purified a trypsin-like cuticle-degrading proteinase from the tobacco hornworm. Interestingly, the enzyme had a molecular mass of 41 kDa. A trypsin from a midgut homogenate of the tobacco hornworm was similar to bovine trypsin in size and was not glycosylated [123].

Insect proteinases vary in glycosylation patterns, which range from no glycosylation to multifarious glycosylation. The procoonases from *Antheraia pernyi* and *A. polyphemus* contain mannose and glucosamine carbohydrates, while another from *A. mylitta* has no glycosylation [108].

Aspartyl proteinases

An aspartyl proteinase from *T. castaneum* was similar to mammalian aspartyl proteinases in lacking sensitivity to ionic compounds [34]. However, whereas mammalian lysosomal cathepsin D is a glycoprotein, the beetle aspartic proteinase is non-glycosylated.

Cysteine proteinases

Some coleopteran insects differ from other insects and from mammals in that thiol proteinases are more abundant in the larval gut, which is moderately acidic. Cysteine proteinases have been isolated and described from the common bean beetle, *Acanthoscelides obtectus* [124], with the following characteristics: maximal activity between pH 5 and pH 7, enhanced activity in the presence of cysteine and dithiothreitol, and inhibition by such compounds as *p*-chloro-mercuribenzoic acid and E-64.

A cathepsin-like enzyme isolated from *M. domestica* eggs is involved in the degradation of yolk proteins during embryogenesis [46]. This enzyme is stored in oocytes as a 55 kDa proenzyme, and a drop in pH triggers the cysteine-proteinase processing of this enzyme to a mature 41 kDa form, similar to the activations of pepsinogen and propapain.

Proteasome proteinases

The proteasome is a complex of numerous subunits and is involved in the degradation of peptides and proteins. An insect proteasome was characterized from *M. sexta* body wall [125]. The insect proteinase complex exhibited five peptidase activities, similar to that observed with proteasomes isolated from lobster and bovine tissues, but differing in specific activity. Substrate hydrolysis by *M. sexta* proteasome was stimulated by SDS similar to that observed with the bovine, but differing from the lobster proteasome, which was not affected. The insect and lobster proteasomes exhibited greater heterogeneity among their subunits than did vertebrate proteasomes, possibly because of more extensive post-translational modifications of the arthropod proteasomes. This discrepancy could also reflect the fact that the small size of the insects necessitates the use of multiple tissues for proteasome preparation.

Other insect proteinases

Insect exopeptidases produce free amino acids, which can then be transported into enterocytes. Larval aminopeptidase N of the silkworm, *B. mori*, was isolated and investigated for its potential role in amino acid transport [126]. Although previous evidence indicated that aminopeptidase N may be linked to Na^+-dependent amino acid cotransport in bovine renal epithelia, no support was obtained for a functional relationship between silkworm aminopeptidase N and amino acid transport. Moreover, structural differences between mammalian and insect aminopeptidase N were proposed. Although most mammalian brush border aminopeptidase N activity can be removed from vesicle preparations by papain, suggesting linkage through a hydrophobic peptide, aminopeptidase N from silkworm brush border membrane vesicles was not affected by papain treatment. Instead, phosphatidylinositol-specific phospholipase C was effective in releasing a portion of the activity from brush border membrane vesicles, indicating that the enzyme is likely to be anchored to the membrane by a glycosyl phosphatidylinositol linkage.

An aminopeptidase involved in cuticle degradation has been purified from the molting fluid of the tobacco hornworm [127]. The enzyme appears to exist as a hexamer of 39 kDa subunits. A carboxypeptidase from *M sexta* brain [128] was found to be comparable to vertebrate carboxypeptidase E in size, glycolysation, pH optimum, kinetics, and stimulation by cobalt. However, unlike bovine carboxypeptidase E, but similar to human carboxypeptidase E, the insect enzyme was not inhibited by *p*-(chloromercuri)benzenesulfonate.

Molecular genetics of insect proteinases

Availability and interpretation of nucleotide sequences

Databases now contain nearly 200 nucleotide sequences of cloned cDNA or genes that encode insect proteinases. More precisely, one should say that the sequences encode putative proteinases, because in the vast majority of cases, the proteins have not been isolated and studied directly themselves. Establishing a one-to-one correspondence between a cDNA or gene and a protein, and verifying the protein's functional properties are clearly important. A nucleotide sequence and an inferred amino acid sequence can place a putative protein in a family, based on sequence similarity, but sequence alone does not assign a function to that putative protein with certainty. Observing a set of catalytic residues in a sequence, such as the catalytic triad in putative serine proteinases, does not in itself guarantee that the putative protein is an enzyme, because amino acid substitutions at many other positions could reduce or even preclude enzyme activity. Of course, in most cases, the putative proteinases will in fact turn out to be enzymes, but demonstrating enzyme activity directly, for instance by expression in cultured insect cells, should be a goal for all putative proteinases identified from nucleotide sequences. Expression and isolation would also allow kinetic characterization to establish an enzymes' specificity and pattern of inhibition by synthetic and proteinaceous inhibitors.

Sequences from families other than the chymotrypsin/trypsin family

We intend to focus our attention on members of the chymotrypsin/trypsin family, but first we will identify some of the known sequences that encode putative proteinases in other families. As a space saving measure, in this portion and the rest of this section we will use database accession numbers rather than literature citations unless we have previously given a literature citation for the work.

Genes that encode furins, eukaryotic members of the subtilisin family of serine proteinases, have been documented in *Drosophila* [X59384, L33831], *Spodoptera frugiperda* [Z68888], and *A. aegypti* [L46373]. These likely function as proprotein processing enzymes. cDNAs that encode cysteine proteinases that are members of the papain family have recently been cloned in two coleopteran species [19, 47], in *Drosophila* [D31970], and in *S. peregrina* [99 and accession number D16533]. Aspartyl proteinases (related to pepsin) are encoded by cDNAs that have recently been isolated from the German cockroach, *B. germanica* [U28863], in which the putative proteinase is a major allergen, and from *A. aegypti* [M95187]. cDNAs that encode proteasome proteinases of both the T1a and T1b families have been cloned from *Drosophila* [X70304, X15497, X52319, M57712, U00790]. cDNAs for an insulin-degrading metalloproteinase and a putative ubiquitin-specific proteinase have been cloned from *D. melanogaster* [M58465, X99211].

Sequences from the chymotrypsin/trypsin family

Just as is true in vertebrates, the chymotrypsin/trypsin family of serine proteinases is complex in insects. As we have seen in previous sections, there is considerable complexity just among the serine proteinases of the gut. Not only do separate enzymes exist with specificities that mirror those of vertebrate trypsin, chymotrypsin, and elastase, but several genes for trypsin and chymotrypsin can exist within a given species. It is also apparent that within other, larger proteins, for instance of the hemolymph, domains occur that encode members of the chymotrypsin/trypsin family just as they do in vertebrates.

To provide a framework for thinking about the complexity of insect serine proteinases, we have constructed a dendrogram or phylogenetic tree (Fig. 1). We restricted our attention to full-length sequences (that is, we did not include sequences of PCR products) and, of the full-length sequences, we chose a representative subset. We aligned the sequences starting with the sequence IVGG, which forms the N-terminus of many of the mature serine proteinases, and with the corresponding sequences in other members of the family, and extending to the C-terminus. The alignment (see http://bru.usgmrl.ksu.edu/proteinases) and unrooted tree were created with the CLUSTAL W multiple alignment program [129]. In addition to insect proteinases, we have included two sequences of mite allergens (members of the chymotrypsin/trypsin family), serine proteinase domains from two horseshoe crab coagulation factors, and several human serine proteinases (trypsin, chymotrypsin, elastase, tissue plasminogen activator, and thrombin).

The tree, perhaps not surprisingly, is highly branched. The branching is clearly not a simple reflection of phylogenetic relationships among insect species since sequences from an individual species occur in several branches. For instance, sequences from *D. melanogaster* occur in all of the major branches of the tree. *Early* and *late* trypsins from *A. aegypti* occur on two distinct branches. Enzymes that have been described as chymotrypsin-like predominate in the extremity of one of the branches whereas trypsin-like enzymes occur in several of the branches. Thus, it appears that some of the branching among trypsin genes occurred as early as the separation of the chymotrypsin branch. The several *Drosophila* serine proteinase domains in gene products that are required for development all occur in one region of the tree, along with the horseshoe crab coagulation enzymes and, interestingly, all of the human serine proteinases. The segregation of the human enzymes into one branch suggests that the divergence of the various human serine proteinase genes occurred after the ancient split of the evolutionary lines leading to insects, on the one hand, and vertebrates, on the other.

A particularly interesting insect sequence is that of the putative elastase-like enzyme from *M. sexta.* This sequence contains a threonine residue at the position of the active-site serine residue in members of the chymotrypsin/trypsin family. It is not yet clear whether the protein product is in fact an enzyme: attempts to obtain enzyme activity in recombinant protein have been unsuccessful (Wells MA, personal communication). Threonine is the nucleophilic attacking group in proteasome proteinases, so it is not unreasonable for it to function in that capacity in members of the chymotrypsin/trypsin family. We would note that in databases there are numerous sequences of PCR products that are clearly parts of chymotrypsin/trypsin homologs and that contain threonine at the position of the active-site serine.

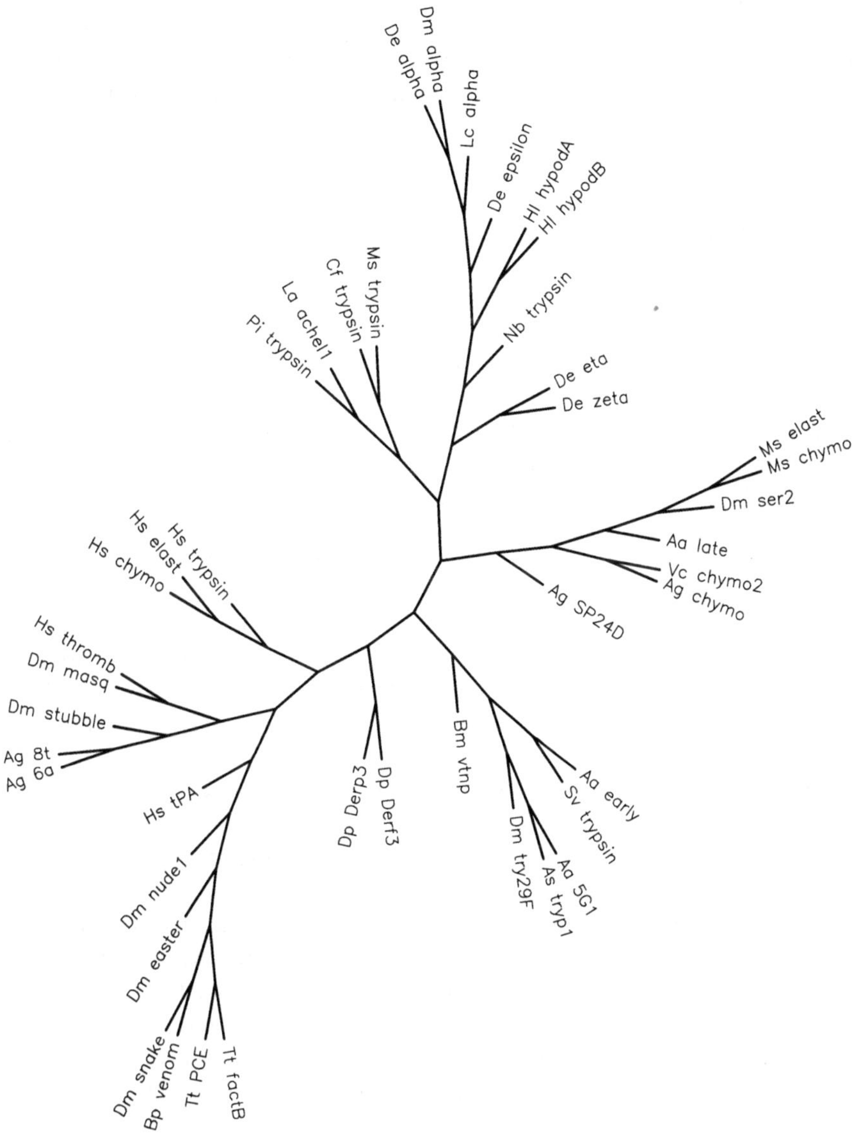

Figure 1. Dendrogram (unrooted tree) for 42 insect, mite, horseshoe crab and human serine proteinases. The tree was constructed using CLUSTAL W [129] with correction for multiple substitution, and plotted using the drawtree program in Felsenstein's PHYLIP package without the branch-length option. The first two letters in the name of each sequence are the initials of the genus and species of the organism. Accession numbers for each sequence are listed below and, except where indicated, are taken from the GenEMBL database.

Another sequence of particular interest is the serine-proteinase-like domain encoded by the *D. melanogaster masquerade* gene [100]. This domain has clear sequence similarity to members of the chymotrypsin/trypsin family yet contains a glycine at the position that corresponds, in serine proteinases, to the active-site serine. As pointed out by Murugasu-Oei et al. [100], this domain is clearly not an enzyme. Their most interesting observation reinforces a point we made above: members of the chymotrypsin/trypsin family (or other families that contain proteinases) are not necessarily enzymes.

Summary and prospectus

To what extent are proteinases in insects similar biochemically and physiologically to proteinases in more thoroughly studied organisms, particularly mammals?

Based on our review of the current literature, we can state that insect proteinases are clearly recognizable, both collectively and individually, as analogous or homologous to their vertebrate counterparts. There is an overall complexity within the insect proteinases that is comparable to that in vertebrates, and individual insect proteinases are all readily identified as members of well recognized protein families. Intracellular processes and intracellular proteinases appear entirely comparable between insects and mammals. Physiological differences involving proteinases are likely to be restricted to extracellular processes. For instance, the hemolymph of insects appears to have primitive and simplified systems analogous to vertebrate coagulation and complement systems. Extracellular specializations are needed in insects, for instance for dissolution of cocoons or digestion of egg-yolk protein. In these cases, existing proteinase genes of the chymotrypsin/trypsin family or the papain family appear to have been recruited and adapted physiologically. In the case of the digestion of egg-yolk protein, different species of insects have selected proteinases from different families.

A flexibility at the genetic level is also apparent in the proteinases used for digestion in the gut in insects. We suggest the following generalized picture. For potential use in protein digestion in the gut, any insect species has genes that encode proteinases of several families (the chymotrypsin/trypsin family, the papain family, the pepsin family, and the metalloproteinase family) and of at least four mechanistic classes (there may also be a threonine proteinase class as well as the more typical classes). There are very likely several, or many, genes from each of the families. Which genes are used (that is, which proteinases are selected from among the palette that is available) is unique to each insect species, or at least the relative abundance of the enzymes varies widely among insect species. Certain generalizations may hold. For instance, coleopterans may rely more heavily on cysteine proteinases, whereas lepidopterans may rely more on serine proteinases. The composition of the rather complex mixture of digestive proteinases can, in individual insects, respond to changes in diet, adding a physiological flexibility to the genetic adaptation of a species.

Although all known insect proteinases (or putative proteinases) can be readily assigned to well known protein families, the functional properties of individual insect proteinases are unique and can differ substantially from the properties of vertebrate digestive proteinases or from other insect proteinases. This can make something as fundamental as assigning a mechanistic class difficult without detailed study of a purified enzyme. Among what appear to be true serine proteinases, there is widespread variation in the stabilities, pH optima, isoelectric points, and even susceptibility to inhibitors that seem quite strict in their reactions with vertebrate proteinases. This functional variability of insect proteinases is evidently part of the adaptation of insects to their individual food sources, which has been a major influence in the evolution of insect species. Among the digestive proteinases, aspartyl and cysteine proteinases have been studied much less than the serine proteinases, both biochemically and genetically. Increased attention to the aspartyl and cysteine digestive proteinases is clearly needed.

Given the economic importance of insects in both agriculture and medicine, and given their evolutionary position as advanced invertebrates, the continued study of insect proteinases is obviously of great importance. To a large extent, insects must be examined species by species since each will be adapted to its own food source and ecological niche. With reagents such as antibodies and cloned DNAs that have been developed in laboratories around the world, the prospects for rapidly advancing our knowledge of proteinases in many insect species is excellent. A major thrust of such work should be the study of the enzymes themselves. Most often this will be by expression of their encoding genes or cDNAs, probably in cultured insect cells. Purified proteinases are needed to understand their unique features and idiosyncracies, to identify inhibitors of them that might be used as insect growth retardants, and to help select mutant inhibitors with desired characteristics for use in transgenic plants and microorganisms. Proteinases are also potentially useful directly as bioinsecticides [130].

Individual insects will and should be chosen for studies of their proteinases based on their economic or medical importance. In addition, it would be appropriate to select several species as model systems. The best choices for this would appear to be well developed genetic systems (for instance, *Drosophila*, the red flour beetle, and the silkworm) or those that offer the advantage of large size (for example, the tobacco hornworm).

There is every reason to believe that we are on the verge of a period of great advances in studying and understanding insect proteinases. The next several years should be truly exciting as new enzymes are studied and their functional adaptations elucidated.

Acknowledgments
Work in our laboratories related to the material in this chapter was supported in part by the Kansas Agricultural Experiment Station, the Agricultural Research Service, and grants from Pioneer Hi-Bred International, NIH, The Rockefeller Foundation, The American Heart Association, US Israel BARD, and the USDA National Research Initiatives Competitive Grants Program. This is contribution number 99-494-B of the Kansas Agricultural Experiment Station.
We thank Walter R. Terra for his very helpful comments on the manuscript.

References

1 Reeck GR, Kramer KJ, Baker JE, Kanost MR, Fabrick JA, Behnke C (1997) Proteinase inhibitors and resistance of transgenic plants to insects. *In*: N Carozzi, M Koziel (eds): *Advances in Insect Control: The Role of Transgenic Plants.* Taylor and Francis, London, 157–183
2 Wilson EO (1992) *The Diversity of Life*. Harvard University Press, Cambridge, MA, 131–162
3 Harvey TL, Wilde GE, Kofoid KD (1997) Designation of a new greenbug, biotype K, injurious to resistant sorghum. *Crop Sci* 37: 989–991

4 Beeman RW, Stuart JJ, Haas MS, Friesen KS (1996) Chromosome extraction and revision of linkage group 2 in *Tribolium castaneum. J Hered* 87: 224–232
5 Goldsmith MR (1995) New directions in the molecular genetics of the silkworm. *In*: E Ohnishi, H Sonobe, SY Takahashi (eds): *Recent Advances in Insect Biochemistry and Molecular Biology*. University of Nagoya Press, Nagoya, 383–405
6 Applebaum SW (1985) Biochemistry of digestion. *In*: GA Kerkut, LI Gilbert (eds): *Comprehensive Insect Physiology Biochemistry and Pharmacology*, vol 4. Pergamon Press, New York, 279–311
7 Terra WR (1988) Physiology and biochemistry of insect digestion: An evolutionary perspective. *Brazilian J Med Biol* 21: 675–734
8 Terra WR (1990) Evolution of digestive systems of insects. *Annu Rev Entomol* 335: 181–200
9 Terra WR, Ferreira C (1994) Insect digestive enzymes: properties, compartmentalization and function. *Comp Biochem Physiol* 109B: 1–62
10 Lehane MJ, Muller HM, Crisanti A (1996) Mechanisms controlling the synthesis and secretion of digestive enzymes in insects. *In*: M Lehane, PF Billingsley (eds): *Biology of the Insect Midgut*. Chapman and Hall, London, 195–205
11 Billingsley PF, Lehane MJ (1996) Structure and ultrastructure of the insect midgut. *In*: MJ Lehane, PF Billingsley (eds): *Biology of the Insect Midgut*. Chapman and Hall, London, 3–30
12 Tellam RL (1996) The peritrophic matrix. *In*: MJ Lehane, PF Billingsley (eds): *Biology of the Insect Midgut*. Chapman and Hall, London, 86–114
13 Terra WR, Ferreira C, Jordão BP, Dillon RJ (1996) Digestive enzymes. *In*: MJ Lehane, PF Billingsley (eds): *Biology of the Insect Midgut*. Chapman and Hall, London, 153–194
14 Terra WR, Ferreira C, Baker JE (1996) Compartmentalization of digestion. *In*: MJ Lehane, PF Billingsley (eds): *Biology of the Insect Midgut*. Chapman and Hall, London, 206–235
15 Cheeseman MF, Pritchard G (1984) Spatial organization of digestive processes in an adult carabid beetle, *Scaphinotus marginatus* (Coleoptera: Carabidae). *Can J Zool* 62: 1200–1203
16 Cheeseman MT, Gillott C (1987) Organization of protein digestion in adult *Calosoma calidum* (Coleoptera: Carabidae). *J Insect Physiol* 33: 1–8
17 Ferreira C, Capella AN, Sitnik R, Terra WR (1994) Digestive enzymes in midgut cells, endo- and ectoperitrophic contents, and peritrophic membranes of *Spodoptera frugiperda* (Lepidoptera) larvae. *Arch Insect Biochem Physiol* 26: 299–313
18 Peterson AM, Fernando GJP, Wells MA (1995) Purification, characterization and cDNA sequence of an alkaline chymotrypsin from the midgut of *Manduca sexta. Insect Biochem Molec Biol* 25: 765–774
19 Matsumoto I, Emori Y, Abe K, Arai S (1997) Characterization of a gene family encoding cysteine proteinases of *Sitophilus zeamais* (maize weevil), and analysis of the protein distribution in various tissues including alimentary tract and germ cells. *J Biochem* 3:464–476
20 Pauluat A (1996) Try29F, a new member of the *Drosophila* trypsin-like proteinase gene family, is specifically expressed in the posterior embryonic midgut. *Gene* 172: 245–247
21 Lane NJ, Dallai R, Ashhurst DE (1996) Structural macromolecules of the cell membranes and the extracellular matrices of the insect midgut. *In*: MJ Lehane, PF Billingsley (eds): *Biology of the Insect Midgut*. Chapman and Hall, London, 115–150
22 Wolfson JL, Murdock LL (1990) Diversity in digestive proteinase activity among insects. *J Chem Ecol* 16: 1089–1102
23 Ward C (1975) Resolution of proteinases in the keratinolytic larvae of the webbing clothes moth. *Aust J Biol Sci* 28: 1–23
24 Baker JE (1981) Resolution and partial characterization of the digestive proteinases from larvae of the black carpet beetle. *In*: G Bhaskaran, S Friedman, JG Rodriguez (eds): *Current topics in insect endocrinology and nutrition*. Plenum, New York, 283–315
25 Christeller JT, Shaw BD, Gardiner SE, Dymock J (1989) Partial purification and characterization of the major midgut proteases of grass grub larvae (*Costelytra zealandica*, Coleoptera: Scarabaeidae). *Insect Biochem* 19: 221–231
26 Gillikin JW, Bevilacqua S, Graham JS (1992) Partial characterization of digestive tract proteinases from western corn rootworm larvae, *Diabrotica virgifera. Arch Insect Biochem Physiol* 19: 285–298
27 Michaud D, Bernier-Vadnais N, Overney S, Yelle S (1995) Constitutive expression of digestive cysteine proteinase forms during development of the Colorado potato beetle, *Leptinotarsa decemlineata* Say (Coleoptera: Chrysomelidae). *Insect Biochem Molec Biol* 25: 1041–1048
28 Christeller JT, Laing WA, Markwick NP, Burgess EPJ (1992) Midgut protease activities in 12 phytophagous lepidopteran larvae: dietary and protease inhibitor interactions. *Insect Biochem Molec Biol* 22: 735–746
29 Valaitis AP (1995) Gypsy moth midgut proteinases: purification and characterization of luminal trypsin, elastase and the brush border membrane leucine aminopeptidase. *Insect Biochem Molec Biol* 25: 139–149
30 Houseman JG, MacNaughton WK, Downe AER (1984) Cathepsin B and aminopeptidase activity in the posterior midgut of *Euschistus euschistoides* (Hemiptera: Pentatomidae) *Can Entomol* 116: 1393–1396
31 Murdock LL, Brookhart G, Dunn PE, Foard DE, Kelley S, Kitch L, Shade RE, Shukle RH, Wolfson JL (1987) Cysteine digestive proteinases in Coleoptera. *Comp Biochem Physiol* 87B: 783–787
32 Silva CP, Xavier-Filho J (1991) Comparison between the levels of aspartic and cysteine proteinases of the lar-

val midguts of *Callosobruchus maculatus* (F.) and *Zabrotes subfasciatus* (Boh.) (Coleoptera: Bruchidae). *Comp Biochem Physiol* 99B: 529–533
33 Lemos FJA, Terra WR (1991) Properties and intracellular distribution of a cathepsin D-like proteinase active at the acid region of *Musca domestica* midgut. *Insect Biochem* 21: 457–465
34 Blanco-Labra A, Martinez-Gallardo NA, Sandoval-Cardoso L, Delano-Frier J (1996) Purification and characterization of a digestive cathepsin D proteinase isolated from *Tribolium castaneum* larvae (Herbst). *Insect Biochem Molec Biol* 26: 95–100
35 Eguchi M, Kuriyama K (1985) Purification and characterization of membrane-bound alkaline proteases from midgut tissue of the silkworm, *Bombyx mori*. *J Biochem* 97: 1437–1445
36 Peterson AM, Barillas-Mury CV, Wells MA (1994) Sequence of three cDNAs encoding an alkaline midgut trypsin from *Manduca sexta*. *Insect Biochem Molec Biol* 24: 463–471
37 Sreedharan SK, Verma C, Caves LSD, Brocklehurst SM, Gharbia SE, Shah HN, Brocklehurst K (1996) Demonstration that 1-trans-epoxysuccinyl-L-leucylamido-(4-guanidino)butane (E-64) is one of the most effective low Mr inhibitors of trypsin-catalyzed hydrolysis. Characterization by kinetic analysis and by energy minimization and molecular dynamics simulation of the E-64-beta-trypsin complex. *Biochem J* 316:777–786
38 Lee MJ, Anstee JH (1995) Endoproteases from the midgut of larval *Spodoptera littoralis* include a chymotrypsin-like enzyme with an extended binding site. *Insect Biochem Molec Biol* 25: 49–61
39 Oppert B, Kramer KJ, Johnson D, Upton SJ, McGaughey WH (1996) Luminal proteinases from *Plodia interpunctella* and the hydrolysis of *Bacillus thuringiensis* CryIA© protoxin. *Insect Biochem Molec Biol* 26: 571–583
40 Novillo C, Castañera P, Ortego F (1997) Inhibition of digestive trypsin-like proteases from larvae of several lepidopteran species by the diagnostic cysteine protease inhibitor E-64. *Insect Biochem Molec Biol* 27: 247–254
41 Michaud D, Faye L, Yelle S (1993) Electrophoretic analysis of plant cysteine and serine proteinases using gelatin-containing polyacrylamide gels and class-specific proteinase inhibitors. *Electrophoresis* 14:94–98
42 Johnson DE, Brookhart GL, Kramer KJ, Barnett BD, McGaughey WH (1990) Resistance to *Bacillus thuringiensis* by the Indianmeal moth, *Plodia interpunctella*: comparison of midgut proteinases from susceptible and resistant larvae. *J Invertebrate Pathol* 55: 235–244
43 Liang C, Brookhart G, Feng GH, Reeck GR, Kramer KJ (1991) Inhibition of digestive proteinases of stored grain Coleoptera by oryzacystatin, a cysteine proteinase inhibitor from rice seeds. *FEBS Lett* 278: 139–142
44 Chen MS, Johnson B, Wen L, Muthukrishnan S, Kramer KJ, Morgan TD, Reeck G (1992) Rice cystatin: bacterial expression, purification, cysteine proteinase inhibiting activity and insect growth suppressing activity of a truncated form of the protein. *Protein Express Purif* 3: 41–49
45 Oppert B, Morgan TD, Culbertson C, Kramer KJ (1993) Dietary mixtures of cysteine and serine proteinase inhibitors exhibit synergistic toxicity toward the red flour beetle, *Tribolium castaneum*. *Comp Biochem Physiol* 105C: 379–385
46 Ribolla PEM, De Bianchi AG (1995) Processing of procathepsin from *Musca domestica* s. *Insect Biochem Molec Biol* 25: 1011–1017
47 Behnke C, Fabrick J, Shen Z, Czapla T, Kramer KJ, Reeck GR (1999) Studies on digestive enzymes of the corn rootworm; *in preparation*
48 Briegel H, Lea AO (1975) Relationship between protein and proteolytic activity in the midgut of mosquitoes. *J Insect Physiol* 21: 1597–1604
49 Broadway RM, Duffey SS (1986) The effect of dietary protein on the growth and digestive physiology of larval *Heliothis zea* and *Spodoptera exigua*. *J Insect Physiol* 32: 673–680
50 Lehane MJ, Blakemore D, Williams S, Moffatt MR (1995) Regulation of digestive enzyme levels in insects. *Comp Biochem Physiol* 110B: 285–289
51 Noriega FG, Shah DK, Wells MA (1997) Juvenile hormone controls early trypsin gene transcription in the midgut of *Aedes aegypti*. *Insect Molec Biol* 6: 63–66
52 Barillas-Mury C, Wells MA (1993) Cloning and sequencing of the blood meal-induced late trypsin gene from the mosquito *Aedes aegypti* and characterization of the upstream regulatory region. *Insect Molec Biol* 2: 7–12
53 Barillas-Mury C, Noriega FG, Wells MA (1995) Early trypsin activity is part of the signal transduction system that activates transcription of the late trypsin gene in the midgut of the mosquito, *Aedes aegypti*. *Insect Biochem Molec Biol* 25: 241–246
54 Noriega FG, Barillas-Mury C, Wells MA (1994) Dietary control of late trypsin gene transcription in *Aedes aegypti*. *Insect Biochem Molec Biol* 24: 627–631
55 Müller H-M, Catteruccia F, Vizioli J (1995) Constitutive and blood meal-induced trypsin genes in *Anopheles gambiae*. *Exp Parasitol* 81: 371–385
56 Moffatt M, Blakemore D, Lehane MJ (1995) Studies on the synthesis and secretion of trypsin in the midgut of *Stomoxys calcitrans*. *Comp Biochem Physiol* 110B: 291–300
57 Blakemore D, William S, Lehane MJ (1995) Protein stimulation of trypsin secretion from the opaque zone midgut cells of *Stomoxys calcitrans*. *Comp Biochem Physiol* 110B: 301–307
58 Santos CD, Ribeiro AF, Terra WR (1986) Differential centrifugation, calcium precipitation and ultrasonic disruption of midgut cells of *Erinnyis ello* caterpillars. Purification of cell microvilli and inferences concerning

secretory mechanisms. *Can J Zool* 64: 490–500
59 Jordão BP, Terra WR, Ribeiro AF, Lehane MJ, Ferreira C (1996) Trypsin secretion in *Musca domestica* larval midguts: a biochemical and immunocyothcmical study. *Insect Biochem Molec Biol* 26: 337–346
60 Lemos FJA, Terra WR (1992) Soluble and membrane-bound forms of trypsin-like enzymes in *Musca domestica* larval midguts. *Insect Biochem Molec Biol* 22: 613–619
61 Ferreira C, Bellinello GL, Ribeiro AF, Terra WR (1990) Digestive enzymes associated with the glycocalyx, microvillar membranes and secretory vesicles from midgut cells of *Tenebrio molitor* larvae. *Insect Biochem* 20: 839–847
62 Jordão BP, Lehane MJ, Terra WR, Ribeiro AF, Ferreira C (1996) An immunocytochemical investigation of trypsin secretion in the midgut of the stablefly, *Stomoxys calcitrans*. *Insect Biochem Molec Biol* 26: 445–453
63 Lecadet MM, Dedonder R (1966) Les protéases de *Pieris brassicae*. II. Spécificité. *Bull Soc Chim Biol* 48: 661–691
64 Forcada C, Alcacer E, Garcera MD, Martinez R (1996) Differences in the midgut proteolytic activity of two *Heliothis virescens* strains, one susceptible and one resistant to *Bacillus thuringiensis* toxins. *Arch Insect Biochem Physiol* 31: 257–272
65 Keller M, Sneh B, Strizhov N, Prudovsky E, Regev A, Koncz C, Schell J, Zilberstein A (1996) Digestion of δ-endotoxin by gut proteases may explain reduced sensitivity of advanced instar larvae of *Spodoptera littoralis* to CryIC. *Insect Biochem Molec Biol* 26: 365–373
66 Oppert BS, Kramer KJ, Johnson DE, MacIntosh SC, McGaughey WH (1994) Altered protoxin activation by midgut enzymes from a *Bacillus thuringiensis* resistant strain of *Plodia interpunctella*. *Biochem Biophys Res Commun* 198: 940–947
67 Oppert B, Kramer KJ, Beeman RW, Johnson D, McGaughey WH (1997) Proteinase-mediated insect resistance to *Bacillus thuringiensis* toxins. *J Biol Chem* 272: 23473–23476
68 Raikhel AS, Dhadialla TS (1992) Accumulation of yolk proteins in insect oocytes. *Annu Rev Entomol* 37: 217–251
69 Yamamoto Y, Takahashi SY (1993) Cysteine proteinase from *Bombyx* eggs: role in programmed degradation of yolk proteins during embryogenesis. *Comp Biochem Physiol* 196B: 35–45
70 Ikeda M, Sasaki T, Yamashita O (1990) Purification and characterization of proteases responsible for vitellin degradation of the silkworm, *Bombyx mori*. *Insect Biochem* 20: 725–734
71 Ikeda M, Yaginuma T, Kobayashi M, Yamashita O (1991) cDNA cloning, sequencing and temporal expression of the protease responsible for vitellin degradation in the silkworm, *Bombyx mori*. *Comp Biochem Physiol* 99B: 405–411
72 Indrasith LS, Sasaki T, Yamashita O (1988) A unique protease responsible for selective degradation of a yolk protein in *Bombyx mori*. *J Biol Chem* 263: 1045–1051
73 Cho WL, Deitsch KW, Raikhel AS (1991) An extraovarian protein accumulated in mosquito oocytes is a carboxypeptidase activated in embryos. *Proc Natl Acad Sci USA* 88: 10821–10824
74 Kageyama T, Takahashi SY (1990) Purification and characterization of a cysteine proteinase from silkworm eggs. *Eur J Biochem* 193: 203–210
75 Medina M, Leon P, Vallejo CG (1988) *Drosophila* cathepsin B-like proteinase: a suggested role in yolk degradation. *Arch Biochem Biophys* 263:355–363
76 Liu X, McCarron RC, Nordin JH (1996) A cysteine protease that processes insect vitellin. Purification and partial characterization of the proenzyme. *J Biol Chem* 271:33344–33351
77 Yamamoto Y, Takimoto K, Izumi S, Toriyama-Sakurai M, Kageyama T, Takahashi SY (1994) Molecular cloning and sequencing of cDNA that encodes cysteine proteinase in the eggs of the silkmoth, *Bombyx mori*. *J Biochem (Tokyo)* 116: 1330–1335
78 Takahashi SY, Yamamoto Y, Shionoya Y, Kageyama T (1993) Cysteine proteinase from the eggs of the silkmoth, *Bombyx mori*: identification of a latent enzyme and characterization of activation and proteolytic processing *in vivo* and *in vitro*. *J Biochem (Tokyo)* 114: 267–272
79 Giorgi F, Yin L, Cecchettini A, Nordin JH (1997) The vitellin-processing protease of *Blatella germanica* is derived from a pro-protease of maternal origin. *Tissue Cell* 29: 293–303
80 Potempa J, Korzus E, Travis J (1994) The serpin superfamily of proteinase inhibitors: structure, function, and regulation. *J Biol Chem* 269: 15957–15960
81 Kawabata S, Muta T, Iwanaga S (1996) The clotting cascade and defense molecules found in the hemolymph of the horseshoe crab. *In*: K Söderhäll, S Iwanaga, GR Vasta (eds): *New Directions in Invertebrate Immunology*. SOS Publications, Fair Haven, NJ, 255–283
82 Muta T, Miyata T, Misumi Y, Tokunaga F, Nakamura T, Toh Y, Ikehara Y, Iwanaga S (1991) Limulus factor C: an endotoxin-sensitive serine protease zymogen with a mosaic structure of complement-like, epidermal growth factor-like, and lectin-like domains. *J Biol Chem* 266: 6554–6561
83 Seki N, Muta T, Oda T, Iwaki D, Kuma K, Miyata T, Iwanaga S (1994) Horseshoe crab (1,3)-β-D-glucan-sensitive coagulation factor G: a serine protease zymogen heterodimer with similarities to β-glucan-binding proteins. *J Biol Chem* 269: 1370–1374
84 Muta T, Hashimoto R, Miyata T, Nishimura H, Toh Y, Iwanaga S (1990) Proclotting enzyme from horseshoe crab hemocytes. CDNA cloning, disulfide locations, and subcellular localization. *J Biol Chem* 265: 22426–22433

85 Muta T, Oda T, Iwanaga S (1993) Horseshoe crab coagulation factor B: a unique serine protease zymogen activated by cleavage of an Ile-Ile bond. *J Biol Chem* 270: 892–897
86 Kanost MR, Jiang H (1996) Proteinase inhibitors in invertebrate immunity. *In*: K Söderhäll, S Iwanaga, GR Vasta (eds): *New Directions in Invertebrate Immunology*. SOS Publications, Fair Haven, NJ, 155–173
87 Söderhäll K, Cerenius L, Johansson MW (1996) The prophenoloxidase activating system in invertebrates. *In*: K Söderhäll, S Iwanaga, GR Vasta (eds): *New Directions in Invertebrate Immunology*. SOS Publications, Fair Haven, NJ, 229–253
88 Yoshida H, Ashida M (1986) Microbial activation of two serine enzymes and prophenoloxidase in the plasma fraction of hemolymph of the silkworm, *Bombyx mori*. *Insect Biochem* 16: 539–545
89 Aspán A, Sturtevant J, Smith VJ, Söderhäll K (1990) Purification and characterization of a prophenoloxidase activating enzyme from crayfish blood cells. *Insect Biochem* 20: 709–718
90 Dohke K (1973) Studies on prophenoloxidase-activating enzyme from cuticle of the silkworm, *Bombyx mori*. II. Purification and characterization of the enzyme. *Arch Biochem Biophys* 157: 210–221
91 Chosa N, Fukumitsu T, Fujimoto K, Ohnishi E (1997) Activation of prophenoloxidase A_1 by an activating enzyme in *Drosophila melanogaster. Insect Biochem Molec Biol* 27: 61–68
92 Jiang H, Kanost MR (1997) Characterization and functional analysis of 12 naturally occurring reactive site variants of serpin-1 from *Manduca sexta*. *J Biol Chem* 272: 1082–1087
93 Katsumi Y, Kihara H, Ochiai M, Ashida M (1995) A serine protease zymogen in insect plasma. Purification and activation by microbial cell wall components. *Eur J Biochem* 228: 870–877
94 Ashida M, Sasaki T (1994) A target protease activity of serpins in insect hemolymph. *Insect Biochem Molec Biol* 24: 1037–1041
95 Coustau C, Rocheleau T, Carton Y, Nappi AJ, Ffrench-Constant RH (1996) Induction of a putative serine protease transcript in immune challenged *Drosophila*. *Dev Comp Immunol* 20: 265–272
96 Sidén-Kiamos I, Skavdis G, Rubio J, Papagiannakis G, Louis C (1996) Isolation and characterization of three serine protease genes in the mosquito *Anopheles gambiae*. *Insect Molec Biol* 5: 61–71
97 Dimopoulos G, Richman A, Della Torre A, Kafatos FC, Louis C (1996) Identification and characterization of differentially expressed cDNAs of the vector mosquito, *Anopheles gambiae*. *Proc Natl Acad Sci USA* 93: 13066–13071
98 Kurata S, Saito H, Natori S (1992) Purification of a 29-kDa hemocyte proteinase of *Sarcophaga peregrina Eur J Biochem* 204: 911–914
99 Takahashi N, Kurata S, Natori S (1993) Molecular cloning of cDNA for the 29 kDa proteinase participating in decomposition of the larval fat body during metamorphosis of *Sarcophaga peregrina* (flesh fly). *FEBS Lett* 334: 153–157
100 Murugasu-Oei B, Balakrishnan R, Yang X, Chia W, Rodrigues V (1996) Mutations in masquerade, a novel serine-proteinase-like molecule, affect axonal guidance and taste behavior in *Drosophila*. *Mech Dev* 57: 91–101
101 Belvin M, Anderson KV (1996) A conserved signaling pathway: The *Drosophila* Toll-Dorsal pathway. *Annu Rev Cell Dev Biol* 12: 393–416
102 Smith C, DeLotto R (1992) A common domain within the proenzyme regions of the *Drosophila* snake and easter proteins and *Tachypleus* proclotting enzyme defines a new subfamily of serine proteases. *Protein Sci* 1: 1225–1226
103 Kafatos FC, Tartakoff AM, Law JH (1967) Cocoonase I. Preliminary characterization of a proteolytic enzyme from silk moths. *J Biol Chem* 242:1477–1487
104 Kafatos FC, Law JH, Tartakoff AM (1967) Cocoonase II. Substrate specificity, inhibitors, and classification of the enzyme. *J Biol Chem* 242: 1488–1494
105 Kafatos FC (1972) The cocoonase zymogen cells of silk moths: a model of egg cell differentiation for specific protein synthesis. *Curr Top Dev Biol* 7: 125–191
106 Berger E, Kafatos FC, Felsted RL, Law JH (1971) Cocoonase III. Purification, preliminary characterization, and activation of the zymogen of an insect protease. *J Biol Chem* 246: 4131–4137
107 Felsted RL, Kramer KJ, Law JH, Berger E, Kafatos FC (1973) Cocoonase IV. Mechanism of activation of procoonase from *Antheraea polyphemus*. *J Biol Chem* 248: 3021–3028
108 Kramer KJ, Felsted RL, Law JH (1973) Cocoonase V. Structural studies on an insect serine protease. *J Biol Chem* 248: 3021–3028
109 Sakal E, Applebaum SW, Birk Y (1988) Purification and characterization of *Locusta migratoria* chymotrypsin. *Int J Peptide Protein Res* 32: 590–598
110 Sakal E, Applebaum SW, Birk Y (1989) Purification and characterization of trypsins from the digestive tract of *Locusta migratoria*. *Int J Peptide Protein Res* 34: 498–505
111 Giebel W, Zwilling R, Pfleiderer R (1971) The evolution of endopeptidases. XII. The proteolytic enzymes of the honeybee (*Apis mellifica* L.) *Comp Biochem Physiol* 38B: 197–210
112 Baumann E (1990) Isolation and partial characterization of a chymotrypsin-like endoprotease from cockroach intestinal system. *Insect Biochem* 20: 761–768
113 Milne R, Kaplan H (1993) Purification and characterization of a trypsin-like digestive enzyme from spruce budworm (*Choristoneura fumiferana*) responsible for the activation of δ-endotoxin from *Bacillus thuringiensis*. *Insect Biochem Molec Biol* 23: 663–673

114 Bernardi R, Tedeschi G, Ronchi S, Palmieri S (1996) Isolation and some molecular properties of a trypsin-like enzyme from larvae of European corn borer *Ostrinia nubilalis* Hübner (Lepidoptera: Pyralidae). *Insect Biochem Molec Biol* 26: 883–889
115 Ahmad Z, Saleemuddin M, Siddi M (1980) Purification and characterization of three alkaline proteases from the gut of the larva of army worm, *Spodoptera litura. Insect Biochem* 10: 667–673
116 Tong NT, Imhoff JM, Lecroisey Keil B (1981) Hypodermin A, a trypsin-like neutral proteinase from the insect *Hypoderma lineatum. Biochim Biophys Acta* 658: 209–219
117 Jany KD, Haug H, Ishay J (1978) Trypsin-like endopeptidases from the midguts of the larvae from the hornets of *Vespa orientalis* and *Vespa crabro. Insect Biochem* 8: 221–230
118 Reeck GR, Neurath H (1972) Pancreatic trypsinogen from the African lungfish. *Biochemistry* 15:503–510
119 Levinsky H, Birk Y, Applebaum SW (1977) Isolation and characterization of a new trypsin-like enzyme from *Tenebrio molitor* L. larvae. *Int JPeptide Protein Res* 10: 252–264
120 Sasaki T, Hishida T, Ichikawa K, Asari S (1993) Amino acid sequence of alkaliphilic serine protease from silkworm, *Bombyx mori*, larval digestive juice. *FEBS Lett* 320: 35–37
121 Brookhart GL, Kramer KJ (1990) Proteinases in molting fluid of the tobacco hornworm, *Manduca sexta. Insect Biochem.* 20: 467–477
122 Samuels RI, Charnley AK, Reynolds SE (1993) A cuticle-degrading proteinase from the moulting fluid of the tobacco hornworm, *Manduca sexta. Insect Biochem Molec Biol* 23: 607–614
123 Miller JW, Kramer KJ, Law JH (1974) Isolation and partial characterization of the larval midgut trypsin from the tobacco hornworm, *Manduca sexta*, Johannson (Lepidoptera: Sphingidae). *Comp Biochem Physiol* 48B: 117–129
124 Wieman KF, Nielsen SS (1988) Isolation and partial characterization of a major gut proteinase from larval *Acanthoscelides obtectus* Say (Coleoptera, Bruchidae). *Comp Biochem Physiol* 89B: 419–426
125 Haire MF, Clark JJ, Jones MEE, Hendil KB, Schwartz LM, Mykles DL (1995) The multicatalytic proteinase (proteasome) of the hawkmoth, *Manduca sexta*: Catalytic properties and immunological comparison with the lobster enzyme complex. *Arch Biochem* 318: 15–24
126 Parenti P, Morandi P, McGivan JD, Consonnic P, Leonardi G, Giordana B (1997) Properties of the aminopeptidase N from the silkworm midgut (*Bombyx mori*). *Insect Biochem Molec Biol* 27: 397–403
127 Samuels RI, Charnley AK, Reynolds SE (1993) An aminopeptidase from the moulting fluid of the tobacco hornworm, *Manduca sexta. Insect Biochem Molec Biol* 23: 615–620
128 Stone TE, Li JP, Bernasconi P (1994) Purification and characterization of the *Manduca sexta* neuropeptide processing enzyme carboxypeptidase E. *Arch Insect Biochem Physiol* 27: 193–203
129 Thompson JD, Higgins DG, Gibson TJ (1994) CLUSTAL W: improving the sensitivity of progressive multiple sequence alignment through sequence weighting, positions-specific gap penalties and weight matrix choice. *Nucleic Acid Res* 22: 4673–4680
130 Purcell JP, Greenplate JT, Duck NB, Sammonds RD, Stonard RJ (1994) Insecticidal activity of proteinases. *FASEB J* 8: A1372

Proteases: New Perspectives
V. Turk (ed.)

Alzheimer's disease and proteinases

Shoichi Ishiura

Department of Life Sciences, Graduate School of Arts and Sciences, The University of Tokyo, 3-8-1 Komaba, Meguro-ku, Tokyo 153-8902, Japan

Introduction

Alzheimer's disease (AD) is the most common form of senile dementia. The disease is characterized by a progressive cognitive decline caused by a loss of neurons from particular regions of the cerebral cortex, accompanied by the presence of amyloid deposition. The core, a central deposit of extracellular amyloid fibrils, is surrounded by dystrophic neurites with activated microglias and reactive astrocytes. Amyloid β protein (Aβ) is a major component of senile plaques and plays a crucial role in the pathogenesis of AD [1–3]. Aβ is a 39–43 amino acid peptide (Fig. 1) produced by abnormal proteolytic processing of the large membrane-bound amyloid precursor protein (APP). Cleavage of APP normally releases a large, soluble N-terminal polypeptide (sAPP). This cleavage occurs within the Aβ region by a putative "α-secretase" (see Fig. 1) and a small 10 kDa C-terminal fragment remains in the membrane. The α-secretory cleavage is known to be due to conventional secretory processing or normal processing and this cleavage precludes further Aβ formation. On the other hand, intact Aβ is produced by abnormal processing that involves the excision of the Aβ region by sequential cutting by putative "β-" and "γ-" secretases. The processing of APP takes place in the trans-Golgi network (TGN) and involves the selective hydrolysis of the substrate from the membrane and the subsequent release of a soluble fragment. Many membrane proteins, such as Fas-ligand, CD43, CD44, L-selectin etc., are now reported to be released from the membrane in this manner. This review will focus on the characteristics of such "secretases" or "sheddases (membrane protein convertases)" involved in the processing of integral membrane proteins, especially APP and presenilins associated with early onset familial AD [4].

α-secretase

The enzyme(s) responsible for α-secretory cleavage has not been isolated. The deletion mutant that removes the α-secretory cleavage site Lys16-Leu17 (numbering from the N-terminus of Aβ) does not prevent cleavage [5]. Therefore, α-secretory cleavage is not sequence-specific, but the secretase cleaves APP at a set distance, 12 amino acids, from the proposed plasma membrane. Hooper and Turner reviewed the available data concerning the site of cleavage within Aβ by α-secretase and found a bulky hydrophobic residue such as Leu, Val, or Phe in the P1'

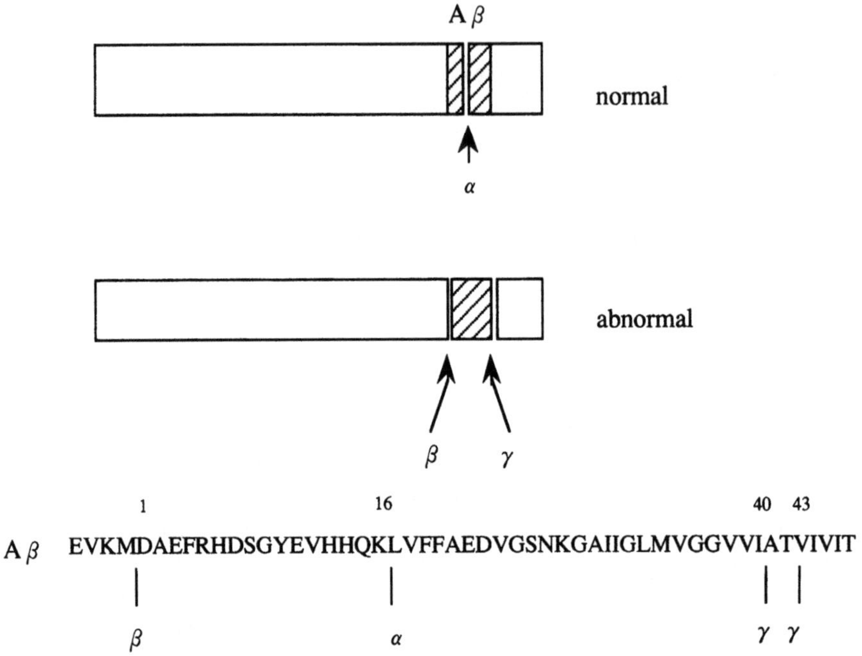

Figure 1. Cleavage sites on APP.

position. They suggested that this specificity is reminiscent of mammalian endopeptidase-24.11 (EC3.4.24.11) and other membrane metalloendopeptidases [6].

A large group of membrane proteins undergo cleavage with the release of their ectodomains into the extracellular medium. Membrane-anchored growth factors such as transforming growth factor α (TGFα), cell adhesion molecules, and APP are included in this group. Some Chinese hamster ovary (CHO) cell clones are unable to cleave membrane-anchored TGFα as well as APP in response to tetradecanoylphorbol acetate (TPA) which normally stimulates TGFα and APP α-cleavage. This indicates that TGFα secretion and APP α-secretory cleavage share the same secretory machinery [7].

Recently, a metalloproteinase was identified as a component involved in membrane-bound tumor-necrosis factor-α (TNFα) release [8, 9]. The protease, disintegrin, cleaves precursor TNFα and inactivation of the putative proteinase gene in T cells causes a marked decrease in TNFα release. The amino acid sequence deduced from the cDNA revealed that this enzyme belongs to the adamalysin family of zinc-binding metalloproteinases (metzincin). Since all mammalian adamalysins (fertilin-α, meltrin-α) so far characterized are involved in embryonic development or cell-cell interaction, the properties of disintegrin in aged cells will be intriguing.

Several lines of evidence suggest that α-secretase activity is modulated by metal ions and metalloproteinase inhibitors. α-Secretory cleavage is enhanced by the addition of bivalent cations (Ca, Mg, Mn) at neutral pH, and inhibited by metal chelators such as EDTA and 1,10-phenanthroline [10]. Disruption of membrane integrity and solubilization of APP by 0.05%

Triton X-100 completely inhibits α-secretory cleavage. These results suggest that α-secretase is an integral membrane protein and that the protein machinery requires the membrane to cleave APP. Iron also enhances the secretion of sAPP and iron chelation inhibits α-cleavage [11]. Although there is no direct evidence for the involvement of iron in the proteolytic system, the above results suggest a metalloproteinase-directed cleavage process.

APP α-secretory cleavage is enhanced by serum withdrawal, TPA treatment [12], and the addition of unsaturated fatty acids [13], and suppressed by C-kinase inhibitors [13] and cholesterol [14]. Since no proteinases have been reported to be affected by these compounds, the α-secretory machinery may be complex and contain many proteins of the signal transduction pathway.

Using synthetic fluorogenic peptide substrates (succinyl-His-His-Gln-MCA, succinyl-His-Gln-Lys-MCA) that harbor the α-secretory cleavage site, we found that only one enzyme hydrolyzes these substrates. The enzyme isolated from rat liver, bovine brain, and human brain was identified as cathepsin B [15]. Purified cathepsin B also cleaves full-length APP at, or close to, the α-secretory cleavage site [16]. Recent evidence that only cathepsin B among cathepsins B, D, and E has the ability to cleave the Lys16-Leu17 bond [17] favors the above possibility, but the data imply only that cathepsin B has the same substrate specificity as the putative α-secretase. Knockout mutants of the cathepsin B gene would ultimately clarify the involvement of this enzyme in α-secretory processing.

The recent finding that proteasome inhibitors selectively reduce the TPA-sensitive α-secretion of APP is also interesting [18]. This indicates that a high molecular mass ATP-dependent protease, the proteasome, participates in APP processing. We have not yet determined whether the well-known inhibitor lactacystin inhibits only the proteasome or some other enzyme as well. In addition, the TGN localization of the proteasome has not been confirmed. Therefore, the involvement of a proteasome in the sAPP secretion process is a problem for future study.

β-secretase

The first proteolytic cleavage in the abnormal amyloidogenic pathway is catalyzed by a putative β-secretase (Fig. 1) which generates the shorter N-terminal sAPP lacking Aβ16 and leaves the C-terminal membrane-bound 100 amino acids (C100). Experiments in which the C-terminal cytoplasmic domains are deleted indicate that the primary stage in Aβ production is β-secretory cleavage [19]. Once C100 is produced, Aβ is promptly secreted.

Candidate β-secretase(s) have been identified, using small synthetic peptides, as proteasome [20–22], metalloprotease [23], gelatinase A [24], and cathepsins [25]. Schoenlein et al. purified a 100 kDa metalloprotease from human brain as β-secretase and determined its activity with Z-Val-Lys-Met-MCA and KTEEISEVKM-pNA; however, the purified enzyme hydrolyzes Arg-MCA 100-times faster than either of these substrates [23]. The protease has the ability to cleave full-length APP to produce a 19 kDa C-terminal fragment which, however, is not the β-secretory cleavage product. Extracellular gelatinase A, once considered to be an α-secretase, has also been reported to have β-secretase activity. However, the cleavage site in synthetic TEEISEVKMDAEFR-NH_2 is shown to be in the vicinity of the β-cleavage site, the Glu-Val bond. When the α-secretase peptide YEVHHQKLVFFAED-NH_2 is incubated with gelatinase A under the same conditions, no breakdown is observed [24]. These results suggest that gelatinase A may not be involved in the intracellular processing of APP.

It has been reported that familial Alzheimer's disease with the Swedish KM to NL mutation shows an increased production of Aβ *in vivo* and *in vitro*. Using a nonapeptide substrate with or without the mutation, Brown et al. evaluated the cleavage efficiency of cathepsin D, G, and thimet oligopeptidase (EC.3.4.24.15) [25]. They purified the β-secretase activity that cleaves between the Met and Asp residues of SEVKMDAEF and identified it as thimet oligopeptidase. When NL replaces KM, as in the Swedish Alzheimer family, cleavage was not increased. On the other hand, cathepsin D dramatically increases the efficiency of cleavage after the introduction of the NL substitution at the cleavage site. These authors speculate that cathepsin D is involved in Aβ generation. However, recent evidence that cathepsin D-knockout mice show normal APP processing and Aβ production rule out cathepsin D as a crucial component of the secretory machinery [26].

Although there is no concrete evidence for the identification of β-secretase, interesting findings have been reported. Zhao et al. produced APP transgenic mice using the neuron-specific enolase promoter or a glial fibrillary acidic protein gene [27]. Efficient β-secretase cleavage was observed only in neurons, not in astrocytes. This suggests either the minute content of β-secretase in glial cells or an abundance of β-secretase-activating factor(s) in neuronal cells.

An irreversible serine protease inhibitor with broad specificity, AEBSF (4-(2-aminoethyl)-benzenesulfonyl-fluoride HCl), inhibits the production of Aβ in both neural and non-neural human cell lines [28]. The inhibitor has been shown to inhibit β-secretase. In addition, an increase in intracellular full-length APP and sAPP has been observed after AEBSF treatment. This also suggests that an uncharacterized serine protease(s) is capable of non-specifically digesting APP or sAPP in the cell.

As shown above, a major obstacle in identifying secretases is the disappearance of secretase activity in cell-free systems. Mok et al. recently prepared rat brain Golgi compartment by using a discontinuous sucrose gradient [29]. Since these fractions may contain endogenous APP and secretase machinery, they incubated the fraction at 37° C. They discovered the breakdown of APP and the concomitant increase in Aβ-containing C-terminal fragments in the Golgi-rich fraction and speculated that the amyloidogenic fragment was generated in the Golgi apparatus by a metalloprotease. The results do not provide sufficient evidence to indicate that the "Golgi" protease is a β-secretase because they did not show the direct production of Aβ in their system and because contamination by a lysosomal enzyme such as cathepsin B [16] might yield the same result.

The results of another reconstitution experiment also suggest the production of Aβ in the trans-Golgi network [30]. Xu et al. used permeabilized neuroblastoma cells expressing full-length APP harboring the Swedish double mutation and showed that a TGN-enriched fraction from these cells produces Aβ at 37° C under standard conditions. This strongly suggests the Golgi localization of β-and γ-secretases, and that Aβ is generated intracellularly. The addition of GTPγS did not affect γ-secretase cleavage, indicating that vesicular budding is not involved in γ-secretase activity.

γ-secretase

As shown in Figure 1, a putative γ-secretase cleaves APP at the membrane-spanning domain. Therefore, γ-secretase has been thought to be a membrane-bound enzyme. γ-secretase activity, as detected by Aβ-formation from the amyloidogenic C-terminal fragment [31], is inhibited by the dipeptide-aldehyde calpain inhibitor MDL28170. Calpain is not a vesicular or Golgi enzyme and this inhibitor appears not to be specific for calpain, so there is little possibility that calpain cleaves APP at the γ-cleavage site.

Analyses of Aβ length in cultured cells revealed a series of Aβ of various lengths (Aβ40, Aβ42, and Aβ43). Many investigators have demonstrated that longer Aβs aggregate rapidly and are deposited in amyloid plaques even at an early stage, suggesting that Aβ ending at residues 42 and 43 is critical in amyloidogenesis. Therefore, it is tempting to postulate that the Aβ40- and Aβ42-generating γ-secretases are distinct, and that the longer Aβ-forming γ-secretase is activated in Alzheimer's brain. Klafki et al. employed the calpain inhibitors Cbz-Leu-Leu-Leu-CHO and calpeptin to distinguish these two activities in cells expressing APP with the Swedish mutation [32]. Calpain inhibitors increased Aβ42 formation, while the generation of the shorter Aβ40 was significantly reduced after treatment. The authors concluded that the cleavages of APP at Aβ residues 40 and 42 are differentially sensitive to calpain inhibitors, and that at least two distinct γ-secretases are involved in the processing. Again, a major obstacle to this theory is the cytosolic localization of various calpain isoforms. We cannot rule out the possible involvement of an unidentified membrane-bound calpain in γ-secretory cleavage since a recently found novel calpain species appears to be attached to the cell membrane [33].

Several years ago, we purified a candidate Aβ42 γ-secretase using the peptide substrates Z-Val-Ile-Ala-MCA and Suc-Ile-Ala-MCA harboring an Aβ42 γ-secretase cleavage site. The enzyme was identified as a neuron-specific prolyl oligopeptidase [21]. We also identified a novel Mg-stimulated prolyl oligopeptidase from the membrane fraction of rat liver [34]. Since the cytosolic prolyl oligopeptidase is a typical peptidase that cannot hydrolyze peptides larger than 30 amino acids, the membrane-bound enzyme may require an unidentified factor for processing larger APP or C100.

Tjernberg et al. reported that Aβ, the protease-resistant core, can be produced from C100 by non-specific proteolysis [35]. They treated C100 with proteinase K and showed that the dissociated, baculovirus-expressed C100 is easily degraded by proteinase K, whereas a significant production of Aβ was observed from membrane-bound intact C100 or polymerized fibrillar C100. This suggests that a compartment with high proteolytic activity is capable of generating Aβ non-specifically.

Aβ-degrading protease

Since Aβ is a minor catabolic product of APP and is responsible for the aberrant pathological process, Aβ should be rapidly degraded into amino acids in normal cells. A portion of Aβ is secreted into the culture medium and is degraded with a half-life of 10–12 h by extracellular proteinase(s) [36]. Inhibitor studies were conducted and the results suggest that multiple proteinases are involved in the degradation of Aβ and no selective inhibitor is able to prevent Aβ degradation. On the other hand, the Kunitz domain of sAPP770 (or sAPP751) effectively inhibits Aβ degradation. Therefore, a trypsin-type serine protease may be the main catabolic enzyme and Kunitz protease inhibitor-containing sAPP may retard the clearance of Aβ *in vivo*.

Cathepsin D is also suspected to be an Aβ degrading protease because of its broad amino acid specificity [37]. The recent finding that a microglial metalloprotease digests extracellular Aβ is also intriguing [38] as it suggests a role for the secreted protease in Aβ catabolism. The Aβ degrading activity is strongly inhibited by 1,10-phenanthroline and partially by EDTA. Activated microglia, which are consistently found in mature senile plaques but less frequently in presenile plaques, may release a digestive protease. Our observation that microglia contain 100-times more lysosomal enzyme activity [39] also favors the possibility that microglia (or brain macrophages) are involved in the inflammatory process in AD, and that anti-inflammatory drugs will modulate APP metabolism [13].

The furin family of proprotein processing enzymes was tested for the ability to process APP because the localization of furin and secretase appears to be very close. Co-transfection of APP695 with various members of the furin family of processing proteinases indicates that furin, PACE 4, PC1/3, PC2, PC4, and PC5/6 do not enhance the secretase reaction [40].

Ubiquitin-dependent degradation of APP-related proteins

ATP-ubiquitin/proteasome-dependent proteolysis has been regarded as one of the major pathways of regulated protein clearance in mammalian cells. This complex system is shown to be mainly involved in antigen presentation, quality control of unfolded proteins in the endoplasmic reticulum (ER), and some stress responses. The intracellular accumulation of ubiquitin conjugates in neurofibrillary tangles is a unique feature of neurodegenerative diseases. However, ubiquitinated tau does not degrade rapidly, suggesting that ubiquitination may play roles other than as a marker for protein degradation in the brain under pathological conditions.

To know whether APP can act as a substrate in a ubiquitin-dependent pathway, full-length APP and soluble sAPP were purified and added to reticulocyte fraction II [41]. In this system, only soluble APP is degraded *in vitro* by the ubiquitin/proteasome pathway. The question remains as to how extracellular protein becomes accessible to the ubiquitin/proteasome system. The same authors also indicate that Aβ inhibits the ubiquitin-dependent degradation of lysozyme [42] as well as proteasome activity itself.

Endoproteolytic cleavage of presenilin

The majority of early-onset autosomal dominant cases of familial Alzheimer's disease can be accounted for by a mutation in Golgi proteins of unknown function, presenilin-1 (PS1) and presenilin-2 (PS2) [43]. PS1 and PS2 share amino acid sequence identity and are predicted to contain nine transmembrane (TM) segments (Fig. 2). The vast majority of mutations in PS1 occur between the sixth and seventh TMs. After translation, full-length PS1 undergoes rapid proteo-

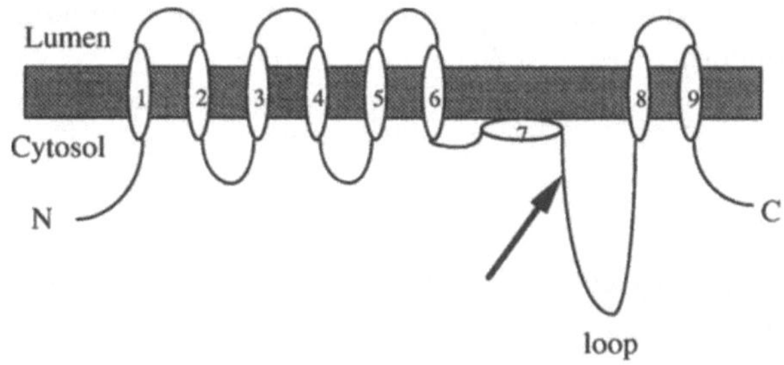

Figure 2. Orientation of presenilin-1 in ER membrane. The arrow shows the approximate site of endoproteolytic cleavage.

lysis in the cytoplasmic loop between TMs 7 and 8. This results in the production of an N-terminal 27 kDa PS1 fragment and a 16 kDa C-terminal fragment. PS1 also has a strong tendency to form high-molecular-mass aggregates [44]. We cannot conclude that these 100-250 kDa SDS-resistant aggregates are conjugates with ubiquitin or complexes formed with APP as shown for PS2 [45]. It has been reported that fibroblasts derived from patients with PS1 mutations produce more and larger Aβ42(43) forms of amyloid. This suggests that the mutations result in changes in the cellular metabolism of APP or trafficking of these proteins.

PS2 is also proteolytically cleaved into two stable peptides, an N-terminal 34 kDa and a C-terminal 20 kDa fragment [46]. PS2 is poly-ubiquitinated *in vivo*, and the degradation of PS2 is inhibited by lactacystin. These results suggest that PS2 normally undergoes endoproteolytic cleavage, and is degraded via the ubiquitin/proteasomal pathway. Brefeldin A and lactacystin treatment result in a greater accumulation of high molecular weight poly-ubiquitinated PS2. This indicates that the proteasomal degradation of PS2 occurs in a pre-Golgi compartment, probably in the ER. It should be noted that the cleavage of PS into two fragments is not inhibited by lactacystin. Therefore, it is unlikely that the proteasome carries out the endoproteolytic cleavage, but ubiquitin/proteasome may be involved in quality control of PS in the ER.

Racemization of Asp residues in Aβ and D-Asp specific endopeptidase

Racemization and/or the isomerization of Asp residues in Aβ affect the aggregation properties of Aβ. Tomiyama et al. reported that synthetic Aβ containing D-Asp shows the earliest increase in turbidity and appears as a smear in SDS-PAGE [47], indicating that racemization at Asp_{23} accelerates peptide aggregation. Substantial levels of D-Asp occur in brain tissue, especially in neurons, and D-amino acids may have some as yet unknown biological function in mammalian brain [48]. Therefore, the mechanism for the repair or degradation of racemized proteins may be disrupted in Alzheimer's disease.

We identified a D-aspartyl endopeptidase in the mammalian mitochondrial fraction that is highly specific for D-Asp [34]. This endopeptidase digests D-Asp_7-containing Aβ10 but not

Aβ10 containing L-Asp. These data suggest that this protease is important in metabolizing D-Asp-containing peptides.

Future studies

Despite our lack of knowledge concerning the identity of AD-related proteases (secretases), it is clear that the inhibition (or activation) of these proteases may offer new therapeutic methods. Inhibitors of β- and γ-secretases may prevent the deposition of Aβ in brain, whereas an activator of α-secretase may inhibit the accumulation of Aβ with the same efficiency. To design a specific inhibitor of Aβ production may also be useful in the treatment of AD.

Protease inhibitors can act strongly on their target proteases *in vitro*, but are often ineffective when administered *in vivo* due to poor penetration into the cell. Therefore, transfection of mammalian cells with cDNA encoding APP or presenilins provides a good tool to search for secretase inhibitors. A potent and selective inhibitor may be an ideal therapeutic tool for AD in the future.

Acknowledgments
I thank the Ministry of Education, Science, Sports and Culture, and the Ministry of Health and Welfare, Japan, for financial support of the work on Alzheimer's disease APP secretases.

References

1 Haass C, Selkoe DJ (1993) Cellular Processing of β-amyloid precursor protein and the genesis of amyloid β-peptide. *Cell* 75: 1039–1042
2 Iversen LL, Mortishire-Smith R, Pollack SJ, Shearman MS (1995) The toxicity *in vitro* of β-amyloid protein. *Biochem J* 311: 1–16
3 Yankner BA (1996) Mechanisms of neuronal degeneration in Alzheimer's disease. *Neuron* 16: 921–932
4 Hooper NM, Karran EH, Turner AJ (1997) Membrane protein secretases. *Biochem J* 321: 265–279
5 Maruyama K, Kametani F, Usami M, Yamao-Harigaya W, Tanaka K (1991) "Secretase", Alzheimer amyloid protein precursor secreting enzyme is not sequence specific. *Biochem Biophys Res Commun* 179: 1670–1676
6 Hooper NM, Turner AJ (1995) Specificity of the Alzheimer's amyloid precursor protein α-secretase. *Trend Biol Sci* 20: 15–16
7 Arribas J, Massague J (1995) Transforming growth factor-α and β amyloid precursor protein share a secretory mechanism. *J Cell Biol* 128: 433–441
8 Black RA, Rauch CT, Kozlosky CJ, Peschon JJ, Slack JL, Wolfson MF, Castner BJ, Stocking KL, Reddy P, Srinivasan S et al. (1997) A metalloproteinase disintegrin that releases tumor-necrosis factor-α from cells. *Nature* 385, 729–733
9 Moss ML, Jin SLC, Milla ME, Burkhart W, Carter HL, Chen WJ, Clay WC, Didsbury JR, Hassler D, Hoffman CR et al. (1997) Cloning of a disintegrin metalloproteinase that processes precursor tumor-necrosis factor-α. *Nature* 385, 733–736
10 Roberts SB, Ripellino JA, Ingall KM, Robakis NK, Felsenstein KM (1994) Non-amyloidogenic cleavage of the β-amyloid precursor protein by an integral membrane metalloendopeptidase. *J Biol Chem* 269: 3111–3116
11 Bodovitz S, Falduto MT, Frail DE, Klein WL (1995) Iron levels modulate α-secretase cleavage of amyloid precursor protein. *J Neurochem* 64: 307–315
12 Kinouchi T, Sorimachi H, Maruyama K, Mizuno K, Ohno S, Ishiura S, Suzuki K (1995) cPKα and nPKε, but not δ, increase APP secretion from PKC cDNA transfected fibroblasts treated with TPA. *FEBS Lett* 364: 203–206
13 Kinouchi T, Ono Y, Sorimachi H, Ishiura S, Suzuki K (1995) Arachidonate metabolites affect the secretion of an N-terminal fragment of Alzheimer's disease amyloid precursor protein. *Biochem Biophys Res Commun* 209: 841–849
14 Racchi M, Baetta R, Salvietti N, Franceschini G, Paoletti R, Fumagalli R, Govoni S, Trabucchi M, Soma M

(1997) Secretory processing of amyloid precursor protein is inhibited by increase in cellular cholesterol content. *Biochem J* 322: 893–898
15 Tagawa K, Kunishita T, Maruyama K, Yoshikawa K, Kominami E, Tsuchiya T, Suzuki K, Tabira T, Sugita H, Ishiura S (1991) Alzheimer's disease amyloid β-clipping enzyme (APP secretase): identification, purification, and characterization. *Biochem Biophys Res Commun* 177: 377–387
16 Tagawa K, Maruyama K, Ishiura S (1992) Amyloid β/A4 precursor protein (APP) processing in lysosome. *Ann N Y Acad Sci* 674: 129–137
17 Mackey EA, Ehrhard A, Moniatte M, Guenet C, Sorokine O, Heintzelmann B, Nay C, Remy J, Wagner J, Danzin C et al. (1997) A possible role for cathepsins D, E, and B in the processing of β-amyloid precursor protein in Alzheimer's disease. *Eur J Biochem* 244: 414–425
18 Marambaud P, Chevallier N, Barelli H, Wilk S, Checler F (1997) Proteasome contributes to the α-secretase pathway of amyloid precursor protein in human cells. *J Neurochem* 68: 698–703
19 Maruyama K, Kawamura Y, Asada H, Ishiura S, Obata K (1994) Cleavage at N-terminal site of Alzheimer amyloid β/A4 protein is essential for its secretion. *Biochem Biophys Res Commun* 202: 1517–1523
20 Ishiura S, Tsukahara T, Tabira T, Sugita H (1989) Putative N-terminal splitting enzyme of amyloid A4 peptide is the multicatalytic proteinase, ingensin, which is widely distributed in mammalian cells. *FEBS Lett* 257: 388–392
21 Ishiura S, Tsukahara T, Tabira T, Shimizu T, Arahata K, Sugita H (1990) Identification of a putative amyloid A4-generating enzyme as a prolyl endopeptidase. *FEBS Lett* 260: 131–134
22 Ishiura S, Nishikawa T, Tsukahara T, Momoi T, Ito H, Suzuki K, Sugita H (1990) Distribution of Alzheimer's disease amyloid A4-generating enzymes in rat brain tissues. *Neurosci Lett* 115: 329–334
23 Schoenlein C, Loeffler J, Huber G (1994) Purification and characterization of a novel metalloprotease from human brain with the ability to cleave substrates derived from the N-terminus of the β-amyloid protein. *Biochem Biophys Res Commun* 201: 45–53
24 LePage RN, Fosang AJ, Fuller S, Murphy G, Evin G, Beyreuther K, Masters C, Small DH (1995) Gelatinase A possesses a β-secretase-like activity in cleaving the amyloid protein precursor of Alzheimer's disease. *FEBS Lett* 377: 267–270
25 Brown AM, Tummolo DM, Spruyt MA, Jacobsen JS, Sonnenberg-Reines J (1996) Evaluation of cathepsin D and G and EC 3.4.24.15. as candidate β-secretase proteases using peptide and amyloid precursor protein substrates. *J Neurochem* 66: 2436–2445
26 Saftig P, Peters C, von Figura K, Craessaerts K, Van Leuven F, De Strooper B (1997) Amyloidogenic processing of human amyloid precursor protein in hippocampal neurons devoid of cathepsin D. *J Biol Chem* 271: 27241–27244
27 Zhao J, Paganin L, Mucke L, Gordon M, Refolo L, Carman M, Sinha S, Oltersdorf T, Lieberburg I, McConlogue L (1996) β-secretase processing of the β-amyloid precursor protein in transgenic mice is efficient in neurons but inefficient in astrocytes. *J Biol Chem* 271: 31407–31411
28 Citron M, Diehl TS, Capell A, Haass C, Teplow DB, Selkoe DJ (1996) Inhibition of amyloid β-protein production in neural cells by the serine protease inhibitor AEBSF. *Neuron* 17: 171–179
29 Mok SS, Evin G, Li QX, Smith AI, Beyreuther K, Masters CL, Small DH (1997) A novel metalloprotese in rat brain cleaves the amyloid precursor protein of Alzheimer's disease generating amyloidogenic fragments. *Biochemistry* 36: 156–163
30 Xu H, Sweeney D, Wang R, Thinakaran G, Lo ACY, Sisodia SS, Greengard P, Gandy S (1997) Generation of Alzheimer β-amyloid protein in the trans-Golgi network in the apparent absence of vesicle formation. *Proc Natl Acad Sci USA* 94: 3748–3752
31 Higaki J, Quon D, Zhong Z, Cordell B (1995) Inhibition of β-amyloid formation identifies proteolytic precursors and subcellular site of catabolism. *Neuron* 14: 651–659
32 Klafki HW, Abramowski D, Swoboda R, Paganetti PA, Staufenbiel M (1996) The carboxy termini of β-amyloid peptide 1-40 and 1-42 are generated by distinct γ-secretase activities. *J Biol Chem* 45: 28655–28659
33 Suzuki K, Sorimachi H, Yoshizawa T, Kinbara K, Ishiura S (1995) Calpain: novel family members, activation, and physiological function. *Biol Chem Hoppe-Seyler* 376: 523–529
34 Ishiura S, Mabuchi Y, Urakami-Manaka Y, Isobe K, Tagawa K, Maruyama K, Sorimachi H, Suzuki K (1997) Proteases involved in the processing of the Alzheimer's disease amyloid precursor protein. *In*: Hopsu-Have (ed.): *Proceedings of 11th Int. Congr. Proteolysis.* VSP, Turku
35 Tjernberg LO, Naeslund J, Thyberg J, Gandy SE, Terenius L, Nordstedt C (1997) Generation of Alzheimer Amyloid β peptide through nonspecific proteolysis. *J Biol Chem* 272, 1870–1875
36 Naidu A, Quan D, Cordell B (1995) β-Amyloid peptide produced *in vitro* is degraded by proteinases released by cultured cells. *J Biol Chem* 270:1369–1374
37 Higaki J, Catalano R, Guzzetta AW, Quon D, Nave JF, Tarmus C, D'Orchymont H, Cordell B (1996) Processing of β-amyloid precursor protein by cathepsin D. *J Biol Chem* 271: 31885–31893
38 Qiu WQ, Ye Z, Kholodenko D, Seubert P, Selkoe DJ (1997) Degradation of amyloid β-protein by a metalloprotease secreted by microglia and other neural and nonneural cells. *J Biol Chem* 272: 6641–6646
39 Tagawa K, Yazaki M, Kinouchi T, Maruyama K, Sorimachi H, Tsuchiya T, Suzuki K, Ishiura S (1993) Amyloid precursor protein is found in lysosome. *Gerontology* 39(suppl 1): 24–29
40 Strooper BD, Creemers JWM, Moechars D, Huylebroeck D, Van de Ven WJM, Van Leuven F, Van de Berghe

H (1995) Amyloid precursor protein is not processed by furin, PACE 4, PC1/3, PC2, PC4 and PC5/6 of the furin family of proprotein processing enzymes. *Biochim Biophys Acta* 1246:185–188
41 Gregori L, Bhasin R, Goldgaber D (1994) Ubiquitin-mediated degradative pathway degrades the extracellular but not the intracellular form of amyloid β-protein precursor. *Biochem Biophys Res Commun* 203: 1731–1738
42 Gregori L, Fuchs C, Figueiredo-Pereira M, Van Nostrand WE, Goldgaber D (1995) Amyloid β-protein inhibits ubiquitin-dependent protein degradation *in vitro*. *J Biol Chem* 270, 19702–19708
43 Lamb BT (1997) Presenilins, amyloid-β and Alzheimer's disease. *Nat Med* 3: 28–29
44 De Strooper B, Beullens M, Contreras B, Levesque L, Craessaerts K, Cordell B, Moechars D, Bollen M, Fraser P, StGeorge-Hyslop et al. (1997) Phosphorylation, subcellular localization, and membrane orientation of the Alzheimer's disease-associated presenilins. *J Biol Chem* 272, 3590–3598
45 Weidemann A, Paliga K, Durrwang U, Czech C, Evin G, Masters CL, Beyreuther K (1997) Formation of stable complexes between two Alzheimer's disease gene products: presenilin-2 and β-amyloid precursor protein. *Nat Med* 3: 328–332
46 Kim TW, Pettingell WH, Hallmark OG, Moir RD, Wasco W, Tanzi RE (1997) Endoproteolytic cleavage and proteasomal degradation of presenilin 2 in transfected cells. *J Biol Chem* 272: 11006–11010
47 Tomiyama T, Asano S, Furiya Y, Shirasawa T, Endo N, Mori H (1994) emization of Asp23 residue affects the aggregation properties of Alzheimer amyloid β protein analogues. *J Biol Chem* 269: 10205–10208
48 Schell MJ, Cooper OB, Snyder SH (1997) D-Aspartate localizations imply neuronal and neuroendocrine roles. *Proc Natl Acad Sci USA* 94: 2013–2018

Proteases: New Perspectives
V. Turk (ed.)

Calpains: structure and function of the calpain super family

Yasuko Ono, Hiroyuki Sorimachi and Koichi Suzuki

Institute of Molecular and Cellular Biosciences, University of Tokyo, 1-1-1 Yayoi, Bunkyo-ku, Tokyo 113-0032, Japan

Calpain (EC 3.4.22.17) is a typical cytosolic cysteine protease whose activity is regulated by calcium, an important second messenger of extra-cellular stimuli [1–4]. Since calpain is distributed ubiquitously among cells and tissues of most higher organisms, it has been predicted to play an essential role in cellular functions. One reason that calpain attracts such interest is its way of hydrolyzing substrates; i.e. calpain may modulate, by limited proteolysis, the function of substrate proteins. Although, the physiological function of calpain as a biomodulator remains to be clarified, recent technological developments enable a variety of new approaches. These include finding tissue-specific calpain species, calpain-homologues in lower organisms, and the introduction of new methodology. In this chapter, the state of the art of calpain is summarized.

Historical background of calpain

Calpain was first identified in 1964 as a caseinolytic activity in rat brain [5] and a phosphorylasekinase activating factor in skeletal muscle [6]. In 1978, calpain was recognized as a Ca^{2+}-dependent factor with various activities. Ca^{2+}-dependent "kinase-activating factor" (KAF) in rabbit muscle and "sarcoplasmic factor" (CASF), which hydrolyzes rabbit muscle Z-disc in a Ca^{2+}-dependent manner, were identified as the same activity. "Ca^{2+}-activated factor" (CAF) and "receptor transforming factor" (RTF) were also proposed to describe the activity executed by calpain. Calpain was also discovered as a novel protease that activates protein kinase C [4]. When Ca^{2+}-activated neutral protease (CANP) was purified for the first time in 1978, all the factors described above were identified as being identical [7]. Further studies showed that the enzyme was a cysteine protease, from which the term calpain originates.

With refinements in purification methods, a second, minor protease activity with distinct Ca^{2+}-sensitivity was discovered. It then became clear that there are at least two isoforms, m-calpain and μ-calpain, requiring mM and μM amounts of Ca^{2+} for proteolytic activity, respectively [8, 9]. Each species is composed of an unique, large catalytic 80 kDa (80K) subunit and a common, small regulatory 30 kDa (30K) subunit [1, 10]. The term "calpain", originally proposed in 1981 [11], stands for Ca^{2+}-dependent cysteine protease. It has been adopted as the recommended term since 1991 [12].

Molecular structure of calpain

The cDNA sequence of calpain was first elucidated in 1984 for chicken 80K [10]. Since then, studies of the structure of calpain subunits at the amino acid sequence level have progressed extensively. 80K and 30K can be divided into four (I, II, III and IV) and two (IV' and V) structural domains, respectively (Fig. 1). The domain structure of calpain is generally defined in relevance to them [13–16].

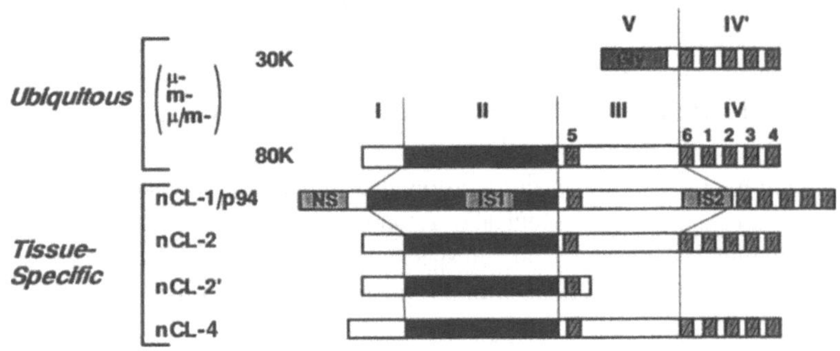

Figure 1. Schematic domain structure of calpain family members in mammls. I, II, III, and IV represent the four calpain domains. II is the cysteine protease domain, and IV and IV' is the Ca^{2+}-binding domain. EF hand motifs including newly identified ones (5 and 6) are represented by hatched boxes. Gly: *Glycine*-rich region in domain V. NS, IS1, IS2: p94-specific insertions.

Domain II shows similarities to the protease domains of other typical cysteine proteases such as papain and cathepsins B, H, and L [17]. Domain IV is a calmodulin-like Ca^{2+}-binding domain consisting of four EF-hand structures [18, 19]. No amino acid sequence showing similarity to either domain I or domain III has been identified so far. The functions of domains I and III remain unclear. However, autocatalytic processing of the N-terminal region of 80K and the subsequent change in Ca^{2+}-sensitivity suggest that domain I is involved in the regulation of activity [2, 4, 20].

30K also has a Ca^{2+}-binding domain similar to that in the 80K domain IV at its C-terminal. This region is referred to as domain IV'. The N-terminal half of 30K is domain V. This domain has a hydrophobic region with repeating glycine residues [21]. The interaction with phospholipids, which is essential for the activation of calpain *in vivo*, is considered to be the function of domain V.

It is generally accepted that calpain exists as a heterodimer of 80K and 30K. The association of these two subunits is supposed to be mediated by the hydrophobic interaction of domain IV and domain IV' [22]. Some lines of evidence proved the binding of domain IV and 30K. The role of 30K has been regarded as that of a regulator of the 80K catalytic subunit. However, various results recently obtained have enlarged the concept of 30K, and the mechanism of the interaction of 80K and 30K is under investigation [23–27].

Constitution of the calpain family

cDNA cloning has shown that there are many 80K species comprising what is called the "calpain family". In vertebrates, five molecular species other than μ- and m-calpains have been identified in our laboratory to date [28–30]. Recently, the existence of calpain homologues have been reported in *Drosophila melanogaster* [31–33], *Caenorhabditis elegans* [34–36], and a Platyhelminth, *Schistosoma mansoni* [37].

Criteria for the classification of family species

Comparisons of these molecular species suggest that the calpain family members can be classified into two groups by either one of two criteria, i.e. mode of expression or molecular structure [38].

According to the former criterion, calpain can be divided into two classes, ubiquitous and tissue-specific calpains. All μ-, μ/m-, and m-calpains are ubiquitously expressed. They exist as dimers of 80K and 30K, and are also called conventional calpains. Calpain 80K homologues found in invertebrates are tentatively classified as ubiquitous, although the existence of a regulatory molecule corresponding to vertebrate 30K is not well investigated. Known tissue-specific calpains at present include p94 (= nCL-1) in skeletal muscle, nCL-2(2') in stomach, and nCL-4 in digestive organs. Since neither p94 nor nCL-2(2') interacts with 30K, tissue-specific calpains seem to differ from ubiquitous calpains in their modes of expression as well as their state of existence.

Another classification system categorizes calpains as "typical calpains" and "atypical calpains", imitating the case of protein kinase C [39]. This nomenclature is based on the integrity of the domain structure of the molecule in comparison with μ- and/or m-calpains, i.e. conventional calpains. Sol from *Drosophila*, and p70, p71, p92, and Tra-3 from *C. elegans* are categorized as atypical calpains according to this criterion. These species, which lack the Ca^{2+} binding domain IV, do not strictly fit the term "calpain", which originally meant calcium-dependent cysteine protease [11]. However, the protease domains of these molecules show a higher similarity to calpain domain II than to any other cysteine protease. Mammalian counterparts for these atypical species have not yet been identified. It is noteworthy that mammalian nCL-2 and CalpA from *Drosophila*, which have the typical domain structures of calpain, generate the atypical species nCL-2' and CalpA', respectively, by alternative splicing [29, 33].

The conventional calpains, μ- and m-calpains

Most of the biochemical and enzymological characterizations of calpain concern the conventional calpains. The μ- and m-calpains are similar in most molecular characteristics, except for the calcium concentration required for activity [8, 9]. This discrepancy in Ca^{2+}-sensitivity is attributed to 80K, since the same 30K is common to both species.

Mechanism of the activation of calpain

The involvement of Ca^{2+} in its activation emphasises calpain as a mediator of the Ca^{2+} signaling cascade. However, the precise activation mechanism *in vivo* remains obscure. In this context, the term "activation" means "hydrolysis of substrates". Although an increase in Ca^{2+} concentration initiates the activation of calpain *in vitro*, it now appears that other factors are involved *in vivo*, including modifiers of Ca^{2+}-sensitivity and an endogenous inhibitor, calpastatin. To date, at least three factors have been identified as regulators of calpain Ca^{2+}-sensitivity; autolysis, phospholipids, and 30K.

Autolytic transition

The autocatalytic sensitization to Ca^{2+} was first observed for chicken calpain in 1981 [20]. In the presence of sufficient Ca^{2+}, both 80K and 30K are converted to their autolyzed forms by cleavage of their N-termini. After autolysis, the Ca^{2+} concentration required for hydrolyzing substrates is lower by about one order of magnitude. However, whether this autolytic processing is necessary for activation remained unclear. In 1995, 80K dissociated from 30K was identified as being fully active as a protease with a Ca^{2+}-sensitivity identical to that of the autolyzed form of the native enzyme. Thus, it was proved that autolysis causes the shift in Ca^{2+}-sensitivity, but not the activation of calpain [26].

The sites of autolysis are located in domain I of 80K and domain V of 30K. As for domain V, phospholipid binding is supposed to be one of its functions [21]. It remains unclear how the removal of the N-terminal polypeptides from calpain raises Ca^{2+}-sensitivity.

The effects of phopholipids

Intracellular Ca^{2+} concentrations hardly exceed 10^{-6} M, which means that the Ca^{2+}-requirements for activation observed *in vitro*, 10^{-4} to 10^{-5} M and 10^{-3} to 10^{-4} M Ca^{2+} for μ- and m-calpain, respectively, are never reached. The search for novel factors in the *in vivo* activation mechanism of calpain originates from the finding that phosphatidylinositol (PI) lowers the Ca^{2+} concentration required for calpain autolysis. This led to the hypothesis that calpain activation occurs at cell membranes composed of this phospholipid [2].

The effects of various phospholipids on the *in vitro* hydrolysis of fodrin by μ-calpain have been extensively examined. The results revealed that phosphatidylinositol polyphosphate (PIP_n) promotes fodrinolysis by μ-calpain more potently than phosphatidylinositol (PI). The presence of PIP_n lowers the Ca^{2+} concentration required for autolysis and accelerates the V_{max} of substrate hydrolysis. The positive effect of PIP_n is inhibited by the antibiotic neomycin, which binds strongly to phosphatidylinositol polyphosphate, both *in vitro* and in cells. PIP_n is thus supposed to be an *in vivo* co-factor. One of the interesting features of these findings is the order of the calpain-activating potency of various PIP_n [40]. The number of negative charges on the phosphate groups coupled to inositol seems to have some relevance, i.e. phosphatidylinositol bisphosphate (PIP_2) promotes fodrinolysis more strongly than phosphatidylinositol monophosphate (PIP). PIP_2 enables autolysis of μ-calpain and m-calpain at 10^{-6} to 10^{-7} and 10^{-5} M Ca^{2+}, respectively.

Chaperone-like effect of 30K

As depicted above, 30K was re-identified as unnecessary for the full proteolytic activity of 80K. In other words, 30K plays a regulatory role by lowering the Ca^{2+} requirement of 80K. The physiological significance of the interaction between 30K and 80K is quite complicated. 30K accelerates the renaturation of 80K dissociated from 30K. 30K, at a molar ratio of 1:1, shortens the period required for the recovery of full proteolytic activity of 80K from one month to one week, which is the same as the effect observed for the *E. coli.* chaperone, GroEL [26]. Since 30K is effective only for 80K, 30K can be assumed to be a calpain-specific molecular chaperone. The purification of calpain from biological sources has never identified 80K or its autolyzed form, 76K, as monomers. It is supposed that these derivatives are quite unstable due to their high Ca^{2+} sensitivities and are autocatalytically degraded. Stabilizing 80K by lowering its Ca^{2+}-sensitivity may be a role of 30K.

Calpastatin, an endogenous inhibitor

Calpastatin, like conventional calpain is ubiquitously expressed in various animal cells [41, 42]. The amino acid sequence of calpastatin shows no similarity to other proteinaceous cysteine protease inhibitors. One calpastatin molecule harbors four inhibitory domains. It has been shown that each domain binds to either domain II, domain IV, or domain IV' of calpain [43]. The interaction of calpastatin and calpain is Ca^{2+}-dependent and calpastatin is also a suicide substrate of calpain. Previous studies have shown that calpastatin inhibits calpain in a reversible manner and the mechanism of inhibition release may be involved in cellular function, a process as important as the activation of calpain.

As will be described later, calpastatin does not inhibit the autolysis of p94, a muscle-specific calpain. It is also notable that the preliminarily estimated K_i value of calpastatin for m-calpain is almost one order of magnitude smaller than that for μ-calpain. The fact that μ-calpain and m-calpain share calpastatin, a negative regulator, as well as 30K, a dual regulator, is also significant.

Studies on the interaction of 80K and 30K

Domain IV in 80K and IV' in 30K are considered to be Ca^{2+} binding domains. Previous studies have proposed that 30K binds to domain IV of 80K by hydrophobic interaction through domain IV'. Studies based on the crystal structure of calpain are essential to a full understanding of the function of Ca^{2+} in regulating calpain activity. To obtain abundant material of high quality, various improvements in the expression system and the purification of the enzyme have been made. Recently, recombinant calpains produced in *E. coli* and insect cells have become available and their specific activities are comparable to that of the highly purified species from biological sources [44–46].

The crystal structure has been solved for a homodimer of bacterially expressed domain IV', both with and without Ca^{2+} [27]. The involvement of EF-hand I and II in Ca^{2+} binding was confirmed and a potential binding site, at the junction of domains V and IV' identified. The dimerization of recombinant domain IV' was an unexpected by-product, and the application of this homodimeric interaction of domain IV' to provide a model of the domain IV/IV' heterodimer will be helpful for further studies.

Physiological implications

The post ischaemic activation of μ-calpain is a well-studied and prominent phenomenon. In the nervous system, the transient elevation of intracellular Ca^{2+} during ischaemia has been suggested to result in neuronal death [47]. Spatial and chronological studies on post-ischaemic brain of gerbils were carried out with a specific antibody for the calpain-proteolyzed form of fodrin. The results clearly demonstrate biphasic fodrin proteolysis in the hippocampus, CA3 and CA1, with the delayed phase in CA1 accompanied by major neuronal degeneration [48]. Further analysis revealed the mode of Ca^{2+} mobilization, PIP_2 breakdown, and subsequent calpain activation in the brains of monkeys [49].

As a therapy for neuronal damage associated with excitotoxicity, stroke, traumatic injury, Alzheimer's disease, and so on, inhibitors targeting μ-calpain have been of great concern. Synthetic inhibitors that are more effective and specific to calpain have been shown to reduce fodrinolysis and cognitive abnormalities [50, 51]. These inhibitors in turn will be useful for analyzing calpain function in other cells and/or tissues.

Protein kinase C (PKC) is also an important enzyme for Ca^{2+}-related cellular responses. Calpain was originally detected as an activator of PKC, although it is now clear that the catalytically active fragment of PKC is an intermediate product of the calpain-catalyzed down regulation of PKC. PKC is activated in response to various extracellular stimuli and the activated moieties are subsequently down-regulated. The involvement of calpain in this down-regulatory processing has been observed in some cell lines [52, 53].

Other groups of proteins, such as transcription factors, adhesion-related molecules, receptors, etc, have been shown to be susceptible to calpain [3, 42, 54–57]. The regulatory mechanism of calpain is often discussed based on the interaction between 30K and 80K or Ca^{2+}-sensitivity. Still, the possibility of other regulatory factors cannot be excluded.

The third species of ubiquitous calpains

In chicken muscle four distinct 80Ks are expressed [30]. One is the third ubiquitous calpain, μ/m-calpain. The others correspond to mammalian μ-calpain, m-calpain, and p94. The cDNA sequence obtained in 1984, which had been regarded as chicken m-calpain, was renamed μ/m-calpain, since its sequence and Ca^{2+}-sensitivity are intermediate between those of μ- and m-calpains [10]. The N- and C-terminal sequences of μ/m-calpain are closer to mammalian m-calpain and μ-calpain, respectively, suggesting that the chicken μ/m-calpain gene may originate from a chimera of the μ- and m-calpain genes [30]. In contrast to mammals, chicken m-calpain has not yet been identified at the protein level, while mammalian μ/m-calpain has not been identified at either the cDNA or protein levels.

The tissue-specific calpains

One innovative concept introduced to the study of calpain is that each tissue contains both ubiquitous and tissue-specific calpains with unique functions. The physiological functions of calpain so far identified or predicted are obscure due to its ubiquitous expression. The identification of p94, a skeletal muscle-specific species, was the start of extensive screening for tissue-specific calpains. In mammals, four tissue-specific species, p94 (= nCL-1), nCL-2 and 2', and nCL-4, have so far been discovered in our laboratory.

Skeletal muscle-specific calpain, p94

The name p94 originates from its molecular mass, 94 kDa. p94 is distinct from ubiquitous or conventional calpains in several characteristics. 1) p94 has three inserted sequences, NS, IS1, and IS2. NS and IS1 exist at the N-terminal and in domain II, and show no homology to other sequences reported so far. In the IS2 region, a possible nuclear localization signal (PxKKKKxKP) has been found. 2) p94 degrades autocatalytically so rapidly that it disappears almost immediately. This might explain why, despite having an mRNA level at least ten times higher than those of μ- and m-calpains in skeletal muscle, p94 has never been identified in the μ- and m-calpain fraction purified from mammalian muscle. 3) p94 binds to connectin/titin through IS2, a finding clarified by the yeast two-hybrid system. In this system, p94 does not associate with 30K under conditions sufficient for 30K to bind to 80K of μ- and m-calpains [58]. 4) p94 is responsible for limb-girdle muscular dystrophy type 2A (LGMD2A) [59]. It is tempting to attribute these features to the biological importance of p94, and the correlation between its structure and function has been investigated.

Autolytic activity of p94

The autolysis of p94 is immediate and almost complete, resulting in the disappearance of the protein. The half-life of p94 is less than one hour for the *in vitro* translation product and ten minutes for the partially purified protein from skeletal muscle. Other proteases, including ubiquitous calpains, also undergo autolysis at the N-terminus, although not so extensively as in the case of p94. Another peculiarity of p94 autolysis is that it occurs in the presence of protease inhibitors effective against conventional calpains (1 mg/ml calpastatin, a specific proteinaceous inhibitor for conventional calpains, 1 mM leupeptin, and 1 mM E-64) and a Ca^{2+} chelator (10 mM EGTA) [60]. Ionic strength, pH, and metal ions, which affect the autolysis of other proteases, have no effect on p94.

To date, three factors have been identified as being involved in p94-specific autolysis, IS1, IS2, and connectin/titin. Although the results are preliminary, the responsibility of IS1 for autolysis has been confirmed, while that of IS2 has already been proved. Both regions can be generated by alternative exon usage, and it was once suggested that these insertions determine the rate of autocatalytic degradation. Some lines of evidence suggest the possibility that alternative splicing variants, including those which lack IS1 and/or IS2, are generated from the same p94 gene, however, this requires confirmation. IS2 is also involved in the interaction with connectin/titin [58]. Connectin/titin is a giant elastic protein in striated muscles and p94 seems to be stabilized by association with connectin/titin. Full-length p94 (94 kDa) can be detected by specific antisera in myofibril fractions containing connectin/titin. Considering the short half-life of p94 (less than one hour for the *in vitro* translation product) and that the purification procedures for this fraction require several hours, the autolytic activitiy of p94 is presumably suppressed by binding to connectin/titin. It is very interesting that IS2 plays dual roles in determining the mode of p94 autolysis [60].

p94 and limb girdle muscular dystrophy (LGMD2A)

In 1995, while it was becoming clear that p94 associates with connectin, p94 was identified as the gene responsible for limb girdle muscular dystrophy type 2A (LGMD2A) [59]. This further emphasized the importance of p94 in the muscle system. What made the report so significant was that this was the first indication of the involvement of calpain in an inherited disease, and also the first time that defects in a single enzyme rather than structural proteins were shown to cause muscular dystrophy. Since LGMD2A is an autosomal recessive disorder and the protease domain of p94 is lacking in some mutations found in LGMD2A, p94 protease activity is thought to be lost in patients [61]. As a means of clinical diagnosis, it was proposed that the absence of full-length p94 could provide an index for the loss of p94 protease activity. In fact, muscles from LGMD2A patients were reported to lack a 94 kDa band when subjected to immunoblotting with a p94 specific antibody [62]. As for other mutations, such as missense mutations, our results suggest the loss of p94-catalyzed proteolysis in the skeletal muscle.

p94 in skeletal muscle

When the viewpoint is shifted from p94 to connectin/titin, the question arises as to the physiological meaning of the interaction between p94 and connectin/titin in skeletal muscle. Connectin/titin is a giant protein (ca. 3000 kDa) in charge of muscle ultrastructure and elasticity [63, 64]. The cDNA sequence of human skeletal and cardiac connectin/titin was determined in 1995, and more details about the structure-function relationships of connectin/titin have been reported [65].

Screening of the p94-binding protein by the yeast two-hybrid system identified two distinct loci on the connectin/titin filament, one in the N_2-line and the other at the C-terminal end [30, 66]. The N_2-line is considered to be the cleavage site in full-length α-connectin giving rise to β-connectin. It has been reported that connectin/titin interacts with the M-line proteins, myomesin and M-protein, at the C-terminal region, and overlaps with the C-terminal region of another connectin/titin molecule. Although no direct evidence is currently available, p94 may play some role in these interactions.

It has been demonstrated that antisense mRNAs to m-calpain and p94 specifically inhibit myoblast fusion and myofibrillar integration *in vitro*, respectively [67, 68]. The antisense treatment targeting p94 results in the absence of mature Z-lines and a diffuse distribution of α-actinin, but does not affect A-band formation. Studies on the assembly of connectin/titin into the Z-line and the M-line, corresponding to the N- and C-termini of connectin/titin, respectively, reveal that the proper settling of the N-terminus of connectin/titin is essential for the assembly of the Z-line and further to sarcomeres [69]. Some inconsistencies remain to be clarified; e.g. it has been shown that Z-line assembly precedes that of the M-line in sarcomeric organization. Still, it is noteworthy that the N-terminus of connectin/titin binds to the C-terminus of α-actinin and direct interaction of p94 with these regions has yet to be identified.

Myofibrillogenesis has been studied as an initial step in muscular organization. It is tempting to investigate the role of p94 in the generation and/or regeneration of skeletal muscle [70]. Considering that impairment not only of protease activity but also of connectin-binding was observed for some of the missense mutants of p94 found in LGMD2A, the association between connectin/titin and p94 may be temporally regulated as a means of expressing its protease activity concordantly with differentiation.

Stomach specific calpains n,CL-2 and 2'

Genomic DNA cloning of nCL-2 proved the expression of two alternative splicing products with (= nCL-2) and without (= nCL-2') a Ca^{2+} binding domain [38]. This was the first identification of an atypical calpain in mammalian tissues, and especially of a gene encoding a typical species simultaneously producing an atypical species by alternative splicing. In contrast to p94 in skeletal muscle, the level of mRNA for nCL-2 and nCL-2' in the stomach is roughly the same as that for m-calpain. No structural specificity is found in nCL-2, i.e. nCL-2 does not contain any specific insertions corresponding to NS, IS1, or IS2 in p94. In other words, nCL-2 shows greater similarity to μ- and m-calpains than p94 (58.1% and 60.9%, and 53.2% and 51.0%, respectively). Since the N-terminal sequences are highly homologous to that of m-calpain, nCL-2(2') may undergo autocatalytic activation similar to m-calpain.

Since both nCL-2 and 2' are expressed at almost the same levels, both products seem to contribute equally to the function of the stomach [38]. At present, it is not clear what this phenomenon means. However, any slight difference between nCL-2 and 2' in temporal and/or spatial localization in the tissue, for example, will provide a strategic hint for further investigation.

Interestingly, CalpA, a *Drosophila* homologue of ubiquitous and typical calpain, also produces CalpA' by alternative splicing [33], which lacks a Ca^{2+} binding domain, like nCL-2', but still has domain III. The fact that alternative splicing is conserved during evolution indicates that other calpain species might also have alternative splicing products.

Other tissue-specific species

The search for tissue-specific calpains in other tissues is ongoing. The latest species identified in our laboratory is expressed specifically in intestine as well as stomach. Distinctive features of this molecule include a hydrophobic region found in the N-terminal region and specific localization when expressed in COS cells. The expression pattern is obviously different from that of nCL-2. At the amino acid level, this molecule is equally similar to μ-, μ/m-, m-calpain, p94 and nCL-2, and is assumed to be an ancestral molecule that branched off at an earlier stage. Further characterization of both the biochemical and phylogenic properties of this molecule is ongoing.

Calpain homologues in other organisms

Calpain homologues so far identified in animals other than mammals include atypical species with diverse molecular structures that can hardly be recognized as calpain at a first glance. However, these rather primitive organisms can be useful tools to analyze the functions of calpain *in vivo*. Brief summaries of the characteristics of molecules identified as calpain homologues follow for each organism (Fig. 2).

Drosophila

The first calpain homologue identified in *Drosophila* was a protein encoded by *sol* (small optic lobes) [31], mutations in which result in the absence of certain classes of columnar neurons by degeneration of developing optic lobes. Of the two kinds of *sol* protein product, the larger has a protease domain similar to calpain in its C-terminus. There are six zinc finger motifs in the N-terminal region and no Ca^{2+}-binding region is found. Sol has a most atypical structure, however, the analysis of *sol* mutants reveal that a mutation in the putative protease domain produces a mutant phenotype, and its function is being re-examined.

CalpA, also known as Dm-calpain, is another homologue [32, 33]. It is a typical calpain and its alternative splicing generates an atypical CalpA'. As for CalpA, its involvement in early embryogenesis has been observed. CalpA has a hydrophobic insertion in domain IV that is thought to mediate the binding of the molecule to membranous structures. Since CalpA' lacks domain IV, the localization of these two molecules as well as Ca^{2+}-sensitivity seems to be controlled by alternative splicing.

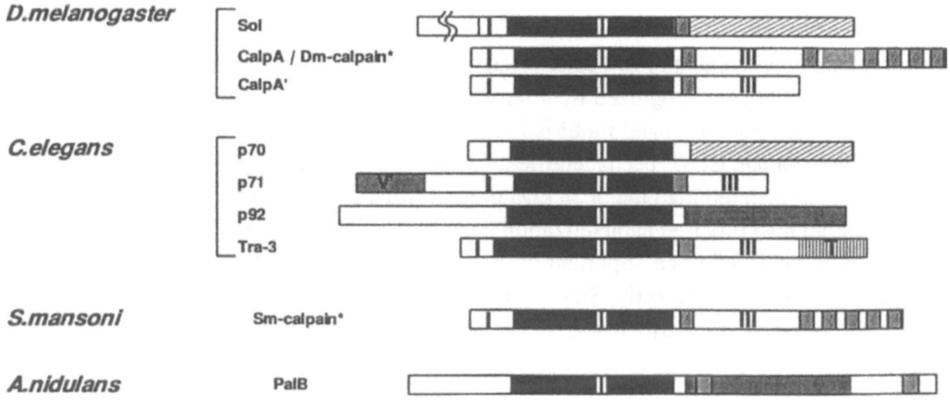

Figure 2. Calpain homologues found in other organism. I, II, III, and V' show similarity to domains I, II, III and V of mammalian calpain 80K and 30K, respectively. EF hand motifs are represented by hatched boxes. Sol and p70, p92 and PalB show similarity to each other in the region adjacent to putative domain II in each molecule. T, domain T; *, Typical calpains.

Caenorhabditis elegans *(*C. elegans*)*

The genome project for *C. elegans* identified three calpain homologues as adjuncts [34, 35] tentatively named p70, p71, and p92, according to their predicted molecular weights. There is another potentially homologous molecule, Ce-CL-1, for which a partial sequence corresponding to a part of domain II is reported. p70, p71 and p92 are atypical calpains. In p71 there is an additional glycine-rich hydrophobic region similar to domain V in 30K that resides in the N-terminal of domain I. Interestingly, this structure can be interpreted as a fusion of domains V, I, II, and III, i. e., all the domains of calpain except the Ca^{2+}-binding domain.

The fifth calpain homologue was identified as a sex determination gene, *tra*-3 [36]. It is also an atypical calpain that lacks domain IV and has an additional 147 residue region named domain T. In *C.elegans*, there are two naturally occurring sexes, the XX hermaphrodite and the XO male. The sex-determination cascade has been genetically studied. The *tra-3* mutation is not lethal, and the potential proteolytic activity of TRA-3 is required to promote female development in XX hermaphrodites, but is not essential for male development in XO males. The substrate candidates for TRA-3 are the products of genes involved in the main sex-detemination pathway, such as tra-2 and fem-1. These proteins contain sequences homologous to the identified substrates of conventional calpains [36, 71, 72].

No typical calpain has been identified so far, however, it is becoming apparent that *C. elegans* is also useful for investigating the *in vivo* functions of calpain. Whether these four or five molecules can be functionally complemented by the calpains found in mammals is of great interest.

Platyhelminths

A typical calpain, Sm-calpain, was discovered in *Schistosoma mansoni*. This molecule was first identified as an antigen recognized by infected host cells [37]. The structure of domain IV is irregular and the fourth EF hand motif is incomplete. Instead, a novel EF hand motif between domain II and domain III has been identified. At the protein level, two isozymes have been purified [73]. It was also reported that a 28 kDa protein, which might correspond to 30K, was copurified, although its precise characterization has not been carried out. The protease activity of Sm-calpain is supposed to be important in the survival strategy of this parasite, i.e. alterations of membrane structure. Recently, Sm-calpain was reported to be an effective vaccine antigen [74]. A relatively low similarity between Sm-calpain and human calpains is taken advantage of in this case.

Fungus

The finding that palB, which is involved in signaling essential for the survival of *Aspergillus nidulans* at alkaline ambient pH, contains an atypical calpain presents a new situation. The sequence similarity between palB and other calpains is restricted to the protease domain and is about 32.3% [75]. This rather weak similarity is understandable considering the evolutionary distance between fungi and mammals, still, the involvement of a calpain homologue in the response of bacteria to ambient changes suggests a primitive model for the signaling cascade in which calpain functions. The final step in the pal signaling cascade is the proteolytic activation of the transcriptional factor, Pac C [76]. However, palB is not responsible for this modification, and a search for candidate substrates is now underway.

Perspectives of calpain

As the number of calpain superfamily species increases, the physiological importance of calpain has enlarged. Extensive *in vitro* experiments have revealed many potential calpain substrates and these suggest the involvement of calpain in many cellular functions. Their modes of existence in the cytosol are not entirely clear, and new questions arise as earlier questions are answered.

The regulatory mechanism for activation, including how to overcome insufficient *in vivo* Ca^{2+} concentrations, has now become clearer. On the other hand, the involvement of other cofactors in the regulation of calpain activity raises other questions. One area of current interest is how calpain isozymes that coexist in cells relate to one another. Conventional calpains, at least, share both 30K and calpastatin as regulators. Although specific substrates for each isozyme have not been identified, the discovery of tissue-specific species strongly suggests functional non-complementarity.

Novel calpain homologues discovered in other organisms suggest another strategic change. Lower animals, with smaller genome sizes, shorter life spans, and ease of genetic manipulation, can provide more direct means for the investigation of calpain. The atypical features of these species are interesting from a phylogenic point of view. Identification of substrates for each homologue may provide clues for clarifying the conserved role of calpain.

Since the cDNA sequence of calpain was first determined, strategic innovations have supported the search for calpains. The expression of recombinant proteins deserving various analyses, and inhibitors with greater specificity for calpain are becoming available, and these have contributed greatly to recent results. The combination of novel techniques and approaches will reveal new aspects of calpain, and disclose its physiological function in greater detail.

References

1 Goll DE, Kleese WC, Okitani A, Kuamamoto T, Cong J (1990) Historical background and current status of the Ca^{2+}-dependent proteinase system. *In*: RL Mellgren, T Murachi (eds): *Intracellular Calcium-dependent Proteolysis*. CRC Press, Boca Raton, 3–24

2 Saido TC, Sorimachi H, Suzuki K (1994) Calpain: new perspective in molecular diversity and physiological-pathological involvement. *FASEB J* 8: 814–822

3 Suzuki K, Sorimachi H, Yoshizawa T, Kinbara K, Ishiura S (1995) Calpain: Novel family members, activation, and physiological function. *Biol Chem Hoppe Seyler* 376: 523–529

4 Sorimahi H, Kimura S, Kinbara K, Kazama J, Takahashi M, Yajima H, Ishiura S, Sasagawa N, Nonaka I, Sugita H et al. (1996) Structure and physiological functions of ubiquitous and tissue-specific calpain species. *Adv Biophys* 33: 101–122

5 Guroff G (1964) A neurtral, calcium-activated proteinase from the soluble fraction of rat brain. *J Biol Chem* 239: 149–155

6 Meyer WL, Fischer EH, Krebs EG (1964) Activation of skeletal muscle phosphorylase kinase b by Ca^{2+}. *Biochemistry* 3: 1033–1039

7 Ishiura S, Murofushi H, Suzuki K, Imahori K (1978) Studies of a calcium-activated neutral protease from chicken skeletal muscle. I. purification and characterization. *J Biochem* 84: 225–230

8 Mellgren RL (1980) Canine cardiac calcium-dependent proteases: resolution of two forms with different requirement for calcium. *FEBS Lett* 109: 129–133

9 Dayton WR, Schollmeyer JV, Lepley RA, Cortes LR (1981) A calcium-activated protease possibly involved in myofibrillar protein turnover. Isolation of a low-calcium-requiring form of the protease, *Biochim Biophys Acta* 659: 48–61

10 Ohno S, Emori Y, Imajoh S, kawasaki H, Kisaragi M, Suzuki K (1984) Evolutionary origin of a calcium-dependent protease by fusion of genes for a third protease and a calcium-binding protein? *Nature* 312: 566–570

11 Murachi T, Tanaka K, Hatanaka M, Murakami T (1981) Intracellular Ca^{2+}-dependent protease (calpain) and its high-molecular-weight endogenous inhibitor (calpastatin). *Adv Enzyme Regul* 19: 407–424

12 Suzuki K (1991) Nomenclature of calcium dependent proteinase. Biomed Biochim Acta 50: 483–484

13 Suzuki K (1987) Calcium activated neutral protease: domain structure and activity regulation. *Trends Biochem Sci* 12: 103–105

14 Suzuki K, Imajoh S, Emori Y, Kawasaki H, Minami Y, Ohno S (1987) Calcium-activated neutral protease and its endogenous inhibitor. *FEBS Lett* 220: 271–277

15 Suzuki K, Ohno S, Emori Y, Imajoh S, Kawasaki H (1987) Calcium-activated neutral protease (CANP) and its biological and medical implications. Prog *Clin Biochem* 5: 44–65

16 Mellgren RL (1988) Calcium-dependent proteases: an enzyme system activated at the cell membrane? *FASEB J* 2: 110–115

17 Berti PJ, Storer AC (1995) Alignment/phylogeny of the papain superfamily of cysteine proteases. *J Mol Biol* 246: 273–283

18 Minami Y, Emori Y, Kawasaki H, Suzuki K (1987) EF hand structure-domain of calcium-activated neutral protease (CANP) can bind Ca^{2+} ions. *J Biochem* 101: 889–895

19 Minami Y, Emori Y, Imajoh-Ohmi S, Kawasaki H, Suzuki K (1988) Carboxyl-terminal truncation and site-directed mutagenesis of the EF hand structure-domain of the small subunit of rabbit calcium-dependent protease. *J Biochem* 104: 927–933

20 Suzuki K, Tsuji S, Ishiura S, Kimura Y, Kubota S, Imahori K (1981) Limited autolysis of Ca^{2+}-activated neutral protease (CANP) changes its sensitivity to Ca^{2+} ions. *J Biochem* 90: 275–278

21 Imajoh S, Kawasaki H, Suzuki K (1986) The amino-terminal hydrophobic region of the small subunit of calcium-activated neutral protease (CANP) is essential for its activation by phosphatidylinositol. *J Biochem* 99: 1281–1284

22 Imajoh S, Kawasaki H, Suzuki K (1987) The COOH-terminal EF hand structure of calcium-activated neutral protease (CANP) is important for the association of subunits and resulting proteolytic activity. *J Biochem* 101: 447–452
23 Saido TC, Nagao S, Shiramine M, Tsukaguchi M, Sorimachi H, Murofushi H, Tsuchiya T, Ito H, Suzuki K (1992) Autolytic transition of μ-calpain upon activation as resolved by antibodies distinguishing between the pre- and post-autolysis forms. *J Biochem* 111: 81–86
24 Saido TC, Nagao S, Shiramine M, Tsukaguchi M, Yoshizawa T, Sorimachi H, Ito H, Tsuchiya T, Kawashima S, Suzuki K (1994) Distinct kinetics of subunit autolysis in mammalian m-calpain autolysis. *FEBS Lett* 346: 263–267
25 Graham-Siegenthaler K, Gauthier S, Davies PL, Elce JS (1994) Active recombinant rat calpain II. Bacterially produced large and small subunits associate both *in vivo* and *in vitro*. *J Biol Chem* 269: 30457–30460
26 Yoshizawa T, Sorimachi H, Tomioka T, Ishiura S, Suzuki K (1995) A catalytic subunit of calpain possesses full proteolytic activity. *FEBS Lett* 358: 101–103
27 Blanchard H, Li Y, Cygler M, Kay CM, Arthur JSC, Davies PL, Elce JS (1996) Ca^{2+}-binding domain VI of rat calpain is a homodimer in solution: Hydrodynamic, crystallization and preliminary X-ray diffraction studies. *Protein Sci* 5: 535–537
28 Sorimachi H, Imajoh-Ohmi S, Emori Y, Kawasaki S, Ohno Y, Minami Y, Suzuki K (1989) Molecular cloning of a novel mammalian calcium-dependent protease distinct from both m- and μ-types. *J Biol Chem* 264: 20106–20111
29 Sorimachi H, Ishiura S, Suzuki K (1993) A novel tissue-specific calpain species expressed predominantly in the stomach comprises two alternative splicing products with and without Ca^{2+}-binding domain. *J Biol Chem* 268: 19476–19482
30 Sorimachi H, Tsukahara T, Okada-Ban M, Sugita H, Ishiura S, Suzuki K (1995) Identification of a third ubiquitous calpain species – chicken muscle expresses four distinct calpains. *Biochim Biophys Acta* 1261: 381–393
31 Delaney SJ, Hayward DC, Barleben F, Fischbach KF, Gabor Miklos GL (1991) Molecular cloning and analysis of small optic lobes, a structural brain gene of *Drosophila melanogaster*. *Proc Natl Acad Sci USA* 88: 7213–7218
32 Emori Y, Saigo K (1994) Calpain localization changes during early embryonic development of *Drosophila*. *J Biol Chem* 269: 25137–25142
33 Theopold U, Pintér M, Daffre S, Tryselius Y, Friedrich P, Nässel DR, Hultmark D (1995) *CalpA*, a *Drosophila* calpain homolog specifically expressed in a small set of nerve, midgut, and blood cells. *Mol Cell Biol* 15: 824–834
34 Waterson R, Martin C, Craxton M, Huynh C, Coulson A, Hillier L, Durbin R, Green P, Shownkeen R, Halloran N et al. (1992) A survey of expressed genes in *Caenorhabditis elegans*. *Nat Genet* 1: 114–123
35 Wilson R, Ainscough R, Anderson K, Baynes C, Berks M, Bonfield J, Burton J, Connell M, Copsey T, Cooper J et al. (1994) 2.2 Mb of contiguous nucleotide sequence from chromosome III of *C. elegans*. *Nature* 368: 32–38
36 Barnes TM, Hodgkin J (1996) The *tra-3* sex determination gene of *Caenorhabditis elegans* encodes a member of the calpain regulatory protease family. *EMBO J* 15: 4477–4484
37 Andresen K, Tom TD, Strand M (1991) Characterization of cDNA clones encoding a novel calcium-activated neutral proteinase from *Schistosoma mansoni*. *J Biol Chem* 266: 15085–15090
38 Sorimachi H, Saido TC, Suzuki K (1994) New era of calpain research: Discovery of tissue-specific calpains. *FEBS Lett* 343: 1–5
39 Nishizuka Y (1992) Intracellular signaling by hydrolysis of phospholipids and activation of protein kinase C. *Science* 258: 607–614
40 Saido TC, Shibata M, Takenawa T, Murofushi H, Suzuki K (1992) Positive regulation of μ-calpain action by polyphosphoinositides. *J Biol Chem* 267: 24585–24590
41 Maki M, Ma H, Takano E, Adachi Y, Lee WJ, Hatanaka M, Murachi T (1991) Calpastatins: biochemical and molecular biological studies. Biomed Biochim Acta 50: 509–516
42 Kawasaki H, Kawashima S (1996) Regulation of the calpain-calpastatin system by membranes. Mol Membr Biol 13: 217–224
43 Takano E, Ma H, Yang HQ, Maki M, Hatanaka M (1995) Preference of calcium-dependent interactions between calmodulin-like domains of calpain and calpastatin subdomains. *FEBS Lett* 362: 93–97
44 Elce JS, Hegadorn C, Gauthier S, Vince JW, Davies PL (1995) Recombinant calpain II: improved expression systems and production of a C105A active-site mutant for crystallography. *Prot Engl* 8: 843–848
45 Meyer SL, Bozyczko-Coyne D, Mallya SK, Spais CM, Bihovsky R, Kaywooya JK, Lang DM, Scott RW, Siman R (1996) Biologically active monomeric and heterodimeric recombinant human calpain I produced using the baculovirus expression system. *Biochem J* 314: 511–519
46 Elce JS, Hegadorn C, Arthur JSC (1997) Autolysis, Ca^{2+} requirement, and heterodimer stability in m-calpin. *J Biol Chem* 272: 11268–11275
47 Seubert P, Lee K, Lynch G (1989) Ischaemia triggers NMDA receptor-linked cytoskeletal proteolysis in hippocampus. *Brain Res* 492: 366–370
48 Saido TC, Yokota M, Nagao S, Yamaura I, Tani E, Tsuchiya T, Suzuki K, Kawashima S (1993) Spatial resolution of fodrin proteolysis in postischaemic brain. *J Biol Chem* 268: 25239–25234

49 Yamashima T, Saido TC, Takita M, Miyazawa A, Yamano J, Miyakawa A, Nishijo H, Yamashita J, Kawashima S, Ono T, Yoshioka T (1996) Transient brain ischaemia provokes Ca^{2+}, PIP_2 and calpain responses prior to delayed neuronal death in monkeys. *Eur J Neurosci* 8: 1932–1944
50 Wang KKW, Yuen PW (1994) Calpain inhibition: an overview of its therapeutic potential. *Trends Pharmacol Sci* 15: 412–419
51 Saatman KE, Murai H, Bautus R, Smith DH, Hayward NJ, Perri BR, Mcintosh TK (1996) Calpain inhibitor AK295 attenuates motor and cognitive deficits following experimental brain injury in the rat. *Proc Natl Acad Sci USA* 93: 3428–3433
52 Eto A, Akita Y, Saido TC, Suzuki K, Kawashim S (1995) The role of the calpain-calpastatin system in thyrotropin-releasing hormone-induced selective down-regulation of a protein kinase C isozyme, nPKCε, in rat pituitary GH_4C_1 cells. *J Biol Chem* 270: 25115–25120
53 Pontremoli S, Melloni E, Sparatore B, Michetti M, Salamino F, Horecker BL (1990) Isozymes of protein kinase C in human neutrophils and their modification by two endogenous proteinases. *J Biol Chem* 265: 706–712
54 Pariat M, Carillo S, Molinari M, Salvat C, Debussche L, Bracco L, Milner J, Piechaczyk M (1997) Proteolysis by calpains: a possible contribution to degradation of p53. *Mol Cell Biol* 17: 2806–2815
55 Bi X, Chen J, Dang S, Wenthold RJ Tocco G, Baudry M (1997) Characterization of calpain-mediated proteolysis of GluR1 subunits of alpha-amino-3-hydroxy-5-methylisoxazole-4-propionate receptors in rat brain. *J Neurochem* 68: 1484–1494
56 Arora AS, De Groen PC, Croall DE, Emori Y, Gores GJ (1996) Hepatocellular carcinoma cells resist necrosis during anoxia by preventing phospholipase-mediated calpain activation. *J Cell Physiol* 167: 434–442
57 Inomata M, Hayashi M, Ohno-Iwashita Y, Tsubuki S, Saido TC, Kawashim S (1996) Involvement of calpain in integrin-mediated signal transduction. *Arch Biochem Biophys* 337: 232–238
58 Sorimachi H, Kinbara K, Kimura S, Takahashi M, Ishiura S, Sasagawa S, Sorimachi N, Shimada H, Tagawa K, Maruyama K, Suzuki K (1995) Muscle-specific calpain, p94, responsible for limbgirdle muscular dystrophy type 2A, associates with connectin through IS2, a p94-specific sequence. *J Biol Chem* 270: 31158–31162
59 Richard I, Broux O, Allamand V, Fougerousse F, Chiannilkulchai N, Bourg N, Brenguier L, Davaud C, Pasturaud P, Roudaut C et al. (1995) Mutations in the proteolytic enzyme calpain 3 cause limb-girdle muscular dystrophy type 2A. *Cell* 81: 27–40
60 Sorimachi H, Toyama-Sorimachi N, Saido TC, Kawasaki H, Sugita H, Miyasaka M, Arahata K, Ishiura S, Suzuki K (1993) Muscle-specific calpain, p94, is degraded by autolysis immediately after translation, resulting in disappearance from muscle. *J Biol Chem* 268: 10593–10605
61 Fardeau M, Eymard B, Mignard C, Tomé FMS, Richard I, Beckmann JS (1996) Chromosome 15-linked limb-girdle muscular dystrophy: clinical phenotypes in reunion island and French metropolitan communities. *Neuromuscular Disord* 6: 447–453
62 Spencer MJ, Tidball JG, Anderson LV, Bushby KM, Harris JB, Passos-Bueno MR, Somer H, Vainzof M, Zatz M (1997) Absence of calpain 3 in a form of limb-girdle muscular dystrophy (LGMD2A). *J Neurol* 146: 173–178
63 Maruyama K (1997) Connectin/titin, giant elastic protein of muscle. *FASEB J* 11: 341–345
64 Labeit S, Kolmerer B, Linke WA (1997) The giant protein titin: emrging roles in physiology and pathophysiology. *Circ Res* 80: 290–294
65 Labeit S, Kolmerer B (1995) Titins: giant proteins in charge of muscle ultrastructure and elasticity. *Science* 270: 293–296
66 Kinabara K, Sorimachi H, Ishiura S, Suzuki K (1997) Muscle-specific calpain, p94, interacts with the extreme C-terminal region of connectin, a unique region flanked by two immunoglobulin C2 motifs. *Arch Biochem Biophys* 342: 99–107
67 Balcerzak D, Poussard S, Brustis JJ, Elamrani N, Soriano M, Cottin P, Ducastaing A (1995) An antisense oligodeoxyribonucleotide to m-calpain mRNA inhibits myoblast fusion. *J Cell Sci* 108: 2077–2082
68 Poussard S, Duvert M, Balcerzak D, Ramassamy S, Brustis JJ, Cottin P, Ducastaing A (1996) Evidence for implication of muscle-specific calpain (p94) in myofibrillar integrity. *Cell Growth Differ* 7: 1461–1469
69 Peckham M, Young P, Gautel M (1997) Constitutive and variable regions of Z-disk titin/connectin in myofibril formation: a dominant negative screen. *Cell Struct Func* 22: 95–101
70 Spencer MJ, Croall DE, Tidball JG (1995) Calpains are activated in necrotic fibers from mdx dystrophic mice. *J Biol Chem* 270: 10909–10914
71 Kuwabara PE, Okkema PG, Kimble J (1992) tra-2 encodes a membrane protein and may mediate cell communication in the *Caenorhabditis elegans* sex determination pathway. *Mol Biol Cell* 3: 461–473
72 Spence AM, Coulson A, Hodgkin J (1990) The product of fem-1, a nematode sex-determining gene, contains a motif found in cell cycle control proteins and receptors for cell-celll interactions. *Cell* 60: 981–990
73 Siddiqui AA, Zhou Y, Podesta RB, Karcz SR, Tognon CE, Strejan GH, Dakaban GA, Clarke MW (1993) Characterization of Ca^{2+}-dependent neutral protease (calpain) from human blood flukes, *Schistosoma mansoni. Biochim Biophys Acta* 1181: 37–44
74 Jankovic D, Åslund L, Oswald IP, Caspar P, Champion C, Pearce E, Coligan JE, Strand M, Sher A, James SL (1996) Calpain is the target antigen of a Th1 clone that transfers protective immunity against *Schistosoma mansoni. J Immunol* 157: 806–814

75 Denison SH, Orejas M, Arst HNJr, (1995) Signaling of ambient pH in *Aspergillus* involves a cysteine protease. *J Biol Chem* 270: 28519–28522
76 Tilburn J, Sarkar S, Widdick DA, Espeso EA, Orejas M, Mungroo J, Peñalva MA, Arst HN Jr (1995) The *Aspergillus* PacC zinc finger transcription factor mediates regulation of both acid- and alkaline-expressed genes by ambient pH. *EMBO J* 14: 779–790

Proteases: New Perspectives
V. Turk (ed.)

Arg-gingipain and Lys-gingipain: a novel class of cysteine proteinases

Kenji Yamamoto, Tomoko Kadowaki and Kuniaki Okamoto

Department of Pharmacology, Kyushu University Faculty of Dentistry, Higashi-ku, Fukuoka 812-8582, Japan

Introduction

Porphyromonas gingivalis, a Gram-negative anaerobic bacterium, is thought to be a major etiologic bacterium associated with several periodontal diseases including chronic adult periodontitis, generalized juvenile periodontitis, periodontal abscess and refractory periodontitis. This bacterium produces various potent virulence factors in both the cell-associated and secretory forms, such as fimbriae, lipopolysaccharides, hemagglutinins and hydrolytic enzymes [1, 2]. Among these factors, proteolytic enzymes have attracted intense interest and led to a vigorous search for agents to prevent or attenuate the virulence of the bacterium, since they are believed to be involved in a wide range of pathologies of progressive periodontal disease [3, 4]. Studies on *P. gingivalis* proteinases started in the middle 1980's when various proteolytic activities, such as trypsin-like activity, glycylprolyl peptidase activity, IgA degrading activity, and collagenolytic activity [5–11], were identified and characterized in different *P. gingivalis* strains. However, pathophysiological functions of such proteinase activities in the progression of the disease have not yet been established. It is also not clear whether the respective activities are attributable to single proteases or whether these enzymes have overlapping functions. Most of the enzymes that have been referred to as trypsin-like enzymes are found to be cysteine proteinases with unusual stimulation by glycine derivatives, apparent inhibition by metal chelators, and activation by thiol-reducing agents in an unusual order of preference [12–16]. Although early attempts to isolate and characterize such trypsin-like proteinases were considered to be completed, it was difficult to compare these enzymes isolated and characterized by different groups in the early studies, since different assay systems were used by each group and since no structural data had been ever presented. More recently, the trypsin-like activity associated with the bacterium was found to be attributable to either Arg-X- or Lys-X-specific cysteine proteinases [17]. These enzymes have now been termed Arg-gingipain (gingipain-R, RGP) and Lys-gingipain (gingipain-K, KGP) on the basis of their peptide cleavage specificity after arginine and lysine residues, respectively. These two enzymes appear to represent a new class of the cysteine proteinase family, since they have novel structural features distinct from other known cysteine proteinases. The present review summarizes the most recent results in the research on RGP and KGP and discusses some important points concerning their pathophysiological roles.

Enzymatic properties of Arg-gingipain and Lys-gingipain

We have recently isolated RGP (formerly, argingipain) and KGP, with both bacterial and collagenolytic activities, from the culture supernatant of the *P. gingivalis* 381 strain [3, 18, 19]. The corresponding proteinases have also been isolated from either the culture supernatant or the cell extracts of various *P. gingivalis* strains by other groups [20–25]. The purified RGP and KGP showed single protein bands on SDS-polyacrylamide gel electrophoresis with apparent molecular masses of near 44 kDa and 51 kDa, respectively. These masses were consistent with those of the native enzymes estimated by gel filtration, indicating that they are composed of single polypeptide chains. Both enzymes have pH optima in the neutral to alkaline range and a requirement for thiol-reducing agents for activation. Analyses of the enzymatic properties revealed several unique features for both enzymes. For example, RGP and KGP have in part the characteristics of serine and/or metalloendopeptidases (Tab. 1). Both enzymes are strongly inhibited by the chloromethylketones of tosyl-L-lysine (TLCK) and tosyl-L-phenylalanine (TPCK) in concentrations of 1 mM, although they are scarcely affected by the general serine protease inhibitors diisopropylfluorophosphate (DFP) and phenylmethylsulfonyl fluoride (PMSF). Metal chelators, such as EDTA, EGTA, and phosphoramidon, are powerful inhibitors for RGP, but not for KGP. Significant differences in susceptibility to other various protease inhibitors are observed between the two enzymes. RGP is intensely inhibited by chymostatin, leupeptin, and antipain, whereas KGP is not sensitive to these inhibitors. It is of special interest to note that internal protease inhibitors, such as cystatins, tissue inhibitors of metalloproteinases (TIMP), and serpins, have no effects on either enzyme activity, suggesting their evasion of normal host defense systems. In addition, RGP and KGP have a narrow specificity for synthetic substrates, limited to peptide bonds containing arginine and lysine residues, respectively, but they can nevertheless unlimetedly degrade various protein substrates, including collagens (types I and IV) and hemoglobin. Both enzymes are also capable of degrading immunoglobulins G and A in a limited degradation manner [3, 18, 19]. RGP also degrades the complement factors C3 and C5 [26, 27]. In this connection, both enzymes have the ability to disrupt the function of polymorphonucear leukocytes, as is evident from their inhibitory effects on the generation of active oxygen species from the activated cells [3, 18]. These findings strongly suggest that both RGP and KGP are closely associated with the disruption of normal host defense mechanisms by *P. gingivalis.*

Structural characterization of Arg-gingipain and Lys-gingipain

Recently, a variety of *P. gingivalis* genes encoding Arg- and Lys-X-specific proteinases have been cloned and sequenced. Eight genes encoding Arg-X-specific proteinases, including *prtH* [28], *agp* (recently renamed *rgpA*) [29], *rgp1* [30], *prpR1* [31], *prtR* [32], *cpgR* [33], *prtP* [34], and *rgpB* [35], have so far been described. A comparison of these nucleotide sequences has been described previously [34, 36]. These genes were found to be highly homologous to one another, with the possible exception of *prtH* and *prtP*, and suggested to be derived from a single *P. gingivalis* genomic locus. The alignment of the deduced amino acid sequences of these

Table 1 Effects of various compounds on the activities of KGP and RGP

Compounds	Concentration	% of Activity remaining	
		RGP	KGP
None	-	100	100
iodoacetic acid	1 mM	33	24
iodoacetamide	1 mM	93	8
leupeptin	50 µg/ml	0	43
E-64	50 µg/ml	4	78
antipain	50 µg/ml	7	84
chymostatin	50 µg/ml	2	115
elastatinal	50 µg/ml	83	89
TPCK	1 mM	5	1
TLCK	1 mM	20	4
PMSF	1 mM	76	94
DFP	1 mM	111	134
EDTA	1 mM	18	110
phosphoramidon	1 mM	46	112
pepstatin	50 µg/ml	112	87
cystatin (egg white)	5.6 µg/ml	-	99
	50 µg/ml	118	-
cystatin S (human)	6.8 µg/ml	-	92
	50 µg/ml	106	-
α1-antichymotrypsin	29 µg/ml	-	98
	50 µg/ml	114	-
TIMP-1	1.3 µg/ml	121	-
TIMP-2	1.0 µg/ml	116	-

The purified enzymes were preincubated with various compounds at the indicated concentrations (37 °C for 5 min) before a 10 min incubation with the respective synthetic substrates. The values are expressed as percentage of the activity determined in the presence of 5 mM cysteine.

gene products is outlined in Figure 1. The amino-terminal regions of these products, which include the prepropeptides and the proteinase active sites, are closely similar to each other, but the carboxy-terminal regions that are thought to be responsible for the hemagglutination differ significantly. The *rgpA* gene lacks a sequence intervening between two direct repeats in the carboxy-terminal domain of the *rgp1* gene. The *cpgR* gene is also missing the carboxy-terminal domain found in other genes. More recently, we have found that *P. gingivalis* possesses two separate genes responsible for RGP activity in its chromosome [37]. The large *rgpA* and the small *rgpB* gene exist on 12.5 and 7.8 kbp *Hind*III chromosomal fragments of the *P. gingivalis* 381 strain. The nucleotide sequence of *rgpB* and the deduced amino acid sequence are essentially identical with those of *rgpA*, except that the carboxy-terminal sequence corresponding to

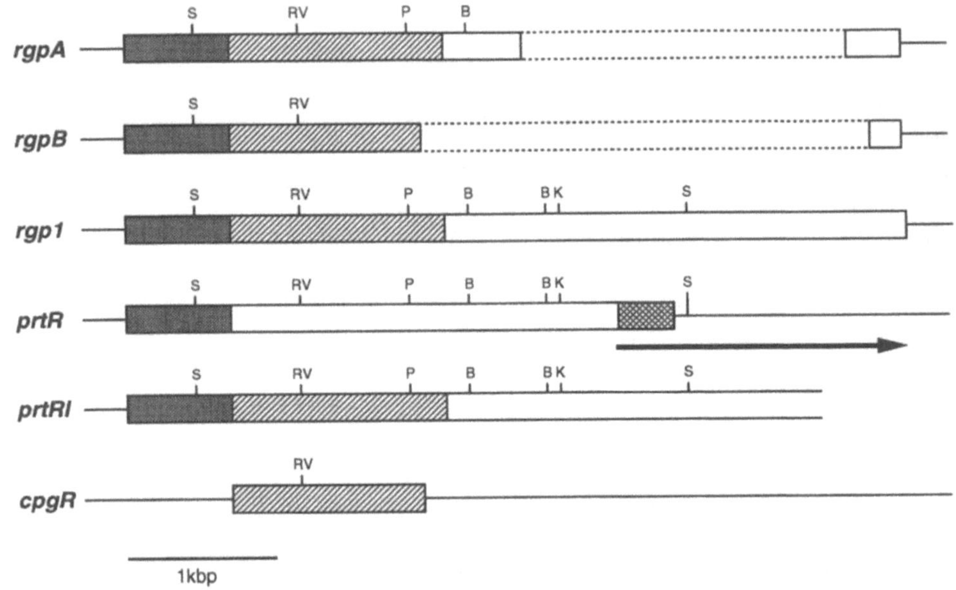

Figure 1. Comparison of the gene structures of Arg-X-specific cysteine proteinases from *Porphyromonas gingivalis* strains. The boxes represent the predicted translated regions encoding the prepropeptides (the closed boxes), the catalytic domains (the hatched boxes), and the carboxy-terminal hemagglutinin domains (the open boxes). The dashed open boxes represent the missing parts in the *rgpA* and *rgpB* genes (381 strain) when compared with the *rgp1* gene (H66 strain). Selected restriction sites are shown above each gene: S, *Sma*I; RV, *Eco*RV; P, *Pst*I; B, *Bam*HI; K, *Kpn*I; RI, *Eco*RI. The 3' end of the *prtR1* gene (W50 strain) is open because the stop codon has not been identified. The arrow shown in the 3' end of the *prtR* gene (W50) (the crossed stripes) indicates the postulated overlapping, but shifted, reading frame corresponding to the 5' region of the *prtH* gene [28].

the sequence Tyr^{654}–Val^{912} of the *rgpA* product [29] is missing in the *rgpB* product [35] (Fig. 1). The deduced amino acid sequences of both *rgpA* and *rgpB*, as well as other Arg-X-specific proteinase-encoding genes, have no significant similarity to the sequences of any known cysteine proteinases. The putative catalytic site sequences of RGP, containing the potential reactive cysteine, histidine, and asparagine residues, also exhibit no homology to those of other known cysteine proteinases. Further, the deduced amino acid sequences of the *rgpA*, *rgp1*, *prtR* and *prpR1* genes indicate that the primary translation products comprise at least four domains: the signal peptide required for the inner membrane transport of the bacterium, the amino-terminal propeptide which is assumed to stabilize the proprotein structure during transport, the catalytic proteinase domain, and the carboxy-terminal hemagglutinin domain. Thus, RGP may belong to a new class of the cysteine proteinase family.

So far, three genes encoding Lys-X-specific proteinases have been described [34, 38, 39]. The alignment of these nucleotide sequences is outlined in Figure 2. These genes are highly homologous to one another, with the exception of significant differences in the carboxy-terminal hemagglutinin domain. Although porphypain has both Arg-X- and Lys-X-specific hydrolyzing activities [25], the nucleotide sequence of its gene (*prtP*) and the deduced amino acid

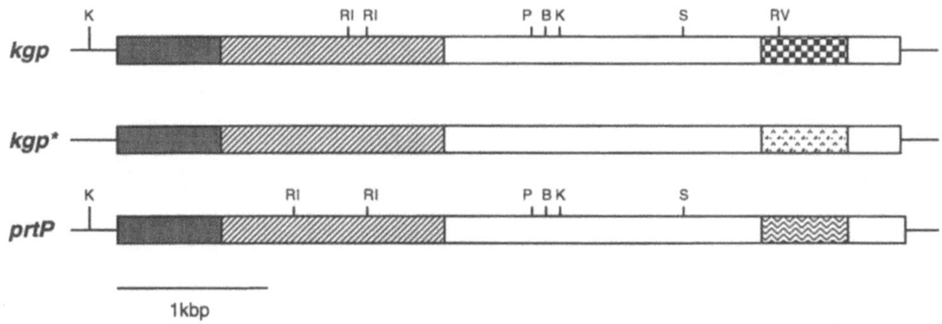

Figure 2. Comparison of the gene structure of Lys-X-specific cysteine proteinases from *Porphyromonas gingivalis* strains. The boxes represent the predicted translated regions, encoding the prepropeptides (the closed boxes), the catalytic domain (the hatched boxes), and the carboxy-terminal hemagglutinin domains (the open boxes). Selected restriction sites are shown above each gene: K, *Kpn*I; RI, *Eco*RI; P, *Pst*I; B, *Bam*HI; S, *Sma*I; RV, *Eco*RV. Significant differences among the three genes observed in the vicinity of the carboxy-terminal ends are marked by different boxes. The asterisk indicates the gene structure of lysine-specific gingipain by Pavloff et al. [39].

sequence are more like the sequences of Lys-X-specific proteinase-encoding genes and their products than those of Arg-X-specific proteinase-encoding genes and their products. The nucleotide sequence of the *kgp* gene determined in our laboratory has a 5169 bp open reading frame encoding 1723 amino acids with a calculated molecular mass of 218 kDa [38]. KGP appears to be similar to RGP in its structural organization, biosynthesis, and maturation processing. The most striking feature of the deduced amino acid sequence of KGP is the presence of three large repeats of homologous sequences, including the two different sequences LKWD(orE)AP and YTYTVYRDGTKI, in the carboxy-terminal domain. A similar multiple homologous sequence is also found in the carboxy-terminal domain of RGP [30]. To compare the structures of the carboxy-terminal domains, as well as the prepropeptides and the proteinase domains, the nucleotide sequences of *kgp* and *rgp1* genes were divided into eight regions on the basis of the similarity in sequence and length between the two genes (Fig. 3). The A region, which includes the prepropeptide and most of the proteinase domain, has no similarity between the two genes, but there is close similarity (more than 94% identity) in nucleotide sequence between the other regions (B, D, E, F, H, and 33 bp after the terminal codon) of both genes. In addition, the D region is essentially identical to the F region. The C region of *kgp* is almost identical to the G region of *rgp1*. Thus, recombinational rearrangement such as transposition or gene conversion has been suggested to occur in the carboxy-terminal region between the two genes. Since each of the two direct repeats is located at each end of the C and G regions, these nucleotide repeats may have assisted in this rearrangement.

Pathophysiological roles of Arg-gingipain and Lys-gingipain

RGP and/or KGP are believed to act as virulence factors of progressive periodontal disease in various ways: (i) by directly degrading structural proteins of the periodontal tissue (e.g. colla-

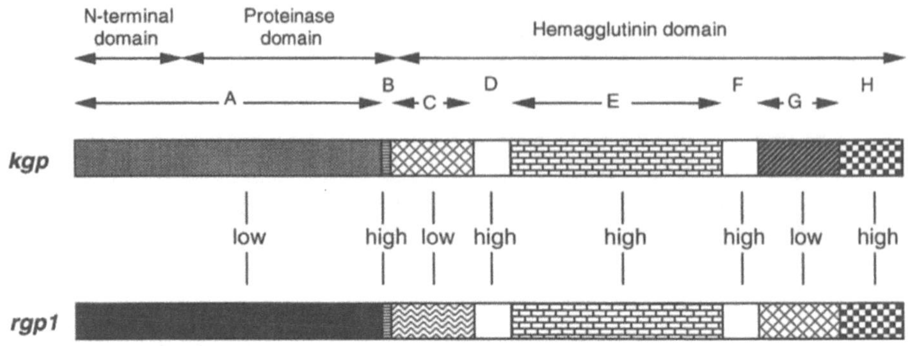

Figure 3. Structural similarity between the *kgp* gene (381) and the *rgp1* gene (H66). Nucleotide sequences of the *kgp* and *rgp1* genes were divided into eight portions based on the similarity in sequences and lengths between the two genes(A, the nucleotide number of the open reading frame of *kgp* 1–2041; B, 2042–2107; C, 2108–2659; D, 2660–2884; E, 2885–4018; F, 4019–4240; G, 4241–4792; H, 4793–5170). The A region includes the sequences encoding the prepropeptide and most of the catalytic domain. The A and G regions of the *kgp* gene are not similar to the corresponding regions of the *rgp1* gene (less than 40% identity). In contrast, the regions of B, D, E, F, and H of the *kgp* gene, as well as 33 bp after the terminal codon, are highly homologous to the corresponding regions of the *rgp1* gene (more than 94% identity).

gens, fibronectin, and fibrinogen) [3, 6, 9, 10, 18, 19, 21–23, 40, 41], (ii) by disrupting host defense mechanisms through (a) suppression of the bactericidal activity of polymorphonuclear leukocytes (PMNs) [3, 18, 19], (b) cleavage of immunoglobulins [3, 11, 18, 19, 26, 42] and complement factors [27, 43–46], and (c) degradation of serum proteinase inhibitors or evasion of inactivation by them [3, 18, 19, 26, 47, 48], (iii) by activating or stimulating the expression of hemagglutinins [37], (iv) by processing and translocating adhesion molecules (e.g. fimbrilin and 75-kDa surface protein) [49], and (v) by inducing or stimulating inflammation through the production of chemical mediators [e.g. kinin and platelet activatig factor (PAF)] [23, 50, 51]. Although the importance of RGP and KGP for the pathogenicity of *P. gingivalis* is thus well understood, it is still not clear as to what extent each enzyme contributes to the entire virulence of the organism. To pursue this issue, RGP- and KGP-deficient mutants have been constructed, by gene disruption using suicide plasmids containing an internal DNA fragment of each gene, and analyzed for their biological properties [3, 37, 52]. The proteolytic activity of RGP on synthetic substrates was markedly eliminated in both the culture supernatants and cell extracts of two single RGP mutants (the *rgpA* or the *rgpB* mutant), but significant activity was still retained in both fractions. The *rgpA rgpB* double mutant, however, showed complete loss of the activity in both its culture supernatant and cell extract. When the protein substrates casein and hemoglobin were used for assay, only 5–10% of the hydrolytic activity found in the culture supernatant of the wild-type strain was detected in the culture supernatant of the double mutant, indicating that RGP is a major extracellular proteinase of *P. gingivalis*. On the other hand, the *kgp* mutant also showed complete loss of the proteolytic activity for KGP.

The *rgpA*, *rgpB kgp* triple mutant showed neither RGP nor KGP activity in both cell extracts and culture supernatants [52]. The culture supernatant of the triple mutant had no proteolytic activity toward gelatin, bovine serum albumin and human immunoglobulin. These results indicate that extracellular protease activity of *P. gingivalis* may be totally attributable to RGP and KGP. As *P. gingivalis* is asaccharolytic and utilizes amino acids and peptides as carbon/energy sources, extracellular proteases are essential for viability of the microorganism. The triple mutant did not grow in a defined medium containing bovine serum albumine as a carbon/energy source [52].

As observed previously, *P. gingivalis* secretes potent virulence factors disrupting the bactericidal function of PMNs [53], and RGP and KGP have been shown to possess this bactericidal activity *in vitro* [3, 18]. Thus, to what extent RGP and KGP contribute to the inhibition of the PMN bactericidal activity was elucidated by use of these mutants. The culture supernatants from both the *rgpA* and the *rgpB* single mutants showed a decrease in the inhibition of PMN bactericidal activity observed with the wild-type strain but still retained this activity. In contrast, this bactericidal activity was almost completely lost in the culture supernatant of the *rgpA rgpB* double mutant. The culture supernatant of the *kgp* mutant exhibited the inhibition of PMN bactericidal activity equivalent to that of the wild-type *P. gingivalis*. Therefore, it has been postulated that production of the virulence factors responsible for disruption of the bactericidal activity of PMNs is attributable to the RGP function.

Protoheme is an essential requirement for the growth of *P. gingivalis*. As protoheme is thought to be produced from erythrocytes in the periodontal pockets, it is particularly important for the organism to agglutinate and lyse erythrocytes in order to survive *in vivo*. As previously described, the carboxy-terminal domains of both *rgp* (but not *rgp*B) and *kgp* gene products contain the sequences identical to the amino-terminal sequences of *P. gingivalis* hemagglutinins. Therefore, experiments to determine to what extent *rgp* and *kgp* genes contribute to hemagglutination of the organism were carried out with the respective mutants [3, 37]. The *rgpA* and *rgpB* single mutants each showed only a small decrease in hemagglutination, as compared with the wild-type strain. Similarly, little or no effect on hemagglutination was observed with the *kgp* mutant. In contrast, the *rgpA rgpB* double mutant exhibited a greater decrease in hemagglutination to that observed with the wild-type strain. These results indicate that most of the hemagglutination activity of *P. gingivalis* depends on the *rgp* genes and their products.

The bacterial colonization in gingival tissues that precedes bacterial penetration and tissue destruction is critical in the pathogenic process of periodontal disease. A variety of the cell surface structures (e.g. fimbriae and vesicles) and potential molecular adhesins (e.g. lectins, hemagglutinins and lipopolysaccharides) are thought to play an important role in attachment of *P. gingivalis* to host cells and matrix proteins. In particular, fimbriae are suggested to act as a key factor facilitating the initial interaction between organism and host [54]. It should be noted that, ultrastructurally, the wild-type *P. gingivalis* strain possesses a number of characteristic curly fimbriae extending radially from the cell surface, while the *rgpA rgpB* double mutant, but not the *rgpA* or *rgpB* single mutants and the *kgp* mutant, showed only a few fimbriae on the cell surface [49]. Western blot analysis with antibodies to fimbrilin, a major subunit protein of fimbriae with a molecular mass of 43 kDa in gels [55], revealed that the double mutant produced a major 45 kDa and a minor 43 kDa band, corresponding to those of the precursor and mature forms of fimbrilin, respectively. Further, hydrolysis of the recombinant prefimbrilin by RGP *in vitro* followed by amino-terminal sequence analysis of the cleavage fragments has shown that the processing of prefimbrilin to the mature fimbrilin is mediated by RGP [55]. Thus, the defect in fimbriation in the *rgp* double mutant may be caused by a deficiency in the processing of prefimbrilin by RGP. Similarly, RGP has been shown to contribute to processing of the precursor to mature forms of the 75-kDa protein, one of the major outer membrane proteins of *P. gingivalis* that is associated with the host-bacterium interaction. It has also been suggested that the precursor of KGP is processed by RGP [55]. Therefore, RGP is involved in processing and translocation of certain cell surface proteins and secretory proteins of *P. gingivalis*.

As described previously, heme is an essential requirement for growth of *P. gingivalis*. When grown in blood agar, *P. gingivalis* stockpiles heme from hemoglobin in blood agar, resulting in black pigmentation of their colonies [56]. Although all the *rgp* mutants show black pigmentation of these colonies, the *kgp* mutant can not form this pigmentation [57]. Therefore, KGP is suggested to be involved in the adsorption and degradation of hemoglobin and in heme accumulation in the organism.

Acknowledgments
Contributions to the experimental work and valuable discussions by Drs. Nakayama K, Yoneda M, Yoshimura F, Misumi Y, and Ikehara Y are gratefully acknowledged. Studies in the laboratory of the author were supported in part by a Grant-in-Aid for Scientific Research from the Ministry of Education, Science and Culture of Japan.

References

1 Mayrand D, Holt SC (1988) Biology of asaccharolytic black-pigmented *Bacteroides* species. *Microbiol Rev* 52: 134–152
2 Grenier D, Mayrand (1993) Proteinases. *In*: HN Shah, D Mayrand, RJ Genco (eds): *Biology of the Species Porphyromonas gingivalis*. CRC Press, Boca Raton, 227–243
3 Yamamoto K, Kadowaki T, Okamoto K, Abe N, Nakayama K (1997) Biological roles of a novel class of cysteine proteinases from *Porphyromonas gingivalis* in periodontal disease progression. *In*: N Katunuma, H Kido, H Fritz, J Travis (eds): *Medical Aspects of Proteases and Protease Inhibitors*. IOS Press, Amsterdam, 139–149
4 Travis J, Pike R, Imamura T, Potempa J (1997) *Porphyromonas gingivalis* proteinases as virulence factors in the development of periodontitis. *J Periodont Res* 32: 120–125
5 Yoshimura F, Nishikata M, Suzuki T, Hoover CI, Newbrun E (1984) Characterization of a trypsin-like protease from the bacterium *Bacteroides gingivalis* isolated from human dental plaque. *Arch Oral Biol* 29: 559–564
6 Toda K, Otsuka M, Ishikawa Y, Sato M, Yamamoto Y, Nakamura R (1984) Thiol-dependent collagenolytic activity in culture media of *Bacteroides gingivalis*. *J Periodont Res* 19: 372–381
7 Abiko Y, Hayakawa M, Murai S, Takiguchi H (1985) Glycylprolyl dipeptidylamino-peptidase from *Bacteroides gingivalis*. *J Dent Res* 64: 106–111
8 Fujimura S, Shibata Y, Nakamura T (1992) Comparative studies of three proteases of *Porphyromonas gingivalis*. *Oral Microbiol Immunol* 7: 212–217
9 Lawson DA, Meyer TF (1992) Biochemical characterization of *Porphyromonas (Bacteroides) gingivalis* collagenase. *Infect Immunity* 60: 1524–1529
10 Sundqvist G, Carlsson SG, Hanstrom L (1987) Collagenolytic activity of black-pigmented *Bacteroides* species. *J Periodont Res* 22: 300–306
11 Frandsen EVG, Reinholdt JR, Kilian M (1987) Enzymatic and antigenic characterization of immunoglobulin A1 proteases from *Bacteroides* and *Capnocytophaga* spp *Infect Immunity* 55: 631–638
12 Tsutsui H, Kinouchi T, Wakano Y, Ohnishi Y (1987) Purification and characterization of a protease from *Bacteroides gingivalis* 381. *Infect Immunity* 55: 420–427
13 Ono M, Okuda K, Takazoe I (1987) Purification and characterization of a thiol-protease from *Bacteroides gingivalis* 381. *Oral Microbiol Immunol* 2: 77–81
14 Otsuka M, Endo J, Hinode D, Nagata A, Maehara R, Sato M, Nakamura R (1987) Isolation and characteri-

zation of protease from the culture supernatant of *Bacteroides gingivalis*. *J Periodont Res* 22: 491–498
15 Sorsa T, Uitto V-J, Suomalainen K, Turto H, Lindy S (1987) A trypsin-like protease from *Bacteroides gingivalis*: partial purification and characterization. *J Periodont Res* 22: 375–380
16 Grenier D, McBride BC (1987) Isolation of a membrane-associated *Bacteroides gingivalis* glycylprolyl protease. *Infect Immunity* 55: 3131–3136
17 Potempa J, Pike R, Travis J (1995) The multiple forms of trypsin-like activity present in various strains of *Porphyromonas gingivalis* are due to the presence of either arg-gingipain or lys-gingipain. *Infect Immunity* 63: 1176–1182
18 Kadowaki T, Yoneda M, Okamoto K, Maeda K, Yamamoto K (1994) Purification and characterization of a novel arginine-specific cysteine proteinase (argingipain) involved in the pathogenesis of periodontal disease from the culture supernatant of *Porphyromonas gingivalis*. *J Biol Chem* 269: 21371–21378
19 Abe N, Kadowaki T, Okamoto K, Nakayama K, Ohishi M, Yamamoto K (1998) Biochemical and functional properties of lysine-specific cysteine proteinase (lys-gingipain) as a virulence factor of *Porphyromonas gingivalis* in periodontal disease. *J Biochem* 123: 305–312
20 Chen Z, Potempa J, Polanowski A, Wikstrom M, Travis J (1992) Purification and characterization of a 50-kDa cysteine proteinase (gingipain) from *Porphyromonas gingivalis*. *J Biol Chem* 267: 18896–18901
21 Sojar HT, Lee J-Y, Bedi GS, Genco RJ (1993) Purification and characterization of a protease from *Porphyromonas gingivalis* capable of degrading salt-solubilized collagen. *Infect Immunity* 61: 2369–2376
22 Bedi GS, William T (1994) Purification and characterization of a collagen-degrading protease from *Porphyromonas gingivalis*. *J Biol Chem* 269: 599–606
23 Scott CF, Whitaker EJ, Hammond BF, Colman RW (1993) Purification and characterization of a potent 70-kDa thiol lysyl-proteinase (lys-gingivain) from *Porphyromonas gingivalis* that cleaves kininogens and fibrinogens. *J Biol Chem* 268: 7935–7942
24 Pike R, McGraw W, Potempa J, Travis J (1994) Lysine- and arginine-specifc proteinases from *Porphyromonas gingivalis*: isolation, characterization, and evidence for the existence of complexes with hemagglutinins. *J Biol Chem* 269: 406–411
25 Ciborowski P, Nishikata M, Allen RD, Lantz MS (1994) Purification and characterization of two forms of a high-molecular weight cysteine proteinase (porphypain) from *Porphyromonas gingivalis*. *J Bacteriol* 176: 4549–4557
26 Grenier D (1992) Inactivation of human serum bactericidal activity by a trypsin-like protease isolated from *Porphyromonas gingivalis*. *Infect Immunity* 60: 1854–1857
27 Wingrove JA, DiScipio RG, Chen Z, Potempa J, Travis J, Hugli TE (1992) Activation of complement components C3 and C5 by a cysteine proteinase (gingipain-1) from *Porphyromonas* (*Bacteroides*) *gingivalis*. *J Biol Chem* 267: 18902–18907
28 Fletcher HM, Schenkein HA, Macrina FL (1994) Cloning and characterization of a new protease gene (*prtH*) from *Porphyromonas gingivalis*. *Infect Immunity* 62: 4279–4286
29 Okamoto K, Misumi Y, Kadowaki T, Yoneda M, Yamamoto K, Ikehara Y (1995) Structural characterization of argingipain, a novel arginine-specific cysteine proteinase as a major periodontal pathogenic factor from *Porphyromonas gingivalis*. *Arch Biochem Biophys* 316: 917–925
30 Pavloff N, Potempa J, Pike RN, Prochazka V, Kiefer MC, Travis J, Barr PJ (1995) Molecular cloning and structural characterization of the arg-gingipain proteinase of *Porphyromonas gingivalis*: biosynthesis as a proteinase-adhesion polyprotein. *J Biol Chem* 270: 1007–1010
31 Aduse-Opoku J, Muir J, Slaney JM, Rangarajan M, Curtis MA (1995) Characterization, genetic analysis, and expression of a protease antigen (PrpRI) of *Porphyromonas gingivalis* W50. *Infect Immunity* 63: 4744–4754
32 Kirszhaum L, Sotiropoulos C, Jackson C, Cleal S, Slakeski N, Reynolds EC (1995) Complete nucleotide sequence of a gene *prtR* of *Porphyromonas gingivalis* W50 encoding a 132 kDa protein that contains an arginine-specific thiol endopeptidase domain and a hemagglutinin domain. *Biochem Biophys Res Commun* 207: 424–431
33 Gharbia SE, Shah HN (1995) Molecular analysis of surface-associated enzymes of *Porphyromonas gingivalis*. *Clin Infect Dis* 20 (suppl. 2): S160–S166
34 Barkocy-Gallagher GA, Han N, Patti JM, Whitlock J, Progulske-Fox A, Lantz MS (1996) Analysis of the *prtP* gene encoding porphypain, a cysteine proteinase of *Porphyromonas gingivalis*. *J Bacteriol* 178: 2734–2741
35 Nakayama K (1997) Domain-specific rearrangement between the two Arg-gingipain-encoding genes in *Porphyromonas gingivalis*: possible involvement of nonreciprocal recombination. *Microbiol Immunol* 41: 185–196
36 Potempa J, Pavloff N, Travis J (1995) *Porphyromonas gingivalis*: a proteinase/gene accounting audit. *Trends Microbiol* 3: 430–434
37 Nakayama K, Kadowaki T, Okamoto K, Yamamoto K (1995) Construction and characterization of arginine-specific cysteine proteinase (Arg-gingipain)-deficient mutants of *Porphyromonas gingivalis*: evidence for significant contribution of arg-gingipain to virulence. *J Biol Chem* 270: 23619–23626
38 Okamoto K, Kadowaki T, Nakayama K, Yamamoto K (1996) Cloning and sequencing of the gene encoding a novel lysine-specific cysteine proteinase (Lys-gingipain) in *Porphyromonas gingivalis*: structural relationship with the arginine-specific cysteine proteinase (Arg-gingipain). *J Biochem* 120: 398–406
39 Pavloff N, Pemberton PA, Potempa J, Chen W-CA, Pike RN, Prochazka V, Kiefer MC, Travis J, Barr PJ (1997)

Molecular cloning and characterization of *Porphyromonas gingivalis* lysine-specific gingipain: a new member of an emerging family of pathogenic bacterial cysteine proteinases. *J Biol Chem* 272: 1595–1600
40 Birkedal-Hansen H, Taylor RE, Zambon J, Barwa PK, Neiders ME (1988) Characterization of collagenolytic activity from strains of *Bacteroides gingivalis. J Periodont Res* 23: 258–264
41 Uitto V-J, Larjava H, Heino J, Sorsa T (1989) A protease of *Bacteroides gingivalis* degrades cell surface and matrix proteins of cultures gingival fibroblasts and induces secretion of collagenase and plasminogen activation. *Infect Immunity* 57: 213–218
42 Grenier D, Mayrand D, McBride BC (1989) Further studies on the degradation of immunoglobulins by black-pigmented *Bacteroides. Oral Microbiol Immunol* 4: 12–18
43 Schenkein HA (1988) The effect of periodontal proteolytic *Bacteroides* species on proteins of the human complement system. *J Periodont Res* 23: 187–192
44 Sundqvist GK, Carlsson J, Hermann BF, Höfling JF, Väätäinen A (1984) Degradtion *in vivo* of the C3 protein of guinea pig complement by a pathogenic strain of *Bacteorides gingivalis. Scand J Dent Res* 92: 14–24
45 Sundqvist GK, Bengtson A, Carlsson J (1988) Generation and degradation of the complement fragment C5a in human serum by *Bacteroides gingivalis. Oral Microbiol Immunol* 3: 103–107
46 Niekrash CE, Patters MA (1986) Assessment of complement cleavage in gingival fluid in humans with and without periodontal disease. *J Periodont Res* 21: 233–242
47 Carlsson J, Herrmann BF, Hofling JF, Sundqvist GK (1983) Degradation of the human proteinase inhibitors alpha-1-antitrypsin and alpha-2-macroglobulin by *Bacteroides gingivalis. Infect Immunity* 43: 644–648
48 Nilsson T, Carlsson J, Sundqvist GK (1985) Inactivation of key factors of the plasma proteinase cascade systems by *Bacteroides gingivalis. Infect Immunity* 50: 467–471
49 Nakayama K, Yoshimura F, Kadowaki T, Yamamoto K (1996) Involvement of arginine-specific cysteine proteinase (Arg-gingipain) in fimbriation of *Porphyromonas gingivalis. J Bacteriol* 178: 2818–2824
50 Hinode D, Nagata A, Ichimiya S, Hayashi H, Morioka M, Nakamura R (1992) Generation of plasma kinin by three types of protease isolated from *Porphyromonas gingivalis* 381. *Arch Oral Biol* 37: 859–861
51 Imamura T, Pike RN, Potempa J, Travis J (1994) Pathogenesis of periodontitis: a major arginine-specific cysteine proteinase from *Porphyromonas gingivalis* induces vascular permeability enhancement through activation of the kallikrein/kinin pathway. *J Clin Invest* 94: 361–367
52 Shi Y, Ratnayake DB, Okamoto K, Abe N, Yamamoto K, Nakayama K (1999) Genetic analyses of proteolysis, hemoglobin binding, and hemagglutination of *Porphyromonas gingivalis*: construction of mutants with a combination of *rgpA*, *rgpB*, *kgp* and *hagA*. *J Biol Chem*; *in press*
53 Yoneda M, Maeda K, Aono M (1990) Suppression of bactericidal activity of human Polymorphonuclear leukocytes by *Bacteroides gingivalis. Infect Immunity* 58: 406–411
54 Isogai H, Isogai E, Yoshimura F, Suzuki T, Kagoya W, Takano K (1988) Specific inhibition of adherence of an oral strain of *Bacteroides gingivalis* 381 to epithelial cells by monoclonal antibodies against the bacterial fimbriae. *Arch Oral Biol* 33: 479–485
55 Kadowaki T, Nakayama K, Yoshimura F, Okamoto K, Abe N, Yamamoto K (1998) Arg-gingipain acts as a major processing enzyme for various cell surface proteins in *Porphyromonas gingivalis. J Biol Chem* 273: 29072–29076
56 Shah HN, Bannett R, Mateen B, Williams RAD (1979) The porphyrin pigmentation of subspecies of *Bacteroides melaninogenicus. Biochem J* 180: 45–50
57 Okamoto K, Nakayama K, Kadowaki T, Abe N, Ratnayake DB, Yamamoto K (1998) Involvement of a lysine-specific cysteine proteinase in hemoglobin adsorption and heme accumulation by *Porphyromonas gingivalis. J Biol Chem* 273: 21225–21231

Proteases: New Perspectives
V. Turk (ed.)

Proteinases in apoptosis

Masanori Tomioka and Seiichi Kawashima

Department of Molecular Biology, Tokyo Metropolitan Institute of Medical Science, 3-18-22 Honkomagome, Bunkyo-ku, Tokyo 113-8613, Japan

Introduction

Apoptosis was defined, by Kerr et al. [1] in 1972, as a new type of cell death differing morphologically from pathological cell death, necrosis. Later, apoptosis was demonstrated to be involved in genetically programmed cell death (PCD) during development and morphogenesis and, after maturation, in tissue remodelling and the immune system. Thus, abnormally accelerated apoptosis results in AIDS and Alzheimers's disease, and defects in apoptosis lead to cancer and autoimmune diseases [2]. Apoptosis, initially defined only as a mode of cell death, is now drawing greater attention in the fields of biology, medical science and pharmacology.

Several proteases have been demonstrated to play very critical roles in apoptosis, either positively or negatively. Early experiments examined the effects of protease inhibitors on apoptosis. Addition of *N*-tosyl-L-phenylalanyl chloromethyl ketone (TPCK) or N^{α}-tosyl-L-lysyl chloromethyl ketone (TLCK) to glucocorticoid-treated cells blocked degradation of nuclear DNA into nucleosome units (DNA ladder formation). Since DNA ladder formation was regarded as a characteristic feature of apoptosis, the result was interpreted as TPCK- or TLCK-sensitive protease(s) degrading linker histones and aiding endonucleases to degrade the internucleosomal DNA domains. Further investigations, however, by genetical approaches using the nematode model, *Caenorhabditis elegans*, showed that proteases are key molecules in the apoptotic machinery and cell survival. In this chapter, we examine proteases that may be involved in apoptosis although their physiological substrates and the implications of substrate proteolysis have yet to be elucidated.

ICE family proteases

Many cell types undergo apoptosis with the accompanying common features of chromatin condensation and DNA ladder formation. Initiation of apoptosis, however, is different depending on the cell type and inducers. The most extensively investigated are the p53-dependent and Fas-dependent systems (Fig. 1). In the former, accumulation of a tumor-suppressing and apoptosis-inducing p53 protein triggers apoptosis. p53 normally turns over very rapidly, but accumulates upon irradiation or administration of anti-tumor drugs through accelerated expression and/or lowered protein degradation. Accumulated p53 acts as a trancription factor of p21, an inhibitor of cyclin-dependent kinase, and Bax, a factor possibly involved in apoptosis. In the latter apop-

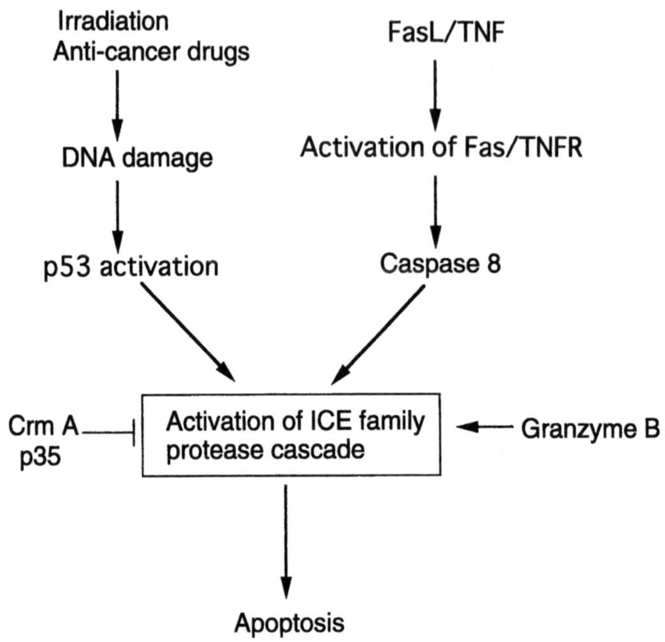

Figure 1. Proposed mechanism of apoptosis through a protease cascade.

totic initiation system where p53 is not involved, activation of a cell surface receptor, Fas, by its ligand or antibody, directly activates an intracellular protease cascade. Thus, the initial events that induce apoptosis are quite different in these two systems, yet common proteases are involved. These are the interleukin-1β-converting enzyme (ICE) and its homologs (ICE family).

ICE and its homologs

ICE was originally found as an intracellular cysteine protease that cleaves an isoform of interleukin-1 (IL-1), IL-1β (an inflammatory cytokine produced by activated macrophages), from its 31 kDa precursor to the 17 kDa mature form after Asp-116 [3, 4]. In 1993, Yuan et al. [6] identified a gene, *ced-3*, that is essential for PCD in *Caenorhabiditis elegans*, and found that its product is a homolog of ICE. This finding suggested that ICE plays an important role in apoptosis and the idea was supported by the fact that overexpression of ICE in fibroblasts induced apoptosis [6]. However, a mouse knock-out model of the ICE gene (ICE–/–) did not cause abnormalities either in mouse development or cellular apoptosis except for slight changes in IL-1β secretion and Fas-induced apoptosis [7, 8]. The involvement of apoptotic proteases similar to ICE was suggested.

At present, ten ICE-like protease species have been identified. These are ICH-1/Nedd-2, CPP32/Yama/apopain, TX/ICH-2/ICErel-II, TY/ICErel-III, Mch2, Mch3/ICE-LAP-

3/CMH-1, MACH/FLICE/Mch5, ICE-LAP6/Mch6, Mch4 and ICE itself [9]. Although these proteases were found in different apoptotic systems, they possess the conserved active site sequence of ICE, QACXG (X = R/Q/G). They are all cysteine proteases with a specificity for Asp residues in substrates. Due to these properties, they are collectively named caspases (cysteine protease with specificity for aspartic acid residues) and numbered according to the sequence of finding. Overexpression of these proteases induces apoptosis but it is not clear why many similar proteases are needed. ICE-like proteases can be divided into three subfamilies, ICE (caspase 1), ICH-1 (caspase 2), and CPP-32 (caspase 3) based on the genetically conserved domain structures (Tab. 1). Since their substrate specificity and inhibitor spectrum share similarities, they may function complementarily during apoptosis among subfamilies. Proteases of the ICE subfamily can be activated autocatalytically, but those of the CPP-32 subfamily cannot activate themselves and need the help of ICE subfamily proteases or granzyme B.

Table 1. ICE family proteases. At this moment, ten species of ICE family protease genes have been identified in mammals and classified into three subfamilies from homology in amino acid sequences. Hatched C represents the active site cysteine residue.

Subfamily	Caspase	Caspase number	Active site	Substrate specificity
ICE subfamily	ICE	1	QACRG	
	TX/ICH-2/ICE$_{rel}$-II	4	QACRG	YVXD-
	TY/ICE$_{rel}$-III	5	QACRG	
CPP32 subfamily	Mch3/ICR-LAP-3/CMH-1	7	QACRG	DXXD-
	CPP32/Yama/apopain	3	QACRG	DXXD-
	Mch2	6	QACRG	DXXD-
	MACH/FLICE/Mch5	8	QACQG	XEXD-
	Mch4	10	QACQG	XEXD-
	ICE-LAP6/Mch6	9	QACQG	–
	Ced-3		QACRG	DXXD-
ICH-subfamily	ICH-1/Nedd-2	2	QACRG	YVXD-

Recently, direct interaction of ICE family proteases with the Fas protein, an apoptosis-inducing molecule, was suggested. The protease demonstrating this interaction was FLICE (FADD-like ICE)/MACH (MOR T1-associated CED-3 homolog). In addition to ICE-like protease activity, FLICE has a domain homologous to FADD (Fas-associating protein with death domain)/MOR T1 (mediator of receptor-induced toxicity 1) that associates with the death domain of the Fas molecule. The discovery of FLICE was a clue for the assignment of signal transduction from apoptosis-inducing factor to ICE family proteases.

Substrates of ICE family proteases

To understand the function of ICE family proteases in apoptosis, the identification of their substrates was essential. The most typical and important substrates of ICE family proteases are the proteases themselves. In cells, ICE family proteases exist as inactive precursors of a single polypeptide chain and are processed autocatalytically to a heterodimer of large and small subunits. ICE family proteases become active by association of these heterodimers to form a tetramer. Through this ICE family protease-specific activation mechanism, the fidelity of apoptotic initiation can be guaranteed. Later, general substrates of ICE family proteases, poly (ADP ribose) polymerase (PARP) as a DNA repairing enzyme and DNA-dependent protein kinase (DNA-PK) as a repairing enzyme of double-strand breaks in DNA, were identified. However, the knock-out mouse of PARP showed normal morphology [10] and this queried whether ICE family proteases inhibit DNA repair during apoptosis. In addition to these, lamin (a nuclear envelope lining protein), actin, Gas 2 (a component of the microfilament system), protein kinase Cδ (PKCδ), SREBP-1 and -2 (sterol regulatory element-binding proteins), and the 70 kDa subunit of U1 small nuclear ribonucleoprotein, are known as substrates of ICE family proteases during induction of apoptosis [9]. However, the implications of the degradation of these substrates has not been elucidated.

Inhibitors of ICE family proteases

Although ICE family proteases are cysteine proteases, their activities are not inhibited by typical cysteine protease inhibitors such as antipain and E-64. Tetrapeptide inhibitors, YVAD-al and YVAD-cmk for ICE and DEVD-al for CPP32, designed on the basis of their substrate specificities, inhibit the respective ICE-like proteases very efficiently. A unique inhibitor of ICE is CrmA, a product of the *crm*A gene encoded by the cowpox virus [11]. *Crm*A gene was identified from the observations that the cells infected with cowpox virus showed a repressed inflammatory response, an accelerated infectivity to other viruses, and especially a repressed secretion of IL-1β. CrmA forms a complex with ICE inhibiting its activity very efficiently ($Ki < 4$ pM). The amino acid sequence of the CrmA active site, LVAD, is similar to that of the cleavage sites for the ICE subfamily proteases. Since CPP32 subfamily proteases require an Asp residue at the P4 position of the substrate, the inhibitory activity of CrmA against these proteases is rather weak. It is also known that apoptosis is inhibited by p35, a virus gene product that is expressed in the early phase after AcMNPV (*Autographa californica* nuclear polyhedrosis virus, a baculovirus) infection [12]. The mechanism of inhibition is quite different from that of CrmA. p35 is first cleaved by ICE family proteases into a 10 kDa and a 25 kDa fragment, which then binds to the two subunits of ICE family proteases to form a stable complex and inhibit the activity. p35 inhibits most ICE family proteases. p35 not only inhibits the apoptosis of virus-infected host cells but also programmed cell death during the development of mammals and nematoda (*Caenorhabditis elegans*), suggesting that its action is evolutionarily conserved and a very critical step in the inhibition of apoptosis.

Proteasome

Proteasome (26S), an intracellular multicatalytic protease complex, is composed of 40–50 heterogenous subunits to give a molecular weight of 2,000 kDa. The proteolytic mechanism of proteasome involves ATP-dependent ubiquitination, with few exceptions. Recently many reports on proteasome have shown that it plays important roles in cell cycle progression and antigen presentation. To identify the function of proteasome, two approaches have been employed; one is the analysis of ubiquitination and the other the use of proteasome inhibitors.

For many years it has been known that the amount of ubiquitin increases on programmed cell death during development and morphogenesis. The expression of ubiquitin also increases during apoptosis induced by γ-irradiation and this apoptotic effect is repressed by antisense ubiquitin mRNA [13]. The effect of ubiquitin expression is now understood to reflect the involvement of the ubiquitin-proteasome system in apoptosis.

More direct evidence for the involvement of proteasome in apoptosis came from studies using proteasome inhibitors. Z-Leu-Leu-leucinal (ZLLLal, MG-132) is a synthetic peptide aldehyde protease inhibitor that was initially found by us to induce neuronal cell differentiation [14]. Although ZLLLal inhibits both proteasome and calpain, another main intracellular protease, its effect on neuronal differentiation is attributed to the inhibition of proteasome activity. Similarly, lactacystin, a microbial metabolite originally isolated as an inducer of neuritogenesis, was later found to be an inhibitor of proteasome. The inhibitors, ZLLLal and lactacystin, did not block apoptosis induced by various stimuli, but rather enhanced apoptosis. Even without apoptosis induction, the proteasome inhibitor by itself induced apoptosis [15]. This suggests that proteasome is preventing cells from inappropriate apoptosis under normal conditions by degradation of intracellular apoptosis inducers which accumulate on inhibition of proteasome, leading to cell death. In the case of infected cells with human papilloma virus (HPV), an apoptosis-inducer p53 protein is actively ubiquitinated and degraded by proteasome to prevent its accumulation and apoptosis [16]. These results demonstrate that proteasome functions as an inhibitor of apoptosis. On the contrary, however, addition of proteasome inhibitors to thymic or neuronal cells did not induce apoptosis but rather inhibited the apoptosis induced by various factors [17, 18]. This effect is similar to that of ICE family proteases suggesting that proteasome can also play a positive role in apoptosis.

Can these contradictory results of proteasome function in apoptosis be explained? So far, the cells in which apoptosis was inhibited are restricted to primary cultured thymic and neuronal cells. Both of these are differentiated post-mitotic cells. In contrast, induction of apoptosis by proteasome inhibitors has been reported only for actively proliferating cells. A clue to resolving the discrepancy was given by Drexler's finding that the cultured cells treated with proteasome inhibitor accumulated at the G1 phase before undergoing apoptosis, and cells at the stationary phase were resistant to proteasome inhibitor-dependent apoptosis [19]. His results suggested that induction of apoptosis by proteasome inhibitors is, at least, dependent on the cell cycle and it is reasonable to hypothesize that the effect of proteasome inhibition on apoptosis is different for dividing cells and non-dividing cells. In non-dividing cells, inhibition of proteasome resulted in inhibition of ICE family proteases, suggesting a role for proteasome upstream of the ICE family proteases.

Calpain

Calpain is an intracellular cysteine protease that is widely distributed in the animal kingdom. Its activity is controlled by Ca^{2+}, which increases on apoptosis. Thus it is speculated to be involved in apoptosis.

Under pathological conditions, short-term brain ischemia and subsequent reperfusion result in neuronal cell death. This cell death demonstrates the morphology of apoptosis and accompanies the degradation of fodrin, a substrate of calpain [20]. In another neurodegenerative disease, Alzheimer's disease, β-amyloid protein deposits in the brain where the amount of active calpain increases. β-amyloid protein is an apoptosis-inducing factor and this induction is blocked by calpain-specific inhibitors [21]. These results suggest the involvement of calpain in apoptosis but the exact mechanism is unclear. Recently, it was shown that calpain-degraded fodrin is further cleaved by ICE family proteases and this cleavage is more closely correlated to apoptosis [22].

Involvement of calpain in apoptosis has also been demonstrated in cultured cells. Addition of Ac-Leu-Leu-norleucinal (ALLNal, calpain inhibitor 1) to thymocytes blocked the irradiation- and glucocorticoid-induced apoptosis [23]. On the other hand, addition of the same inhibitor to human prostate adenocarcinoma cells induced apoptosis [24]. The inhibitor ALLNal used as a calpain inhibitor is also known to inhibit proteasome [25]. As described above, inhibition of proteasome induces apoptosis in dividing cells and inhibits apoptosis in non-dividing cells. Thus, the effect of ALLNal on apoptosis is now regarded as the result of proteasome inhibition. Although no clear evidence has been obtained about the involvement of calpain in apoptosis, a synthetic inhibitor more specific for calpain [26] and an antibody that exclusively recognizes the fodrin fragment generated by calpainolysis [27] have recently been developed. These powerful new tools may elucidate the role of calpain.

Granzymes

Apoptosis is regulated mainly by intracellular factors. However, some apoptosis can be induced directly from the outside of cells as has been demonstrated for the immune system. Cytotoxic T lymphocytes (CTL) and natural killer cells (NK cells) induce apoptosis in their target cells via two different pathways [28]. One is dependent on the Fas protein on the surface of target cells and another is Ca^{2+}- and perforin/granzyme-dependent. In the latter pathway, CTL or NK cells bind to target cells and release perforin stored in intracellular granules. Perforin binds to the surface of target cells in a Ca^{2+}-dependent manner and polymerizes, creating small pores in the membrane. This leads to cell death, but perforin alone is not sufficient to kill the cells. Granzyme is needed as an effector of cell death. Among several granzyme species identified so far, granzyme A (GraA) and granzyme B (GraB) are well characterized.

GraA is a trypsin-like protease that hydrolyzes the synthetic substrate N^{α}-benzyloxycarbonyl-L-lysine thiobenzyl ester. Since co-expression of GraA with perforin induced apoptosis accompanying nuclear DNA fragmentation, GraA was the candidate for apoptosis induction by CTL and NK cells. Later, DNA fragmentation was observed in a GraA-deficient mouse, suggesting that GraA was not an essential effector molecule for DNA fragmentation and apoptosis [29].

GraB was also found to be involved in the induction of DNA fragmentation. GraB is a serine protease like GraA, but is unique in its specificity for Asp residues. The specificity is similar to that of ICE family proteases and, thus, GraB was expected to play an important role in cell killing by CTL. In the GraB-deficient mouse, hematopoiesis occurred normally, but DNA fragmentation and nuclear condensation characteristic of apoptosis were not observed [30]. From these results, it was concluded that GraB is an essential effector in CTL-induced apoptosis of the target cells. How does GraB induce apoptosis in target cells? GraB cleaves its substrates after Asp residues similarly to ICE family proteases. In 1994, it was reported that GraB cleaves and activates CPP32/Yama/apopain, an ICE family protease, leading to apoptosis of target cells [31]. Furthermore, CrmA, a serine protease inhibitor from cowpox virus that inhibits CTL-induced apoptosis, inhibited the activation of CPP32 by GraB. In addition to CPP32/Yama/apopain, GraB was later shown to activate ICE and other members of the CPP32/Yama/apopain subfamily.

Fragmentin 2 is a serine protease with homology to GraB and its introduction into cells with the help of perforin causes uncontrolled activation of a serine/threonine protein kinase (p34cdc2), inducing apoptosis [32]. For apoptosis, dephosphorylation of p34cdc2 is required and phosphorylation of p34cdc2 at Tyr-15 of the ATP-binding site by a nuclear protein kinase, Weel, inhibits the apoptosis induced by GraB. It is not known if the activation of p34cdc2 is located downstream of ICE family proteases in the apoptotic pathway.

Other proteases

In addition to intracellular proteases such as ICE family proteases, proteasome, and calpain, the involvement of other unidentified proteases has been suggested. In the presence of TLCK or TPCK, DNA fragmentation accompanying apoptosis, induced in mouse thymic cells and lymphocytes by glucocorticoid or topoisomerase inhibitors, was blocked [33]. It is difficult to explain this effect by the involvement of the proteases previously mentioned. Since another serine protease inhibitor dichloroisocoumarin blocks apoptosis, the candidate protease is possibly a serine protease. However, TPCK and TLCK inhibit the DNA fragmentation that accompanies apoptosis, but not completely the morphological characteristics of apoptosis [34]. Thus, TPCK- and TLCK-sensitive proteases may be involved in the DNA fragmentation pathway, but they do not play a central role in apoptosis induction.

Perspectives

In this chapter, we have reviewed proteases that are or may be involved in apoptosis, either positively or negatively. Among them, the ICE family proteases are the sole common factor involved in apoptosis induction. Still, there are many questions to be answered. What are the target substrates of these proteases? In which step of apoptosis do proteases function? How are proteasome and calpain involved in apoptotic systems? Dysregulated apoptosis is a cause of serious diseases and controlling apoptosis by protease inhibitors may provide an effective therapeutic treatment. For this purpose, elucidation of the apoptotic mechanism and the mechanism underlying the determination of cell death or survival is required.

Acknowledgements
We express our sincere thanks to Dr. Charles Chalfant for his proof-reading of the manuscript.

References

1 Kerr JF, Wyllie AH, Currie AR (1972) Apoptosis: a basic biological phenomenon with wide-ranging implications in tissue kinetics. *Brit J Cancer* 26: 239–257
2 Thompson CB, Bull HG, Calaycay JR, Chapman KT, Howard AD, Kostura MJ, Miller DK, Molineaux SM, Weidner JR, Aunins J (1995) Apoptosis in the pathogenesis and treatment of disease. *Science* 267: 1456–1462
3 Thornberry NA, Bull HG, Calaycay JR, Chapman KT, Howard AD, Kostura MJ, Miller DK, Molineaux SM, Weidner JR, Aunins J et al. (1992) A novel heterodimeric cysteine protease is required for interleukin-1β processing in monocytes. *Nature* 356: 768–774
4 Cerretti DP, Kozlosky CJ, Mosley B, Nelson N, Ness KV, Greenstreet TA, March CJ, Kronheim SR, Druck T, Cannizzaro LA et al. (1992) Molecular cloning of the interleukin-1β converting enzyme. *Science* 256: 97–100
5 Yuan J, Shaham S, Ledoux S, Ellis HM, Horvitz HR (1993) The *C. elegans* cell death gene *ced-3* encodes a protein similar to mammalian interleukin-1β converting enzyme. *Cell* 75: 641–653
6 Miura M, Zhu H, Rotello R, Hartwieg EA, Yuan J (1993) Induction of apoptosis in fibroblasts by IL-1β converting enzyme, a mammalian homolog of the *C. elegans* cell death gene *ced-3*. *Cell* 75: 653–660
7 Kuida K, Lippke JA, Ku G, Harding MW, Livingston DJ, Su MS-S, Flavell RA (1995) Altered cytokine export and apoptosis in mice deficient in interleukin-1β converting enzyme. *Science* 267: 2000–2003
8 Li P, Allen H, Banerjee S, Franklin S, Herzog L, Johnston C, McDowell J, Paskind M, Rodman L, Salfeld J et al. (1995) Mice deficient in IL-1β converting enzyme are defective in production of mature IL-1β and resistant to endotoxic shock. *Cell* 80: 401–411
9 Whyte M (1996) ICE/CED-3 proteases in apoptosis. *Trends Cell Biol* 6: 245–248
10 Wang ZQ, Auer B, Stingl L, Berghammer H, Haidacher D, Schweiger M, Wagner EF (1995) Mice lacking ADPRT and poly(ADP-ribosyl)ation develop normally but are susceptible to skin disease. *Gene Develop* 9: 509–520
11 Tewari M, Quan LT, O'Rourke K, Desnoyers S, Zeng Z, Beidler DR, Poirier GG, Salvesen GS, Dixit VM (1995) Yama/CPP32β, a mammalian homolog of CED-3, is a CrmA-inhibitable protease that cleaves the death substrate poly(ADP-ribose) polymerase. *Cell* 81: 801–809
12 Clem RJ, Fechheimer M, Miller LK (1991) Prevention of apoptosis by a baculovirus gene during infection of insect cells. *Science* 254: 1388–1390
13 Delic J, Morange M, Magdelenat H (1993) Ubiquitin pathway involvement in human lymphocyte γ-irradiation-induced apoptosis. *Mol Cell Biol* 13: 4875–4883
14 Saito Y, Tsubuki S, Ito H, Kawashima S (1990) The structure-function relationship between peptide aldehyde derivatives on initiation of neurite outgrowth in PC12h cells. *Neurosci Lett* 120: 1–4
15 Shinohara K, Tomioka M, Nakano H, Tone S, Ito H, Kawashima S (1996) Apoptosis induction resulting from proteasome inhibition. *Biochem J* 317: 385–388
16 Hubbert NL, Sedman SA, Schiller JT (1992) Human papilloma virus type 16 E6 increases the degradation rate of p53 in human keratinocytes. *J Virol* 66: 6237–6241
17 Grimm LM, Goldberg AL, Poirier GG, Schwartz LM, Osborne BA (1996) Proteasomes play an essential role in thymocyte apoptosis. *EMBO J* 15: 3835–3844
18 Sadoul R, Fernandez P-A, Quiquerez A-L, Martinou I, Maki M, Schroter M, Becherer JD, Irmler M, Tschopp J, Martinou J-C (1996) Involvement of the proteasome in the programmed cell death of NGF-deprived sympathetic neurons. *EMBO J* 15: 3845–3852
19 Drexler HCA (1997) Activation of the cell death program by inhibitor of proteasome function. *Proc Natl Acad Sci USA* 94: 855–860
20 Blomgren K, Kawashima S, Saido TC, Karlsson J-O, Elmered A, Hagberg H (1995) Fodrin degradation and subcellular distribution of calpains after neonatal rat cerebral hypoxic-ischemia. *Brain Res* 684: 143–149
21 Jordan J, Galindo MF, Miller RJ (1997) Role of calpain- and interleukin-1β converting enzyme-like proteases in the β-amyloid-induced death of rat hippocampal neurons in culture. *J Neurochem* 68: 1612–1621

22 Vanags DM, Porn-Ares MI, Coppola S, Burgess DH, Orrenius S (1996) Protease involvement in fodrin cleavage and phosphatidylserine exposure in apoptosis. *J Biol Chem* 271: 31075–31085
23 Squier MKT, Miller ACK, Malkinson AM, Cohen JJ (1994) Calpain activation in apoptosis. *J Cell Physiol* 159:229–237
24 Zhu W, Murtha PE, Young CYF (1995) Calpain inhibitor-induced apoptosis in human prostate adenocarcinoma cells. *Biochem Biophys Res Commun* 214: 1130–1137
25 Figueiredo-Pereira ME, Banik N, Wilk SJ (1994) Comparison of the effect of calpain inhibitors on two extralysosomal proteinases: the multicatalytic proteinase complex and m-calpain. *J Neurochem* 62: 1989–1994
26 Tsubuki S, Saito Y, Tomioka M, Ito H, Kawashima S (1996) Differential inhibition of calpain and proteasome activities by peptidyl aldehydes of di-leucine and tri-leucine. *J Biochem (Tokyo)* 119: 572–576
27 Saido TC, Yokota M, Nagao S, Yamaura I, Tani E, Tsuchiya T, Suzuki K, Kawashima S (1993) Spatial resolution of fodrin proteolysis in postischemic brain. *J Biol Chem* 268: 25239–25243
28 Takayama H, Kojima H, Shinohara N (1995) Cytotoxic T lymphocytes: the newly identified Fas (CD95)-mediated killing mechanism and a novel aspect of their biological functions. *Adv Immunol* 60: 289–321
29 Ebnet K, Hausmann M, Lehmann-Grube F, Mullbacher A, Kopf M, Lamer M, Simon MM (1995) Granzyme A-deficient mice retain potent cell-mediated cytotoxicity. *EMBO J* 14: 4230–4239
30 Heusel JW, Wesselschmidt RL, Shresta S, Russell JH, Ley J (1994) Cytotoxic lymphocytes require granzyme B for the rapid induction of DNA fragmentation and apoptosis in allogenic target cells. *Cell* 76: 977–987
31 Darmon AJ, Nicholson DW, Bleackley RC (1995) Activation of the apoptotic protease CPP32 by cytotoxic T-cell-derived granzyme B. *Nature* 377: 446–448
32 Shi L, Nishioka WK, Th'ng J, Bradbury EM, Litchfield DW, Greenberg AH (1994) Premature $p34^{cdc2}$ activation required for apoptosis. *Science* 263: 1143–1145
33 Fearnhead HO, Rivett AJ, Dinsdale D, Cohen GM (1995) A pre-existing protease is a common effector of thymocyte apoptosis mediated by diverse stimuli. *FEBS Lett* 357: 242–246
34 Ghibelli L, Maresca V, Coppola S, Gualandi G (1995) Protease inhibitors block apoptosis at intermediate stage: a compared analysis of DNA fragmentation and apoptotic nuclear morphology. *FEBS Lett* 377: 9–14

Proteases: New Perspectives
V. Turk (ed.)

Caspases: cytokine activators and promoters of cell death

Guy S. Salvesen and Henning R. Stennicke

The Program for Apoptosis and Cell Death Research, The Burnham Institute, 10901 North Torrey Pines Road, La Jolla, CA 92037, USA

Programmed cell death

Apoptosis is the morphology that results from triggering programmed cell death (PCD): a mechanism enabling metazoans to eliminate cells that are damaged, mislocated, or have become superfluous (reviewed by [1]). The ability to undergo PCD is inherent in all metazoan nucleated cells and, though originally studied within the context of normal development, PCD plays important roles in mature adults. For example, cell numbers are maintained by a balance between cell proliferation and cell death, with for instance, approximately 10^{11} neutrophils and 10^{10} colon epithelial cells turning over per day in humans. Moreover, abundant evidence implicates dysregulation of PCD in the pathogenesis of many diseases (reviewed in [2, 3]). Inappropriate increases in cell death have been reported in AIDS, neurodegenerative disorders, and ischemic injury, whereas decreases in cell death contribute to cancer, autoimmune diseases, and restenosis.

Cell deaths from engagement of the program result in apoptosis – the ordered dismantling of the cell that results in its "silent" demise, with packaged cell fragments removed by phagocytic cells. A key to PCD is the discovery in many laboratories that, irrespective of the lethal stimulus, death results in the same apoptotic morphology that includes cell and organelle dismantling and packaging, DNA cleavage to nucleosome-sized fragments, and engulfment of the fragmented cell. Apoptosis is therefore distinct from the other type of cell death, necrosis, which is mediated more by acute trauma to a cell, resulting in spillage of potentially toxic and antigenic cellular components into the intercellular milieu [4]. The vital role of apoptosis in normal animal development and its aberrations in pathology led to an intensive search for the components of the program. As is often the case in such novel areas of research, the first clues to the components were discovered by genetic analysis in the model organism *Caenorhabditis elegans*.

Apoptosis is promoted by proteases

During the development of the adult form of *C. elegans*, 131 of the total 1090 somatic cells die with apoptotic morphology [5]. Genetic analysis of mutants in which these cells fail to die correctly identified two "killer" genes, *ced*-3 and *ced*-4, that the worm must express for the cells to die [6]. Similar studies identified a third gene, the "protector" *ced*-9, that acts as an inhibitor of the pathway [7]. Nucleotide sequence data reveal that CED-3 is a homolog of the cysteine protease human interleukin-1β converting enzyme (caspase 1) [8]. Thus, the discovery that CED-3 is a protease may be regarded as a singularly important advance in understanding the biochemistry of programmed cell death because it directly implicates proteins with a defined biochemical activity, i.e. proteases. An understanding of the regulation of mammalian CED-3 homologs, now christened the caspases, therefore has yielded significant insights into the programmed cell death pathway.

Caspases

On the basis that CED-3 is a key regulator of apoptosis in nematodes, it was predicted that a human ortholog would carry out the same function. However, the primordial paradigm of apoptosis observed in the worm is recapitulated with an apparently large degree of complexity in higher animals. The problem here is that humans contain at least ten caspases, complicating the assessment of their individual functions. The caspases are thought to be located in the cell cytosol, where they await activation to transmit their respective functions. Much of the recent work on caspases has focused on how they are regulated, and how the limited proteolytic cleavages now known to be typical of apoptosis relate to caspase activity.

Caspases cleave a number of cellular proteins, and the process is one of limited proteolysis where a small number of cuts, usually only one, are made in inter-domain regions. Sometimes cleavage results in activation of the protein, sometimes in inactivation, but never in degradation since their primary specificity for Asp distinguishes the caspases as among the most specific endopeptidases. It is currently very difficult to determine which caspases are responsible for specific cleavages in whole cells, nevertheless, by combining *in vivo* observations with *in vitro* tests one can recognize substrate specificity within the caspase family. For example, caspase 1 is remarkably specific for the precursors of IL-1β and IL-18 (interferon-γ-inducing factor), making a single initial cut in each pro-cytokine that activates them and allows exit from the cytosol. Ectopic expression of caspase 1 in some cells can result in apoptosis, but a role in developmental PCD is unlikely given the normal phenotype of knockout mice [12]. In stark contrast to this are caspase 3 knockout mice which demonstrate a severe phenotype with early lethality caused by massive brain enlargement. This is presumed to be caused by failure of developmental cell death of neuronal precursors [13].

Whereas caspase 1 (and possibly 4 and 5) are primarily involved in pro-cytokine activation, other caspases are considered to promote pathways to apoptosis. This conclusion is based largely on the following observations: (1) the zymogens are seen to be processed during apoptosis, or *in vitro* models of apoptosis, (2) at least *in vitro*, they cut proteins whose cleavage is associated with apoptotic cell death. Importantly, distinct caspases are implicated in death receptor mediated initiation, and subsequent execution (see Fig. 1). The executioners cleave and inactivate key proteins required for the maintenance of homeostasis (reviewed in [10]), leading to apoptotic collapse and demise of the cell. The advantage of utilizing proteases, rather than conventional signal transduction proteins such as protein kinases, is that they produce irreversible events. Therefore they achieve rapid execution of the cell once commitment is engaged. This does not mean that the caspases are the only enzymes that participate in apoptosis, since nucle-

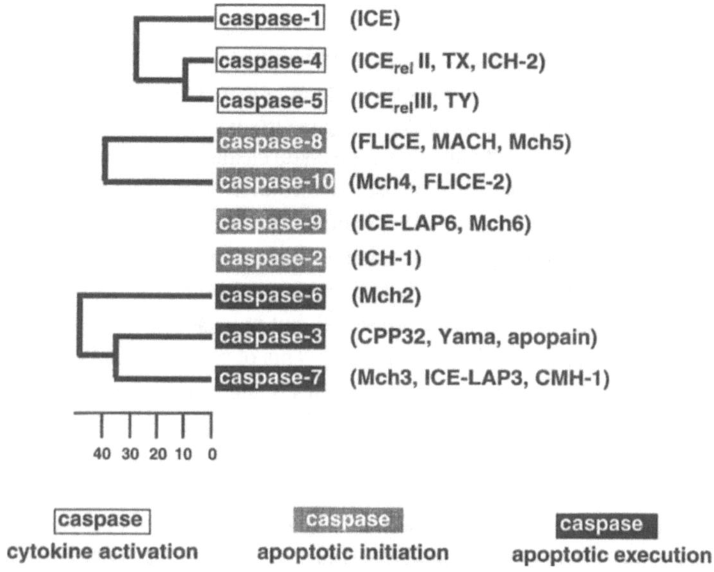

Figure 1. Human caspase family relationships. The name of the family signifies that they are cysteine-dependent aspartate-specific proteases. Membership in the family is based on sequence similarities in the catalytic chains. All of the caspases have now been shown to cause apoptosis when transfected into at least one cell line, and have either proteolytic activity on protein or synthetic peptide substrates. Alignment of the catalytic domains demonstrates evolutionary relationships between the caspases which presumably diverged from a common ancestor in an early metazoan. The scale is in millions of years, with an arbitrary cutoff at 50. The caspases can be divided into three groups based on their expected biological functions (as reviewed in [9–11]).

ases and protein kinases must also participate, but they are absolutely required for the accurate and limited proteolytic events that typify this type of programmed cell death. Why have such a complex network in place to kill a cell when all that is really needed is to turn off a crucial house-keeping gene? Presumably the presence of necrotic cells is dangerous, and the apoptotic response is thus a mechanism to dismantle cells for disposal in a way that does not compromise the rest of the organism.

Mechanisms of activation

Once synthesized, each caspase is maintained in a latent state as a single chain zymogen. In common with other protease zymogens [14], generation of the active form requires limited proteolysis, and for the caspases this results from cleavage in an inter-domain linker segment to give a heterodimeric enzyme, with both chains containing essential components of the catalytic machinery. The activation of the caspases has been found to take place by several different mechanisms which may be summarized as hetero-activation and homo-activation.

Hetero-activation

While there is abundant evidence that the catalytically active form of a caspase consists of two processed chains, the pathways to activation are not clear. The classic way of activating a protease zymogen, i.e. by another protease, clearly occurs with caspases, and this can be termed a "hetero-activation". In this mechanism a single chain caspase zymogen is processed in the inter-chain connector to yield an active protease. Hetero-activators include other caspases, and the highly conserved Asp297 is a major site for activation (Fig. 2). Because of the almost total conservation of this residue throughout the caspase family, and because the caspases are specific for Asp, there is a general consensus that caspases constitute a proteolytic cascade. Definitive data for a true cascade have yet to be presented, but at least caspase 6 and 8 can directly activate the zymogens of caspases 3 and 7 [15–17]. This observation finds a mechanistic explanation in the $S_{2-S}4$ substrate preferences of caspases 6 and 8 discussed later. These resemble the region upstream of Asp297 in caspases 3 and 7, implicating caspases 6 and 8 as activators, and caspases 3 and 7 as executioners in at least a minimal two step cascade that serves to amplify apoptotic signals.

Another Asp-specific protease, the serine protease granzyme B, plays a major role as a physiological activator of caspases. Granzyme B has an extraordinarily limited tissue specificity, being found only in the cytolytic granules of activated cytotoxic T-lymphocytes and natural killer cells. The restricted tissue distribution and Asp-specificity of granzyme B is a wonderful example of adaptation of the mammalian immune system, since this allows cytotoxic cells to

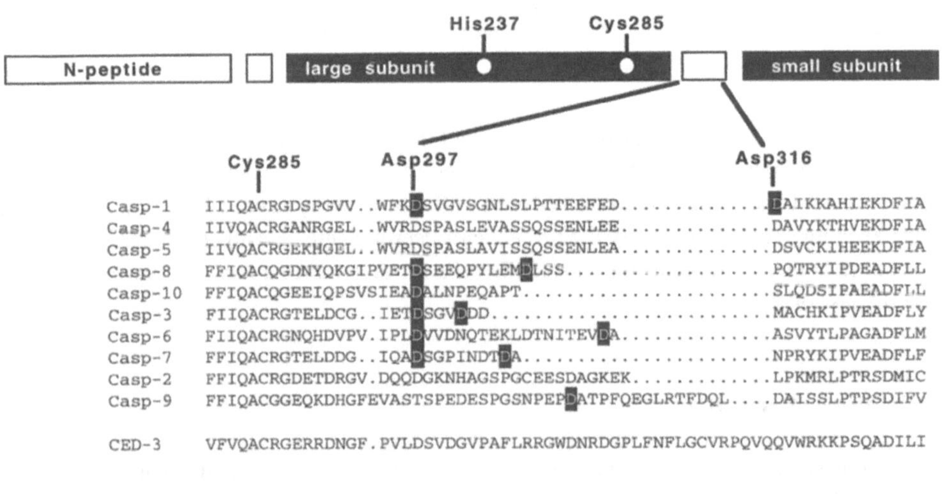

Figure 2. Schematic representation of a processed caspase. The active form of a caspase consists of large (17–20 kDa) and small (9–12 kDa) chains released from a precursor by proteolysis. An N-terminal polypeptide, sometimes known as the pro-domain, is also released from several caspases during activation. The N-terminal region is not conserved between caspases, but the two chains (shown in black) that fold to form the catalytic site contain extensive identities throughout the family. Removal of the pro-peptide does not result in activation. Rather, cleavage between the large and small subunits generates the active protease, and sometimes a small "linker" segment is released. The linker segment is expanded to demonstrate the wide degree of sequence variation between caspases, with outlined Asp residues constituting the major activation sites.

trigger apoptosis in virally infected or transformed cells. Granzyme B has been demonstrated to activate most caspase zymogens *in vitro* [18–21], presumably by cleaving at Asp297 (though the precise cleavage site has only been clearly demonstrated for pro-caspase 7 [22]), thus initiating a massive lethal signal. There is considerable controversy over exactly how granzyme B is delivered from what are essentially secretory vesicles in cytotoxic cells, into the cytoplasm of target cells, but it is probable that activation of the caspases is this proteases major function.

More surprising is the demonstration that proteases without specificity for Asp are able to activate caspase zymogens, at least *in vitro*. For example, the serine protease subtilisin Carlsberg, which has little affinity for synthetic substrates with Asp at P_1, rapidly activated pro-caspase 7 by cleaving at Asp297, and the serine protease cathepsin G activates the same zymogen by cleaving at Gln295 [22]. These observations forced the conclusion that the linker segment between the large and small subunits is an unusually susceptible inter-domain connector designed to be utilized for rapid proteolysis. Normally, within the hierarchy of caspase cascades, zymogen activation takes place at conserved Asp residues, but the proteolytic sensitivity of the caspase inter-domain link may allow non-Asp-specific proteases, such as those from lysosomes or viruses, to engage the apoptotic apparatus under pathological conditions.

Homo-activation

Caspase zymogens also seem to be able to activate themselves by an unusual "homo-activation" mechanism. This activation pathway has been examined in most detail for caspase 1. The processing of caspase 1 is apparently quite complex, and the resulting protein intermediates have different abilities to bind an active site-directed peptide ketone inhibitor [23]. Though this is not really definitive of proteolytic activity, it approximates to it and forces an interesting conclusion [23]. The results of the study concluded that full length pro-caspase 1 (p45) has extremely low intrinsic activity, the form produced by cleavage between the large and small catalytic subunits is moderately active, and full activity is only attained with removal of the N-peptide to give the mature p20/p10 heterotetramer [23]. This does not explain how activation occurs, but it does reveal that intermediates can have activity.

Related to this finding are experiments demonstrating that the purified caspase 1 zymogen undergoes spontaneous activation to the final p20 and p10 catalytic units, when refolded from denaturing solution [27], by a pathway that is dependent on the N-terminal peptide [28]. In another context, precise and essentially full activation to the large and small catalytic units of caspases 3, 6 and 7 occurs when the full length zymogens are expressed in *E.coli* [15]. Since fully processed active caspase 7 is unable to activate its own precursor at any detectable rate [22], a reasonable explanation is that activation of the zymogens takes place spontaneously at the high concentrations achieved during expression in *E.coli*. On the basis of these indirect lines of evidence one can predict that the homo-activation pathway is best explained by the single chain zymogens of some caspases having small amounts of catalytic activity, allowing them to process *in trans* when brought together at high concentration. Indeed, the homo-activation pathway may explain the initiation of caspase signals in the Fas pathway since clustering of caspase 8 zymogens at the cytosolic face of the receptor during oligomerization may achieve a

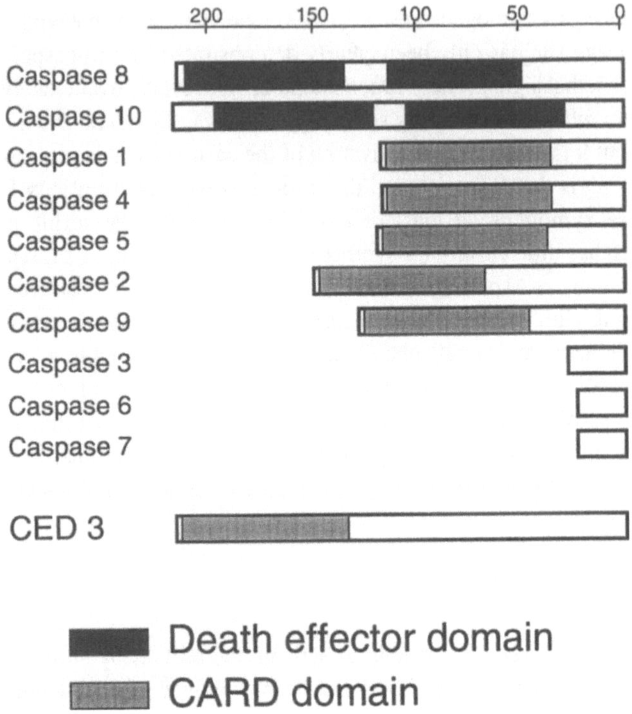

Figure 3. N-peptide domains. The length of the N-peptides varies from 22 amino acid residues for caspases 3 and 6 to over 200 in caspases 8 and 10. Embedded within some of these N-terminal segments are distinct domains that are probably required for specific protein interactions utilized during homo-activation of the caspases. Thus caspases 8 and 10 contain tandem "death effector domains" (DED) required for binding to adaptor proteins during death receptor ligation [20, 24, 25]. Caspases 1, 2, 4, 5 and 9 as well as Ced-3 contain a "caspase recruitment domain" (CARD) which is recognized by sequence similarity [26] and is postulated to be another protein interaction domain required for assembly of activation complexes. Interestingly, the executioner caspases 3, 6 and 7 have very short N-peptides whose function is currently unknown. The scale at the top of the figure indicates the number of amino acid residues in the individual N-peptides.

local concentration sufficient for homo-activation. The molecular explanation of how local clustering of partially active caspase zymogens results in full activation is completely unknown, and provides a fertile area of research.

Regulation by natural caspase inhibitors

Not surprisingly, inhibition of caspases is a strategy adopted by viruses in their attempt to elude the apoptotic response of the cell to the infectious insult. Thus, much of our grasp of the pivotal role played by proteases in apoptosis and cytokine activation comes from our understanding that the poxvirus protein CrmA [29] and the baculovirus protein p35 [30] specifically target caspases. Interestingly, no homologs of p35 are known in mammals, and though mammals possess many homologs of the serpin CrmA, none of the known ones seem to be targeted against caspases. So far the only demonstrated caspase inhibitors endogenous to mammals are members of the "inhibitor of apoptosis" (IAP) family, including the human proteins XIAP, cIAP-1 and cIAP-2. All three of these IAPs seem to specifically target the executioner caspases 3 and 7, and not 1, 6, 8 or 10 [31, 32]. This clearly distinguishes these inhibitors from poxviral CrmA which is an efficient inhibitor of caspases 1 and 8, but very weak with caspases 3 and 7, and from the baculovirus inhibitor p35 which seems to inhibit most caspases with comparable efficiency [30, 33, 34] (Q. Zhou and G. Salvesen, unpublished observations).

Caspase enzymatic characteristics

As demonstrated in the three-dimensional models of caspases 1 and 3 [35–39], active enzyme contains two heterodimers interacting via the small chains, so each molecule contains two active sites. Inspection of these structures shows that the peptide chain of a substrate binds into a surface groove on the enzyme, with substrate side-chains adapting to pockets on the enzyme surface (Fig. 4). The substrate is positioned over the catalytic machinery to allow attack of the scissile bond of the substrate by the enzyme's nucleophilic Cys, with cleavage following a conventional cysteine protease mechanism. The primary substrate pocket (S_1) is probably common to all caspases and gives them their characteristic dominant specificity for Asp. The only other known protease with a similar preference is the serine protease granzyme B which, as discussed above, is a caspase activator. Distinguishing the caspases from each other are preferences in adjacent pockets (S_2–S_4) which are superimposed on the primary specificity.

Extended substrate specificity

All known caspases clearly cleave after Asp residues in synthetic substrates, but the degree of discrimination versus the next best side-chain at this position in most caspases remains uncharacterized. Since the initial studies using caspase 1 revealed no gain in catalysis beyond occupancy of S_4, further experimental studies on substrate specificity of the caspases have concentrated mainly on occupancy of subsites S_2–S_4. Within this range, all known caspases seem to prefer Glu at P_3 and differences in substrate specificities are largely dominated by individual preferences at P_2 and P_4. Based on subsite mapping using combinatorial libraries exploring the P_2–P_4 positions, Thornberry et al. [40] have divided caspases 1-9 and *C.elegans* CED-3 into three specificity groups according to P_4 preferences. Of these, group II is the most distinct with a high degree of selectivity for Asp at this position. Significantly, this group contains the executioner caspases 3 and 7, whose natural substrates cleaved during apoptosis frequently contain Asp at P_4. Groups I and III are less distinct since they both accept branched aliphatic side-chains fairly well, and the key distinction is that group I, including caspases 1, 4 and 5, the presumed cytokine activators (Fig. 1), will also accept aromatic side-chains. Division into these three groups is useful in understanding the natural substrates and activation pathway, but a note

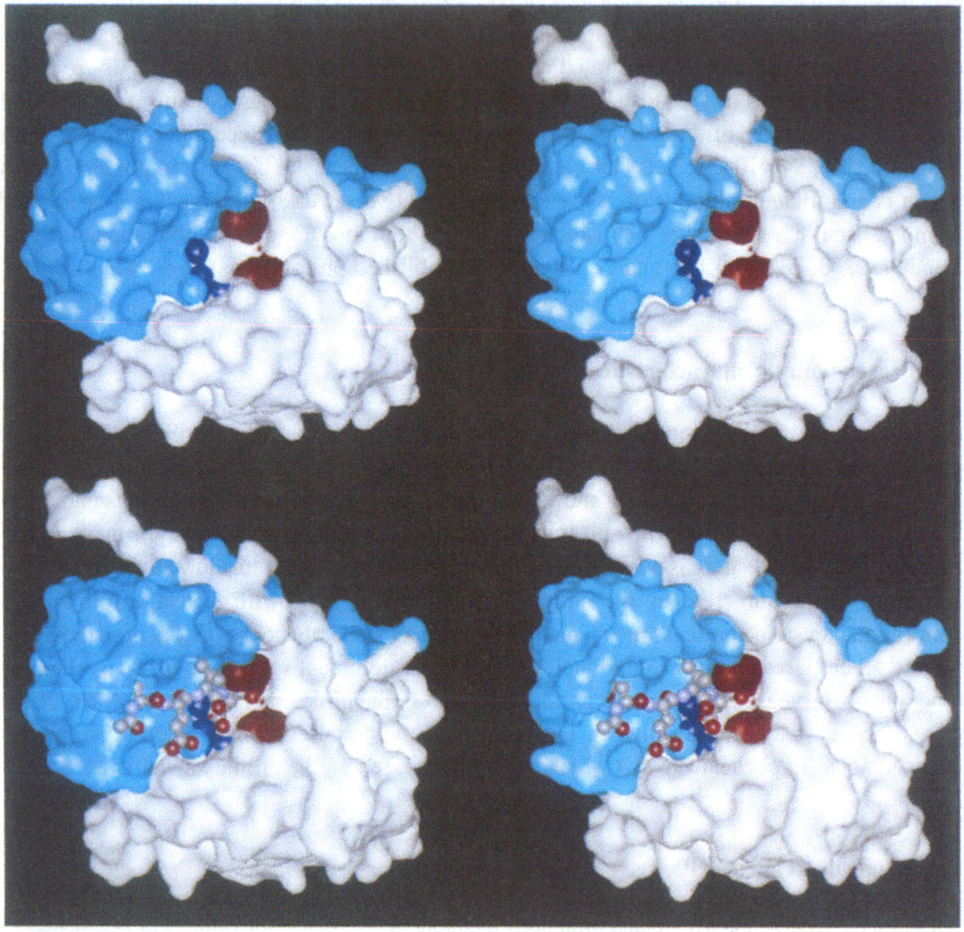

Figure 4. The substrate binding surface. In these stereo pairs of caspase 3 in the standard orientation the S_1 is lined by nitrogen groups on the side chains Arg179, Gln283 and Arg341 (blue) which interacts with the side chain carboxylate of the P_1 Asp residue. The top panel shows the opening to the S_1 pocket and the catalytic His237 and Cys285 (in red) and the bottom panel shows the tetrapeptide aldehyde (Ac-DEVD-CHO) occupying the S_4–S_1 subsites. The patch to the right of the S_1 pocket would be large enough to accommodate P_1' and P_2' if the substrate chain runs in the extended conformation. It is clear from this view that the small subunit (in cyan) contributes to non-prime site substrate interactions and that the large subunit (in white) must contribute to prime site substrate interactions.

of caution is added by Talanian et al. who demonstrated that occupancy of S_5 was necessary for efficient cleavage of substrates by caspase 2 [41].

Based on initial data of the specificity of caspases 1 and 3 for synthetic substrates [42, 43] a number of peptidyl inhibitors have been synthesized, and some have been used by numerous authors in attempts to discriminate the role of individual caspases *in vivo* and in cell free systems. Because caspases are cysteine proteases, reactive groups successful with cysteine proteases of the papain family have been tagged onto simple peptides in an attempt to convey specificity. These reactive groups include reversible aldehydes, and the less reversible halomethyl ketones, various acyloxymethyl ketones (AOMK), and diazomethanes [37, 38, 41, 42, 44–47]. In general, fluoromethyl ketones (FMK) and AOMK derivatives are effective since they show minimal cross reactivity with other proteases, and though the FMK are more widely available from commercial sources, the AOMK are preferred due to greater stability in aqueous solution [44]. FMK and aldehyde derivatives of the tetrapeptide YVAD (based on an early caspase 1 substrate) and DEVD (based on an early caspase 3 substrate) have been utilized by many researchers in attempts to relate specific apoptotic events or cytokine activation to specific caspases. However, as demonstrated in Table 1, it is dangerous to extrapolate very far with these reagents. For example, though the inhibitors containing the YVAD frame are reasonably selective for caspases 1 and 4, those based on DEVD show little selectivity. Table 1 is compiled from various literature sources, and is far from complete since not even one inhibitor has been published with all caspases. Nevertheless, the general trend seems to indicate that tetrapeptide-based inhibitors are unlikely to achieve the specificity required to allow inhibition of individual caspases.

Table 1. Specificity of synthetic inhibitors. Apparent K_is for the reversible inhibitors and apparent inactivation constants for the irreversible ones are compared with different caspases*. Note that a reasonable degree of specificity is achieved only for caspases 1 and 4, with Tyr in P_4. Highly selective inhibitors of the other caspases have yet to be reported. A range of values for an inhibitor signifies that several different values have been published.

	Caspase-1	Caspase-2	Caspase-3	Caspase-4	Caspase-7
Reversible (nM)					
Ac-YVAD-CHO	0.7–6 [a,b]		$1.2–5 \times 10^5$ [a,c]	14 [a]	$0.1–5 \times 10^5$ [a,e]
Ac-WEHD-CHO	0.056 [b]				
Ac-DEVD-CHO	15 [a]	1,750 [d]	0.52–0.8 [a,c]	135 [a]	1.8–35 [a,e]
Ac-VDVAD-CHO		3.5 [d]	1.0 [d]	7.5 [d]	
EVD-CHO	360 [a]		0.65×10^5 [a]	340 [a]	2×10^5 [a]
Irreversible (s^{-1} M^{-1})					
Z-VAD-DCB	7×10^5 [a]		0.13×10^5 [a]	4.5×10^5 [a]	3,600 [a]
Ac-YVAD-CHN_2	0.15×10^5 [a]		<500 [a]	6,300 [a]	<200 [a]

*The table is not a complete compilation of caspase inhibitors since the majority of inhibitors have only been thoroughly characterized with caspase 1. Apparent K_i values and apparent inactivation rate constants have been obtained from [a][37]; [b][47]; [c] [38]; [d] [41]; [e][46].

Biochemical properties of the caspases

Characterization of caspases 3, 6, 7 and 8 reveals that there are only minor differences observed in the pH profiles of the four caspases, and all are maximally active within the pH range and ionic strength of cell cytosols [48]. They all exhibit a bell-shaped pH dependence with optima in the range 6.8–7.2, signifying the existence of one active form of the enzyme with the increase in activity most likely due to the de-protonation of the catalytic Cys residue. In this respect the caspases closely resemble other unrelated cysteine proteases such as papain in their activity pH profiles although the pH-profile is much more narrow than that found for papain [49]. Caspase 3 was found to be active over a broader pH range with an optimum slightly higher than the other three. Though the pH dependence of all four enzymes may be described as a simple bell-shaped curve, there is a faster than expected drop-off in activity at low pH, most clearly observed with caspases 3 and 6. This indicates that more than one group is protonating, possibly another group on the enzyme, or the substrate carboxylate(s).

The influence of transition metal ions on the activity of cysteine proteases has been well established for a long time, being predominantly due to interaction with the catalytic thiol. It is therefore not surprising that the caspases are sensitive to Zn^{2+}, being completely inhibited in the mM range [48], although there are significant differences in their affinity. Caspase 6 is most readily inhibited by Zn^{2+} (completely inactivated by 0.1 mM) and caspase 3 is the least sensitive, requiring more than 1 mM for complete inactivation. The inhibition of caspases by Zn^{2+} may explain the inhibitory action of this metal on apoptosis [50, 51], though the interaction is blocked by thiol compounds [48] and therefore presumably highly dependent on the redox potential of the cell. Ca^{2+} has no effect on the activity of caspases 3, 6, 7 and 8 at concentrations up to 100 mM. Thus, the reported role of Ca^{2+} in apoptosis (see for example [52]), is unlikely to be due to any effect on the caspases.

Three-dimensional structure of caspases

The three-dimensional structures have thus far been determined for caspases 1 (2.5 and 2.6 Å resolution) [35–37] and 3 (2.3 and 2.5 Å resolution) [38, 39]. The two structures are in general quite similar, both demonstrating a molecule composed of two large and two small subunits interacting with inverse two-fold symmetry at the interface of the small subunits. Each symmetry-related unit of the tetramer contains a large five stranded parallel β-sheet with a single anti-parallel strand near the C-terminus; this anti-parallel strand is one of the main contributors in the formation of the tetramer, which is hypothesized to be the active species (Fig. 5).

From a mechanistic point of view the caspase contains a catalytic Cys-His pair, equivalent to that found in all other cysteine proteases, with Cys285 acting as the nucleophile and His237 acting as the general base to abstract the proton from the catalytic Cys, thus promoting the nucleophile. In other cysteine proteases such as papain it is believed that the Cys-His dyad exists as an ion-pair in which the thiol proton of the catalytic Cys has been transferred to the catalytic His prior to substrate binding [53]; this mechanism of action represents one of the major differences between the serine and cysteine proteases. Further comparisons between papain and the caspases reveal distinctive components of catalysis.

The majority of the members of the cysteine proteases contain an Asn side-chain that is believed to orientate the histidine in the Cys-His ion-pair, thereby determining the pH-dependence of the enzyme [54]. A significant deviation from the papain family is demonstrated in the identity of the third component of the catalytic triad, which in caspases is the backbone nitrogen of residue 177. Interestingly, substitution of Asn175 in papain results in a consider-

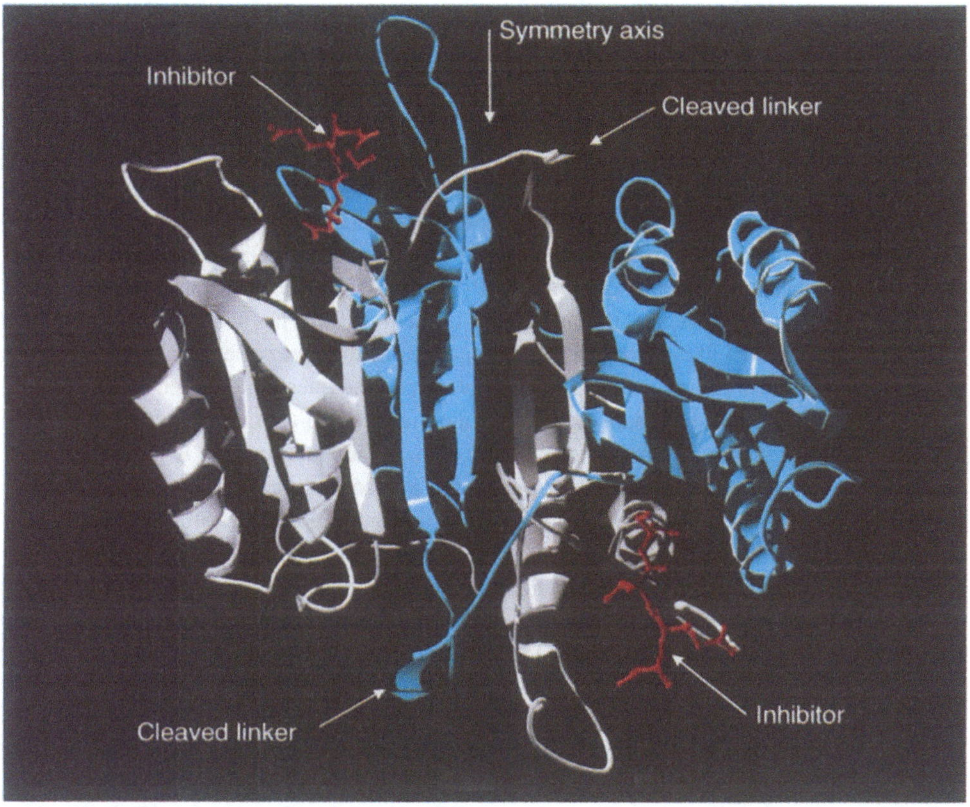

Figure 5. Tetrameric structure of caspase 3. Both the caspase 1 and 3 structures show the caspases existing as tetramers consisting of two large and two small subunits giving rise to a two-fold symmetry of the dimer. As pointed out in the original publications, one of the consequences of this tetrameric structure is that its formation may be achieved by interdigitation of the small subunit with the large subunit of a separate molecule. This is signified in the figure by displaying the precursor molecules in different colors, which places the cleavage sites that resulted in activation close to each other. Published data are still controversial as to whether interdigitation is required for activation, or whether already activated molecules simply associate to form the tetramer.

able narrowing of the pH profile due to the increased distance between the catalytic residues [54]. Indeed, examination of the structure of papain complexed with leupeptin reveals the distance between the catalytic residues to be 3.75 Å [55], which is significantly less than the distance of 5.2 Å found in the caspase:aldehyde structures [35, 39]. Another difference between these proteases is that papain stabilizes the oxyanion that is developed during catalysis by two H-bonds, whereas the caspases utilize a single H-bond. Removal of one of these in papain (Gln19Ala) results in a narrowing of the pH profile, increasing the pK value of the acidic leg by approximately 0.5 pH units [56]. Thus the narrow pH profile of the caspases, compared to papain, may be explained by looser interactions in both the catalytic triad and oxyanion pocket.

Another key feature in the function of any enzyme is the ability to discriminate between substrates. In the caspases this discrimination is primarily based on the recognition of Asp in the P_1 position. In both the structures solved so far the S_1 subsite contains three residues interacting directly with the P_1 Asp–Arg179, Gln283 and Arg341. Together these three residues form the most prominent binding site in the caspases; this binding site is often referred to as the Asp-hole since it shows up as a hole on the surface of the active site (Fig. 4). Of the three residues involved in P_1 binding, Arg179 appear to be the primary hydrogen bond donor in the binding of the negatively charged P_1 residue side-chain, whereas Gln283 and Arg341 each donate a single hydrogen bond. Arg341 also contributes to the binding of the P_3 Glu in the caspase 3 structures as well as backbone interactions with the P_3 residue. The issue of the protonic state of the Arg residues in the S_1 subsite and the side-chains of the P_1 residue has so far not been resolved, however, it seems most likely from the pH dependence of these enzymes that the P_1 Asp is in the deprotonated form during catalysis. This leaves open the question of whether Arg179 and Arg341 interact with the side-chain carboxyl group via a charge-charge interaction or through hydrogen bonding. If the enzyme relies on charge-charge interactions, the Arg residues, or at least one of them, must be protonated, however, due to electrostatic repulsion it seems unlikely that both should be protonated. Alternatively they may both be deprotonated and simply bind the charged side-chain via strong hydrogen bonds between a charged and an uncharged partner as seen in a number of other systems, such as the recognition of the C-terminal carboxyl group of the serine carboxypeptidases [57] and binding of the P_1 side-chain in the glutamic acid specific protease from *Streptomyces griseus* [58].

As discussed previously the caspases do not exhibit any significant P_2 substrate preference and this lack of preference in the S_2 subsite is also evident in the three-dimensional structures of caspases 1 and 3. None of the structures solved so far have shown a well defined binding site for P_2 and, as may be seen in the surface drawing of caspase 3 complexed with Ac-DEVD-CHO, the area surrounding the P_2 Val is open and flat with only a few non-specific interactions (Fig. 4). Thus, the substrate binding mode in the S_2 position of the caspases resembles that of the chymotrypsin-family proteases rather than that of the papain-family proteases.

Distinctions between the caspase 1 and 3 structures

The caspase 1 structure complexed with Ac-YVAD-CHO does not show any significant P_3 interactions. However, when complexed with Ac-DEVD-CHO, caspase 1 forms a hydrogen bond between the P_3 Glu and the guanido group of Arg341 in accordance with the structure of caspase 3 complexed with the same inhibitor.

One of the major differences between caspase 1 and 3 is the size of loop 3, which in caspase 3 contains ten amino acid residues more than in caspase 1. This difference in the size of loop 3 does to some extent explain the difference in the S_4 substrate specificity of these enzymes, i.e. the preference for a large hydrophobic residue in the case of caspase 1 and a small negatively charged residue in the case of caspase 3. In caspase 1 the P_4 residue is accommodated by hydrophobic components of residues Trp340, His342 and Arg383 as well as the backbone of Arg341. Together these residues form a narrow cleft interacting with the hydrophobic

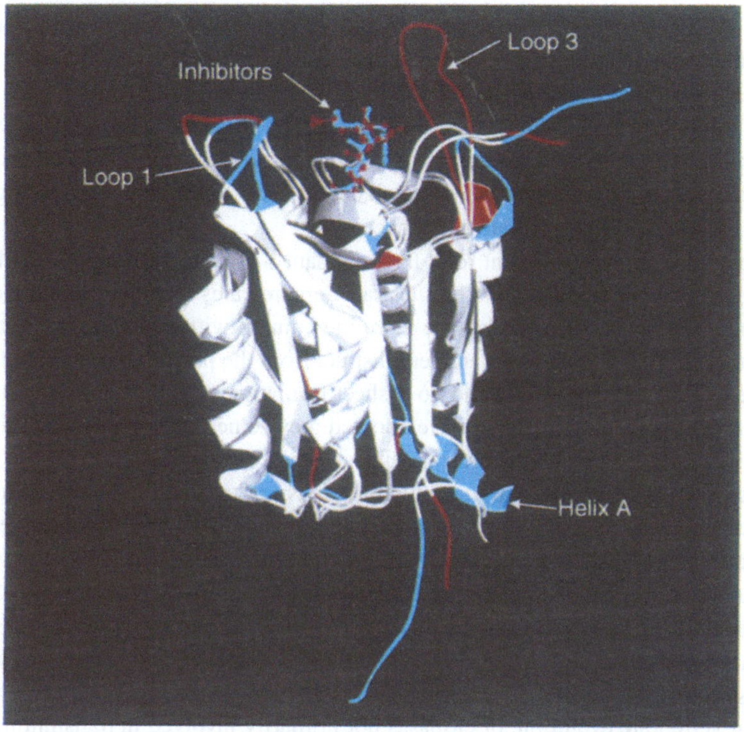

Figure 6. Structural organization of the caspase fold. Both caspases 1 and 3 reveal a conserved fold consisting of a central β-sheet core with strands (in the order 213456) sandwiched between five α-helixes. When superimposed based on the Cα groups the structures of caspases 1 and 3 are found to have a strikingly similar topology. In fact large parts of the molecules can be superimposed with a root mean square deviation of only about 1.3 Å. However, compared to caspase 3 the structure of caspase 1 shows an additional α-helix, which is positioned perpendicular to the other helices at the bottom of the enzyme. Other significant differences are the extended loop 3 found in caspase 3, which is believed to play an important role in the preference for Asp in the S_4 pocket of caspase 3, and loop 1 in caspase 1, which may be defining the prime site specificity of this enzyme. Segments that differ between the two caspases have been colored cyan (caspase 1) and red (caspase 3) while segments superimposing are colored white.

part of the P_4 Tyr, but open to solvent at the position of the hydroxyl group, thus making caspase 1 capable of recognizing a variety of aromatic P_4 substituents. The three-dimensional structure of caspase 1 complexed with Ac-DEVD-CHO indicates that there are no specific interactions between the P_4 Asp residue and the enzyme since there are no well defined electron densities for this part of the structure [37]. In contrast, the recognition of the P_4 in caspase 3 is based on specific interactions between the enzyme and substrate. The P_4 Asp carboxyl group is accommodated by hydrogen bonds to the side-chains of Asn342 and Trp214 as well as the backbone amide of Phe250. This requirement for exact hydrogen bond interactions results in a large degree of discrimination for Asp over Glu in the P_4 position.

Another significant feature related to the enzyme substrate interactions is the influence of peptide backbones on substrate/inhibitor binding. Although the importance of these backbone interactions is difficult to investigate in detail due to the problems associated with changing the structure of the backbone, this is a central part of the binding of the substrate during catalysis. In both the caspase 1:Ac-YVAD-CHO and the caspase 3:Ac-DEVD-CHO structures the backbone of the inhibitor forms an anti-parallel β-sheet like structure with residues Ser339 to Arg341. Similar binding modes are found for the chymotrypsin family of proteases, whereas the papain class of enzymes do not show an elaborate backbone hydrogen bonding network between the enzyme and the substrate backbone.

Though placed on a very different fold, the dominant S_1 specificity and β-sheet like backbone hydrogen-bond interactions with the substrate demonstrates that the overall mode of substrate binding observed with caspases have more in common with serine proteases than other cysteine proteases. Thus the caspases are very distinct from the papain family of proteases that display a predominant P_2 specificity and no β-sheet formation with the substrate. The caspases can be considered as proteases that have adopted a cysteine nucleophile to hydrolyze proteins bound in a more-or-less serine protease-like substrate conformation.

As pointed out earlier the primary specificity for Asp is unusual among mammalian proteases, and this hints at an evolutionary explanation. Since the apoptotic response seems to be of fundamental importance to metazoans one can argue that there would be strong selective pressure against other proteases that may develop Asp specificity and inadvertendly trigger death. Consequently, it is likely that the caspase substrates required to deliver the classic apoptotic effect have evolved alongside the caspases during the course of metazoan radiation to adapt to the caspase primary specificity. Somewhat surprising in this context is the emergence, somewhere during vertebrate evolution, of caspases not primarily involved in transmitting apoptotic signals, as exemplified by caspase 1. Thus the secondary evolutionary drive would be to ensure that cytokine-activating caspases cannot cleave apoptotic substrates, hence the development of a tightly regulated secondary specificity at the S_4 subsite.

References

1 Raff MC (1992) Social controls on cell survival and cell death. *Nature* 356: 397–400
2 Thompson CB (1995) Apoptosis in the pathogenesis and treatment of disease. *Science* 267: 1456–1462
3 Nicholson DW (1996) ICE/CED3-like proteases as therapeutic targets for the control of inappropriate apoptosis. *Nat Biotechnol* 14: 297–301
4 Tomei LD, Cope FO (1991) *Current Communications in Cell and Molecular Biology, vol. 3*, CSHL Press, Cold Spring Harbor
5 Ellis HM, Horvitz HR (1986) Genetic control of programmed cell death in the nematode *C. elegans*. *Cell* 44: 817–829
6 Yuan J, Horvitz HR (1990) The *Caenorhabditis elegans* genes ced-3 and ced-4 act autonomously to cause programmed cell death. *Dev Biol* 138: 33–41
7 Hengartner MO, Ellis RE, Horvitz HR (1992) *Caenorhabditis elegans* gene ced-9 protects cells from programmed cell death. *Nature* 356: 494–499
8 Yuan J, Shaham S, Ledoux S, Ellis HM, Horvitz HM (1993) The *C. elegans* cell death gene ced-3 encodes a protein similar to mammalian interleukin-1β-converting enzyme. *Cell* 75: 641–652
9 Cohen GM (1997) Caspases: the executioners of apoptosis. *Biochem J* 326: 1–16
10 Nicholson DW, Thornberry NA (1997) Caspases: killer proteases. *Trends Biochem Sci* 22: 299–306
11 Salvesen GS, Dixit VM (1997) Caspases: Intracellular signaling by proteolysis. *Cell* 91: 443–446
12 Kuida K, Lippke JA, Ku G, Harding MW, Livingston DJ, Su MSS, Flavell RA (1995) Altered cytokine export and apoptosis in mice deficient in interleukin-1-beta converting enzyme. *Science* 267: 2000–2003
13 Kuida K, Zheng TS, Na S, Kuan C-Y, Yang D, Karasuyama H, Rakic P, Flavell RA (1996) Decreased apoptosis in the brain and premature lethality in CPP32-deficient mice. *Nature* 384: 368–372

14 Neurath H (1989) Proteolytic processing and physiological regulation. *Trends Biochem Sci* 14: 268–271
15 Orth K, O'Rourke K, Salvesen GS, Dixit VM (1996) Molecular ordering of apoptotic mammalian CED-3/ICE-like proteases. *J Biol Chem* 271: 20977–20980
16 Srinivasula SM, Ahmad M, Fernandes-Alnemri T, Litwack G, Alnemri ES (1996) Molecular ordering of the Fas-apoptotic pathway: the Fas/APO-1 protease Mch5 is a CrmA-inhibitable protease that activates multiple Ced-3/ICE-like cysteine proteases. *Proc Natl Acad Sci USA* 93: 14486–14491
17 Muzio M, Salvesen GS, Dixit VM (1997) FLICE Induced Apoptosis in a Cell-free System. Cleavage of caspase zymogens. *J Biol Chem* 272: 2952–2956
18 Duan H, Orth K, Chinnaiyan AM, Poirier GG, Froelich CJ, He W-W, Dixit VM (1996) ICE-LAP6, a novel member of the ICE/Ced-3 gene family, is activated by the cytotoxic T cell protease granzyme B. *J Biol Chem* 271: 16720–16724
19 Quan LT, Tewari M, O'Rourke K, Dixit V, Snipas SJ, Poirier GG, Ray C, Pickup DJ, Salvesen G (1996) Proteolytic activation of the cell death protease Yama/CPP32 by granzyme B. *Proc Natl Acad Sci USA* 93: 1972–1976
20 Fernandes-Alnemri T, Armstrong R, Krebs J, Srinivasula SM, Wang L, Bullrich F, Fritz L, Trapani JA, Croce CM, Tomaselli KJ et al. (1996) *In vitro* activation of CPP32 and Mch3 by Mch4, a novel human apoptotic cysteine protease containing two FADD-like domains. *Proc Natl Acad Sci USA* 93: 7464–7469
21 Talanian RV, Yang X, Turbov J, Seth P, Ghayur T, Casiano CA, Orth K, Froelich CJ (1997) Granule-mediated Killing: Pathways for Granzyme B-initiated Apoptosis. *J Exp Med* 186: 1323–1331
22 Zhou Q, Salvesen GS (1997) Activation of pro-caspase-7 by serine proteases includes a non-canonical specificity. *Biochem J* 324: 361–364
23 Yamin T-T, Ayala JM, Miller DK (1996) Activation of the native 45-kDa precursor form of interleukin-1-converting enzyme. *J Biol Chem* 271: 13273–13282
24 Muzio M, Chinnaiyan AM, Kischkel FC, O'Rourke K, Shevchenko A, Ni J, Scaffidi C, Bretz JD, Zhang M, Gentz R et al. (1996) FLICE, a novel FADD-homologous ICE/CED-3-like protease, is recruited to the CD95 (Fas/APO-1) death-inducing signaling complex. *Cell* 85: 817–827
25 Boldin MP, Goncharov TM, Goltsev YV, Wallach D (1996) Involvement of MACH, a novel MORT1/FADD-interacting protease, in Fas/APO-1- and TNF receptor-induced cell death. *Cell* 85: 803–815
26 Hofmann K, Bucher P, Tschopp J (1997) The CARD domain: a new apoptotic signalling motif. *Trends Biochem Sci* 22: 155–156
27 Ramage P, Cheneval D, Chbei D, Graff P, Hemmig R, Heng R, Kocher HP, Mackenzie A, Memmert K, Revesz L et al. (1995) Expression, refolding, and autocatolytic proteolytic processing of the interleukin-1beta-converting enzyme precursor. *J Biol Chem* 270: 9378–9383
28 Van Criekinge W, Beyaert R, Van de Craen M, Vandenabeele P, Schotte P, De Valck D, Fiers W (1996) Functional characterization of the prodomain of interleukin1b-converting enzyme. *J Biol Chem* 271: 27245–27248
29 Ray CA, Black RA, Kronheim SR, Greenstreet TA, Sleath PR, Salvesen GS, Pickup DJ (1992) Viral inhibition of inflammation: cowpox virus encodes an inhibitor of the interleukin-1β converting enzyme. *Cell* 69: 597–604
30 Bump NJ, Hackett M, Hugunin M, Seshagiri S, Brady K, Chen P, Ferenz C, Franklin S, Ghayur T, Li P, et al. (1995) Inhibition of ICE family proteases by baculovirus antiapoptotic protein p35. *Science* 269: 1885–1888
31 Deveraux Q, Takahashi R, Salvesen GS, Reed JC (1997) X-linked IAP is a direct inhibitor of cell death proteases. *Nature* 388: 300–304
32 Roy N, Deveraux QL, Takahashi R, Salvesen GS, Reed JC (1997) The c-IAP-1 and c-IAP-2 proteins are direct inhibitors of specific caspases. *EMBO J* 16: 6914–6925
33 Bertin J, Mendrysa SM, LaCount DJ, Gaur S, Krebs JF, Armstrong RC, Tomaselli KJ, Friesen PD (1996) Apoptotic suppression by baculovirus P35 involves cleavage by and inhibition of a virus-induced CED-3/ICE-like protease. *J Virol* 70: 6251–6259
34 Xue D, Horvitz HR (1995) Inhibition of the *Caenorhabditis elegans* cell-death protease CED-3 by a CED-3 cleavage site in baculovirus p35 protein. *Nature* 377: 248–251
35 Wilson KP, Black JA, Thomson JA, Kim EE, Griffith JP, Navia MA, Murcko MA, Chambers SP, Aldape RA, Raybuck SA et al. (1994) Structure and mechanism of interleukin-1 beta converting enzyme. *Nature* 370: 270–275
36 Walker NPC, Talanian RV, Brady KD, Dang LC, Bump NJ, Ferenz CR, Franklin S, Ghayur T, Hackett MC, Hammill LD et al. (1994) Crystal structure of the cysteine protease interleukin-1beta-converting enzyme: A (p20/p10)2 homodimer. *Cell* 78: 343–352
37 Margolin N, Raybuck SA, Wilson KP, Chen W, Fox T, Gu Y, Livingston DJ (1997) Substrate and inhibitor specificity of interleukin-1 beta-converting enzyme and related caspases. *J Biol Chem* 272: 7223–7228
38 Mittl PR, Di Marco S, Krebs JF, Bai X, Karanewsky DS, Priestle JP, Tomaselli KJ, Grutter MG (1997) Structure of recombinant human CPP32 in complex with the tetrapeptide acetyl-Asp-Val-Ala-Asp fluoromethyl ketone. *J Biol Chem* 272: 6539–6547
39 Rotunda J, Nicholson DW, Fazil KM, Gallant M, Gareau Y, Labelle M, Peterson EP, Rasper DM, Tuel R, Vaillancourt JP et al. (1996) The three-dimensional structure of apopain/CPP32, a key mediator of apoptosis. *Nat Struct Biol* 3: 619–625

40 Thornberry NA, Rano TA, Peterson EP, Rasper DM, Timkey T, Garcia-Calvo M, Houtzager VM, Nordstrom PA, Roy S, Vaillancourt JP et al. (1997) A combinatorial approach defines specificities of members of the caspase family and granzyme B. *J Biol Chem* 272: 17907–17911

41 Talanian RV, Quinlan C, Trautz S, Hackett MC, Mankovich JA, Banach D, Ghayur T, Brady KD, Wong WW (1997) Substrate specificities of caspase family proteases. *J Biol Chem* 272: 9677–9682

42 Thornberry NA, Bull HG, Calaycay JR, Chapman KT, Howard AD, Kostura MJ, Miller DK, Molineaux SM, Weidner JR, Aunins J et al. (1992) A novel heterodimeric cysteine protease is required for interleukin-1beta processing in monocytes. *Nature* 356: 768–774

43 Nicholson DW, Ali A, Thornberry NA, Vaillancourt JP, Ding CK, Gallant M, Gareau Y, Griffin PR, Labelle M, Lazebnik YA et al. (1995) Identification and inhibition of the ICE/ced-3 protease necessary for mammalian apoptosis. *Nature* 376: 37–43

44 Thornberry NA, Peterson EP, Zhao JJ, Howard AD, Griffin PR, Chapman KT (1994) Inactivation of Interleukin-1β Converting Enzyme by Peptide (Acyloxy)methyl Ketones. *Biochemistry* 33: 3934–3940

45 Prasad CVC, Prouty CP, Hoyer D, Ross TM, Salvino JM, Awad M, Graybill TL, Schmidt SJ, Osifo IK, Dolle RE et al. (1995) Structural and stereochemical requirements of time-dependent inactivators of the interleukin-1β converting enzyme. *Bioorg Med Chem Lett* 5: 315–318

46 Fernandes-Alnemri T, Takahashi A, Armstrong R, Krebs J, Fritz L, Tomaselli KJ, Wang L, Yu Z, Croce CM, Salvesen G et al. (1995) Mch3, a novel human apoptotic cysteine protease highly related to CPP32. *Cancer Res* 55: 6045–6052

47 Rano TA, Timkey T, Peterson EP, Rotonda J, Nicholson DW, Becker JW, Chapman KT, Thornberry NA (1997) A combinatorial approach for determining protease specificities: application to interleukin-1β converting enzyme (ICE). *Chem Biol* 4: 149–155

48 Stennicke HR, Salvesen GS (1997) Biochemical characteristics of caspases 3, 6, 7 and 8. *J Biol Chem* 272: 25719–25723

49 Menard R, Carriere J, Laflamme P, Plouffe C, Khouri HE, Vernet T, Tessier DC, Thomas DY, Storer AC (1991) Contribution of the glutamine 19 side chain to transition-state stabilization in the oxyanion hole of papain. *Biochemistry* 30: 8924–8928

50 Takahashi A, Alnemri ES, Lazebnik YA, Fernandes-Alnemri T, Litwack G, Moir RD, Goldman RD, Poirier GG, Kaufmann SH, Earnshaw WC (1996) Cleavage of lamin A by Mch2 alpha but not CPP32: multiple interleukin 1 beta-converting enzyme-related proteases with distinct substrate recognition properties are active in apoptosis. *Proc Natl Acad Sci USA* 93: 8395–8400

51 Perry DK, Smyth MJ, Stennicke HR, Salvesen GS, Duriez P, Poirier GG, Hannun YA (1997) Zinc is a potent inhibitor of the apoptotic protease, caspase-3. A novel target for zinc in the inhibition of apoptosis. *J Biol Chem* 272: 18530–18533

52 Bian X, Hughes FM Jr, Huang Y, Cidlowski JA, Putney JW Jr (1997) Roles of cytoplasmic Ca^{2+} and intracellular Ca^{2+} stores in induction and suppression of apoptosis in S49 cells. *Amer J Physiol* 272: C1241–C1249

53 Lewis ER, Johnson FA, Shafer JA (1981) Effect of cystein-25 on the ionization of histidine-159 in papain as determined by proton nuclear magnetic resonance spectroscopy. Evidence for a His-159-Cys-25 ion pair and its possible role in catalysis. *Biochemistry* 20: 48–51

54 Vernet T, Tessier DC, Chatellier J, Plouffe C, Lee TS, Thomas DY, Storer AC, Menard R (1995) Structural and functional roles of asparagine 175 in the cysteine protease papain. *J Biol Chem* 270: 16645–16652

55 Schroder E, Phillips C, Garman E, Harlos K, Crawford C (1993) X-ray crystallographic structure of a papain-leupeptin complex. *FEBS Lett* 315: 38–42

56 Menard R, Plouffe C, Laflamme P, Vernet T, Tessier DC, Thomas DY, Storer AC (1995) Modification of the electrostatic environment is tolerated in the oxyanion hole of the cysteine protease papain. *Biochemistry* 34: 464–471

57 Mortensen UH, Remington SJ, Breddam K (1994) Site-Directed Mutagenesis on (Serine) Carboxypeptidase-Y – A Hydrogen Bond Network Stabilizes the Transition State by Interaction with the C-Terminal Carboxylate Group of the Substrate. *Biochemistry* 33: 508–517

58 Stennicke HR, Birktoft JJ, Breddam K (1996) Characterization of the S1 binding site of the glutamic acid-specific protease from *Streptomyces griseus*. *Protein Sci* 5: 2266–2275

Proteases: New Perspectives
V. Turk (ed.)

Lysosomal cysteine proteinases: Structure and regulation

Vito Turk, Gregor Gunčar and Dušan Turk

Department of Biochemistry and Molecular Biology, J. Stefan Institute, Jamova 39, 1000 Ljubljana, Slovenia

Introduction

Cysteine peptidases (EC 3.4.22) are widely distributed enzymes in animals, plants and microorganisms. There are many distinct families of cysteine peptidases (also called proteinases, proteases, proteolytic enzymes) the largest and best known being the papain family [1, 2]. The papain family consists of peptidases ubiquitous in plant and animal cells. Until recently, information about mammalian papain-like cysteine peptidases has been limited only to lysosomal cathepsins B, H, L and S. While human cathepsins B, H and L are expressed in all tissues, cathepsin S is selectively expressed in antigen presenting cells, indicating a specific role in the activation of the MHC-class II pathway. These enzymes have been purified from different tissues and species, including human, characterised, and their complete amino acid sequences determined [3–5]. They are synthesised as glycosylated precursors and then converted to the mature enzymes by limited proteolysis either by other proteinases or by autocatalysis. The proregion serves as a regulator of catalytic activity. The propeptides of cathepsins B and L have been shown to be potent inhibitors of the mature enzymes [6, 7]. The proregion is also required for the proper folding of the protein [8]. Cathepsins B, H, L and S differ in their specificity, cathepsin L and S being endopeptidases, cathepsin H acting as an aminopeptidase and cathepsin B as a dipeptidyl carboxypeptidase [5].

Cathepsin C, later termed dipeptidyl-peptidase I, differs from cathepsins B, H, L and S in its large molecular size, oligomeric structure, retention of a large part of the propeptide in the mature form and requirement of halide ions for activity [9–12]. Using a new molecular biology approach, several human enzymes of the papain-like peptidases have been discovered. Novel proteins with high sequence homology to papain, cathepsin L and S have been cloned, expressed and characterized independently by several groups and referred to variously as cathepsin O [13, 14], cathepsin K [15] and cathepsin O_2 [16]. The name cathepsin K (EC 3.4.22.38) has recently been assigned to them. Cathepsin K is highly expressed in osteoclasts and may play a central role in bone resorption. In contrast, relatively low levels of cathepsins L, S and B were found within osteoclasts [17].

Recently, a novel human cysteine peptidase has been identified from the dbEST databank and named cathepsin W [18, 19]. Cathepsin W expression is restricted to human $CD8^+$, T lymphocytes, suggesting a specific function in the regulation of the cytolytic activity of cytotoxic T cells. Another recently identified and characterised human cysteine proteinase of the papain family, tentatively called cathepsin Z, is widely expressed in normal human tissues and in human cancer cell lines and primary tumours [20].

Three-dimensional structures have been determined by X-ray crystallography for several members of the papain family, starting with papain itself [21]. Among the lysosomal cysteine proteinases, the structures of human [22] and rat [23] cathepsin B, human cathepsin L [24], human cathepsin K [25, 26], and recently porcine cathepsin H [27] and human cathepsin S [28] have been determined. Their structures, together with the crystal structure of cruzipain [29], the major cysteine proteinase present in a protozoan parasite *Trypanosoma cruzi*, show that all papain-like enzymes so far investigated share a common fold. The mature enzymes consist of two domains which separate on the top in a "V" shaped active site cleft, in the middle of which are the catalytic residues Cys 25 and His 159 (papain numbering). The binding sites within the active site cleft provide the structural basis for enzyme specificity which, within this group of enzymes, is relatively broad.

In mammals, lysosomal cysteine proteinases represent an important part of the lysosomal protein degradation system. Proteins are degraded in lysosomes non-selectively, and the resulting end-products, dipeptides and amino acids, diffuse through the lysosomal membrane and are re-used in protein biosynthesis [5, 30]. Lysosomal cathepsins have been shown to be capable of specific processing of other proteins, e.g. hormones [31]. Recent studies have elucidated the role of some cysteine proteinases in degrading the invariant chain (Ii) and regulating the association of processed antigen and major histocompatibility complex (MHC) class II dimers competent for peptide loading [32].

In order to prevent unwanted proteolysis by the proteinases, their potentially highly destructive activity must be regulated [33, 34]. There are two main mechanisms of regulation of cysteine proteinase activity: zymogen activation and proteinase-inhibitor interaction. The crystal structure of procathepsin B has shown that the N-terminal propeptide inhibits the enzyme by blocking access to the already formed active site [35–37]. However, the main means of regulating cysteine proteinase activity is by interaction with their endogenous inhibitors, cystatins [38]. The recently discovered thyroglobulin type-1 domain inhibitors, thyropins, strongly inhibit (rather selectively) cysteine proteinases [39]. In addition, the activity of cysteine proteinases is regulated by pH [40–42].

Disturbance of the normal balance of enzymatic activity may lead to the development and progression of a variety of human diseases, including cancer and metastasis, Alzheimer's disease, pancreatitis, arthritis, gingivitis and periodontitis, inflammation, lung disorders, muscle dystrophy, Chagas disease and trauma [5].

There are numerous excellent papers and reviews on this important group of proteinases. This chapter will focus on current information about the structure, function, regulation and biological role of lysosomal cysteine proteinases.

Biochemical properties of cathepsins

Lysosomal cathepsins B, H, L, S, K and C are well characterized enzymes. They are monomers with low molecular masses in the range of 24–35 kDa, with the exception of cathepsin C, which is a tetramer with a molecular mass of 200 kDa [9]. Cathepsins B, H and L were found to exist in a single-chain and/or a two-chain form with the proportions of the single and two-chain forms varying from species to species. No evidence exists to indicate that the cleavage is the consequence of autoprocessing. In contrast, mature cathepsins S and K are single-chain proteins. In general, the catalytic activity of cathepsins is dependent on a pH below 7.0 and they are unstable at neutral and alkaline pH values. The stability of cathepsin S above pH 7.0 is a distinctive characteristic of this enzyme. Cathepsin K is significantly more stable at pH 6.5 than cathepsin L, but less stable than cathepsin S (reviewed in [43]).

Cathepsins B, H, L, S, K and C are susceptible to the various classes of irreversible inhibitors that have been developed such as peptidyl chloromethane, diazomethane and fluoromethane [44–46]. Selectivity has been obtained by a new type of irreversible cysteine proteinase inhibitor identified as 1-[L-N-(trans-epoxysuccinyl)leucyl]amino-4-guamidinobutane, and named E-64 [47]. Several E-64 derivatives have been synthesized, and one of them, named CA030 (ethylester of epoxysuccinyl – Ile-Pro-OH) proved to be a specific inhibitor of cathepsin B [48].

Numerous protein inhibitors of cysteine proteinases have been described. The most abundant are the cystatins, which are reversible and tight-binding protein inhibitors displaying structural and functional similarities. On the basis of sequence homology, they have been subdivided into three families, stefins, cystatins and kininogens (reviewed in [38, 49, 50]). They differ in their inhibitory specificity against different cysteine proteinases [51, 52]. The decreased affinity of these inhibitors toward cathepsin B may be explained by the steric hindrance caused by the partial occlusion of the active site cleft of cathepsin B, as seen from its crystal structure [22]. Similarly, the decreased affinity for cathepsin H by some cystatins might be explained by competition with the mini-chain residues for the same binding site [27]. This is in agreement with the suggestion that reduced binding of stefins with elongated amino termini is caused by the mini-chain of cathepsin H [53]. Modeling of complexes between stefins and cathepsin H, based on the stefin B-papain complex [54], suggests that the N-terminal trunk of stefin B collides with the mini-chain residues, in particular with the Thr 83 P [27].

The discovery that a fragment of the p41 invariant chain associated with MHC class II strongly inhibits cathepsin L [55] and cruzipain from *T. cruzi* [56] suggested the appearance of a new superfamily of protein inhibitors of cysteine proteinases. This finding was confirmed by the discovery of a new inhibitor, equistatin, isolated from sea anemone *Actinia equina,* which binds tightly and rapidly to cathepsin L ($K_i = 0.051$ nM) and papain ($K_i = 0.57$ nM). This provided the now firmly established basis for a new superfamily of inhibitors, named thyropins [39].

Three-dimensional structures

Papain-like cysteine proteinases share similar sequences [1, 2] as well as 3-dimensional structures [56]. The structural data are strong evidence that they all arise from a common ancestor. The structures of the plant enzyme papain [21], of mammalian cathepsin B [22], cathepsin H [27], cathepsin L [57] and of cruzipain from *T. cruzi* (only the catalytic domain lacking the C-terminal domain) [29] show that these enzymes share a common fold (Fig. 1), similar to cathepsin K and S (not shown). The mature enzymes consist of two domains which separate on the top in a "V" shaped active site cleft, in the middle of which are the catalytic residues Cys

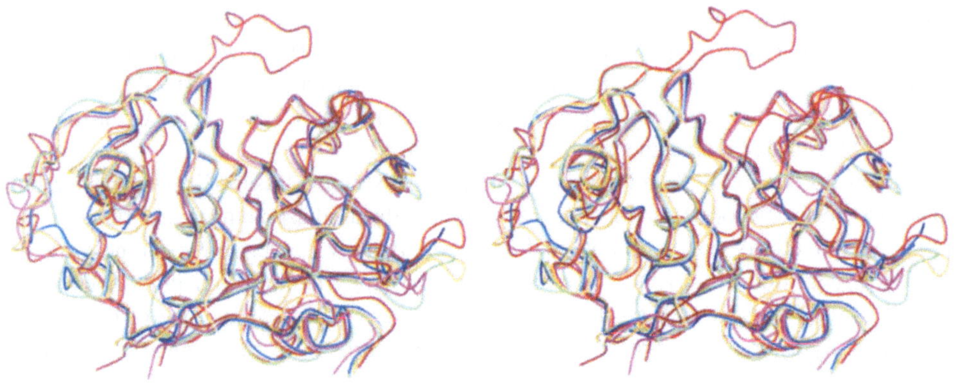

Figure 1. Cα-worm stereo representation of the superimposed cathepsin B (red), cathepsin H (yellow), papain (magenta), cruzipain (green) and a mature part of the procathepsin L (blue). Figure was made using the program MOLSCRIPT [90] and rendered using Raster 3D [91].

25 and His 159 (papain numbering), one from each domain, which form the catalytic site of the enzyme. The domains are referred to as the left (L-) and right (R-) domains according to their positions in the standard view shown in Figure 1. The most prominent feature of the L-domain is the central α helix, whereas the R-domain forms a kind of α β barrel which also includes a shorter α-helical motif. The interactions between the domains have hydrophilic as well as hydrophobic character, and are specific for the particular enzyme. Cys 25 and His 159 form, between pH ~ 3.5 and 8.0, a thiolate-imidazolium ion pair essential for the enzyme activity. Cathepsin C and cruzipain have additional domains attached to these two domains, however, their structures are not known.

The structures of cathepsin K [25, 26], cathepsin L [24, 57] and cathepsin S [28] show a high degree of sequence homology. Cathepsin S shows 57% identity to cathepsins L and K, and 31% identity to cathepsin B [22]. The differences in the hydrophobic S2 and S3 pockets between cathepsin K, cathepsin S and cathepsin L could be important for structure-based design of selective inhibitors.

Most of the enzymes exhibit predominantly endopeptidase activity. The main reason for their differentiation according to their endo- and exopeptidase activity seems to be defined by adding features to the rather well conserved endopeptidase basic structure, as shown for cathepsins K, B and H (Fig. 2a, b, and c, respectively). The structure of human cathepsin B revealed that an insertion of about 20 residues, termed the occluding loop, blocks the active site cleft at the back of its S' site. It has been suggested that the occluding loop residues His 110 and His 111 with their positive charges bind the C-terminal carboxylic group of the P2' residue thus making cathepsin B a carboxydipeptidase [22]. This suggestion was later confirmed with the structure of cathepsin B in a complex with CA030, a substrate-mimicking inhibitor [58]. The occluding loop does not allow cystatin-like inhibitors to bind to cathepsin B as they do to papain and other endopeptidases. Removal of the flexible part of the occluding loop converts the enzyme into an

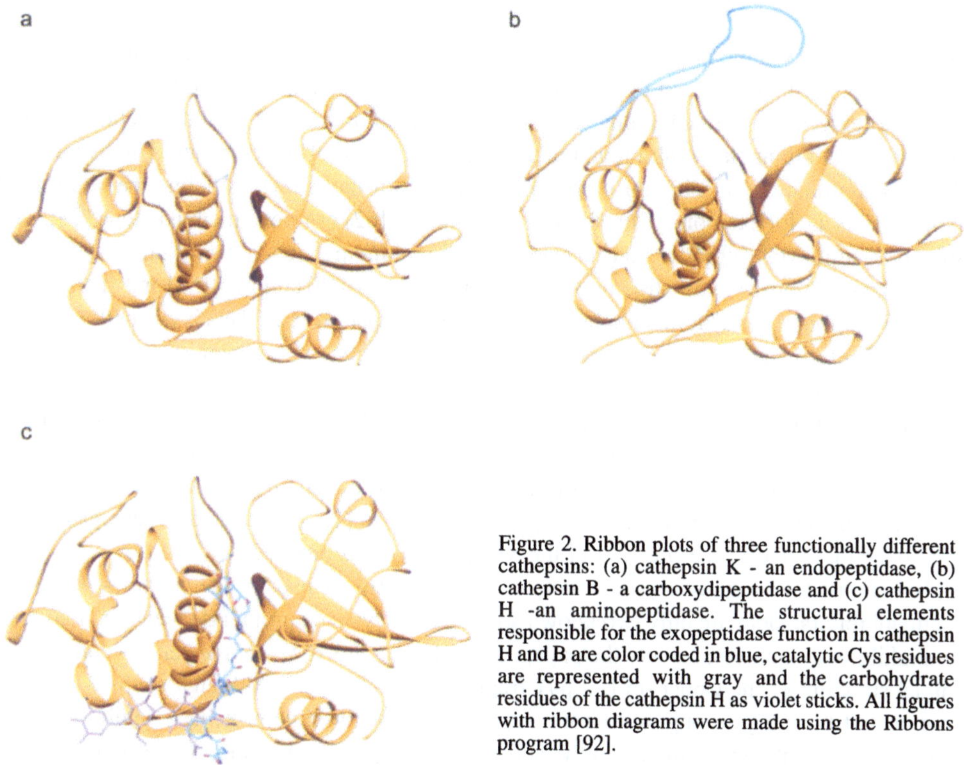

Figure 2. Ribbon plots of three functionally different cathepsins: (a) cathepsin K - an endopeptidase, (b) cathepsin B - a carboxydipeptidase and (c) cathepsin H -an aminopeptidase. The structural elements responsible for the exopeptidase function in cathepsin H and B are color coded in blue, catalytic Cys residues are represented with gray and the carbohydrate residues of the cathepsin H as violet sticks. All figures with ribbon diagrams were made using the Ribbons program [92].

endopepdidase, with no carboxydipeptidase activity, and increases its affinity for the inhibitor cystatin C [59].

Another exopeptidase with known crystal structure, cathepsin H [27], contains an octapeptide called the mini-chain, which binds into the active site cleft (Fig. 2c) thus defining its aminopeptidase activity. The octapeptide EPQNCSAT is attached via a disulphide bond to the body of the enzyme and bound in a narrowed active site cleft in the direction of a bound substrate with the negatively charged carboxylic group of its C-terminal Thr 83P attracting the positively charged N-terminus of a substrate. The mini-chain fills the region that in the related enzymes comprises the non-primed substrate-binding sites from S2 backwards. A carbohydrate chain further tightens the active site cleft, presumably enhancing binding of the mini-chain.

Cathepsin H behaves similarly to cathepsin B in that, besides its exopeptidase activity, it exhibits some endopeptidase activity [60]. Cathepsin H also binds cystatins more strongly than does cathepsin B, but more weakly than endopeptidase cathepsin L and papain [53, 61]. It can be concluded that longer substrates as well as protein inhibitors compete with the mini-chain for the same binding sites.

Substrate specificity of cathepsins

All papain-like cysteine proteinases have the same basic mechanism of action, but their specificity and point of cleavage in the substrate differs from one member of the family to another. The knowledge of substrate binding sites is based on the kinetics and crystal structures of synthetic substrate-mimicking inhibitors bound to the active site of the enzymes. The majority of the crystal structures of inhibitor complexes with papain-like enzymes define the nonprimed binding sites as, for example, with chloro-methyl ketone [62] or fluoro-methyl ketone [29]. The substrate-mimicking inhibitor that revealed the prime binding sites is based on an epoxysuccinyl reactive group as in the structures of a cathepsin B – CA030 complex [58] and of an almost identical inhibitor, CA074, in complex with bovine cathepsin B [63]. This is currently the only source of information concerning the S1' and S2' binding sites in papain-like enzymes. The active-site cleft of cysteine peptidases like papain [21], cathepsin L [24, 57], cathepsin K [25, 26] and cathepsin S [28] is unoccupied and so is able to bind substrates along its full length, whereas the active-site clefts of cathepsins B [22] and H [27], both exopeptidases, are partially filled, thereby limiting the free substrate binding sites. In cathepsin B, a carboxypeptidase, the active-site cleft behind the S2'-binding site is occupied [22, 35, 59] and in cathepsin H, an aminopeptidase, the active site cleft of the S1-binding site is blocked.

In contrast to the many crystal structures of complexes of papain-like enzymes with synthetic inhibitors, there are only two known structures of cysteine proteinase – endogenous protein inhibitor complexes, namely the papain-stefin B complex [64] and the very recently determined structure of the cathepsin L – p41 fragment complex [57]. Although stefin B and p41 fragment structurally belong to different superfamilies of cysteine proteinase inhibitors, their structural elements which bind into the active site cleft are similar (Fig. 3). Cystatins exhibit relatively little specificity in their inhibition of the papain-like enzymes [50, 56]. In contrast the p41 fragment [55] and equistatin [65, 66], as members of the thyropins, discriminate between the different members [39]. There are similarities and differences between the p41 fragment and stefin B, each in inhibitory complexes with cathepsin L and papain, respectively. The wedge shape and three-loop arrangement of the p41 fragment bound to the active site cleft of cathepsin L are reminiscent of the inhibitory edge of cystatins. However, the different fold of the p41 fragment results in additional contacts with the top of the R-domain of the enzymes, which defines the specificity – determining S2 and S1' substrate-binding sites. This enables thyropin inhibitors, in contrast to the non-selective cystatins, to exhibit specificity for their target enzymes. Thus, the p41 fragment discriminates between cathepsins L and S, based on three selective interaction regions. The ability of the p41 fragment to bind to various cathepsins is based on electrostatic and some favourable and unfavourable packing interactions with surface residues on the top of the cathepsin L and S loops embracing the S2-S1' substrate-binding sites [57]. The structures show that the substrate binding sites of papain-like enzymes are located to both sides of the active site cleft. Loops coming out of the core of the R- and L- domains embrace the binding sites. The location of the substrate binding sites beyond S3 and S2' is uncertain. Therefore, the substrate binding sites as defined [67], which encompass the stretch of sites from S4 to S3' on papain, need to be revised, as has been suggested [56].

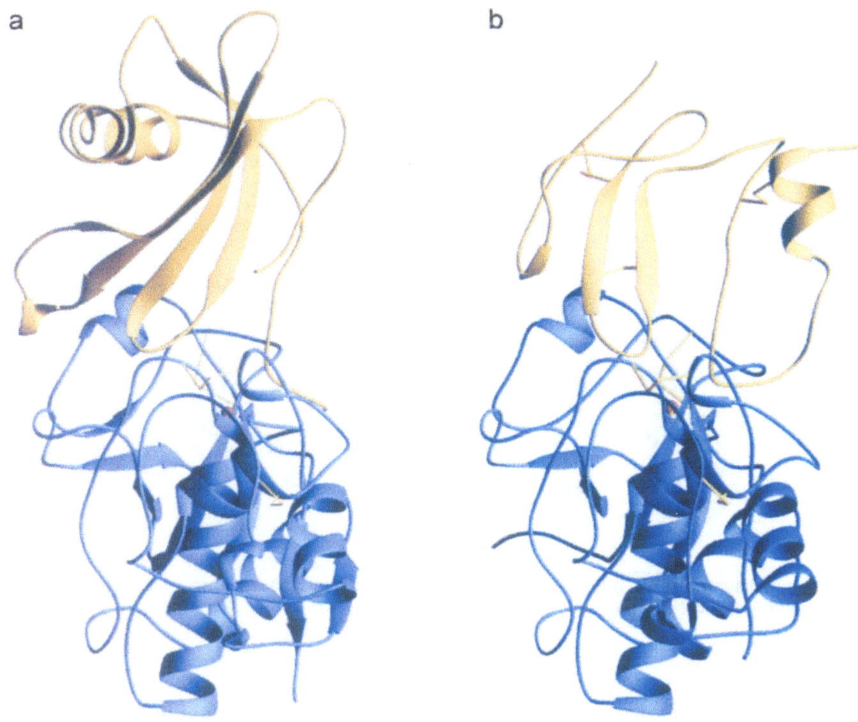

Figure 3. Side view of the papain-stefin B complex (a) and the cathepsin L- p41 fragment complex (b). Papain and cathepsin L are shown as blue and both inhibitors as yellow ribbons. Disulfide bridges are represented with yellow sticks and the sulfur atom of the catalytic cysteine residue as a red ball.

Zymogen activation

Lysosomal cysteine proteinases are synthesized as zymogens and are converted to their active forms by proteolytic cleavage of the N-terminal proregion either by autocatalysis or by other proteinases. Proregions act as inhibitors, blocking the active site cleft [68–70] and are quite selective inhibitors [7, 36, 37]. It has been shown that the proregion of cathepsin L is crucial for the correct folding of the newly synthesized protein and it stabilizes the enzyme when exposed to a higher pH environment [8, 68]. The peptide of cathepsin L has been implicated in the targeting of the enzyme [70].

The crystal structures of procathepsin B [35–37], procathepsin L [71] and procaricain [72] provide evidence that the mature portion of the enzyme is in the active conformation, with the propeptide chain folding on the surface of the enzyme domains and blocking access of substrate to the already formed active site.

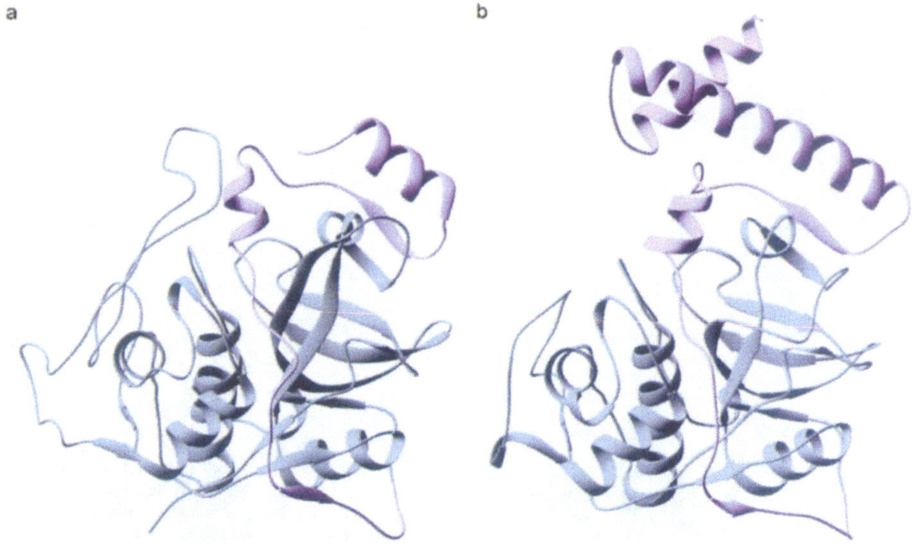

Figure 4. Ribbon diagrams of the procathepsin B (a) and procathepsin L (b) structure. Mature parts of the enzymes are colored in gray and the pro-parts in violet.

In procathepsin B the propeptide chain wraps around the enzyme domains and covers the-active site. It starts as a short α-helix-antiparallel β-strand motif and runs via the kinked region into a short α-helix which ends above the active site (Fig. 4a). The occluding loop, which in mature cathepsin B is positioned above the active site and is responsible for the exopeptidase activity, is raised by this propeptide above the enzyme and stabilized by hydrogen bonds. The propeptide chain continues as a linker peptide along the active site cleft and the interdomain region towards the N-terminal residue of the enzyme, whose position is retained in the mature form. However, the propeptide chain orientation in the substrate-binding site is opposite to that of the bound substrate. This spatial arrangement of the propeptide chain also prevents substrates from binding to the active site cleft.

The interface between the propeptide and the mature enzyme exhibits hydrophilic as well as hydrophobic character, the latter being emphasized by the only Trp residue in the proregion which anchors the propeptide on the hydrophobic surface of the R-domain of the enzyme. The linker peptide forms very few contacts and the only autocatalytic scissile bond lies in the C-terminal part of this region, between Met P56 and Phe P57 [73]. The peptide bond is exposed to the solvent and available for cleavage by another proteinase molecule but is inaccessible to the active site on the same molecule. The autoactivation and autoprocessing of procathepsin B takes place in acidic conditions [73, 74]. It was proposed that a moderate change in pH affects the propeptide acidic residues exposed to the surface, destabilizes the propeptide secondary structure and invokes autoactivation of the enzyme. The loss of the propeptide structure leads to distortion of the propeptide-enzyme contacts. The loss of the interactions causes the dissociation of the propeptide from the enzyme surface, leading to intermolecular proteolytic cleavage and the release of active, mature cathepsin B molecules [35]. A similar mechanism has been proposed for the *in vitro* autoactivation of procaricain [72].

The propeptide of procathepsin L (Fig. 4b) shares little sequence homology with that of procathepsin B, yet the structures show high overall similarity. The main difference between the two structures is at the N-terminus of procathepsin L where the 34 residue longer propeptide of procathepsin L forms an additional α-helix and prolongs the second α-helix, which is structurally homologous to the N-terminal helix in procathepsin B. The propeptide of cathepsin B does not significantly inhibit cathepsin L and is selective for cathepsin B [75], whereas inhibition by the cathepsin L propeptide was found to be selective for cathepsin L [71].

Knowledge of the crystal structures of the three papain-like proenzymes provides further possibilities for investigating the molecular basis for the proregion selectivity and might provide a further basis for the design of more specific inhibitors [76].

Cysteine proteinases and MHC class II function

In general, the majority of protein antigens must be internalized by endocytosis for subsequent processing within intracellular compartments. The processing includes proteolysis and disulphide reduction. During intracellular transport, MHC class II molecules associate with a non-polymorphic type II transmembrane protein termed invariant chain (Ii). Ii is expressed as two isoforms, p31 and p41, derived from alternative splicing [77]. Peptides resulting from limited proteolysis are loaded into the binding groove of MHC class II molecules and presented at the cell surface for recognition by $CD4^+$ T lymphocytes. Proteolysis participates in two critical steps: First, cleavage of the invariant chain (Ii) is required for peptide binding to class II molecules, which depends on its prior removal from the class II peptide binding groove. Secondly, proteolysis of large polypeptides is essential for generating the peptide antigens presented by class II molecules [78].

Different proteinases are involved in the processing of Ii and antigenic peptides. The data suggest that both cysteine and aspartic proteinases may be important in generating antigenic epitopes for presentation by class II molecules (reviewed in [32, 79–81]). However, most reports relate to the involvement of lysosomal cysteine proteinases cathepsin L and S [81–84]. Cathepsin L was found to be necessary for Ii degradation in cortical thymic epithelial cells but not in bone marrow-derived antigen presenting cells, where cathepsin S probably plays an essential role [86, 87].

The discovery of the complex of p41 fragment with cathepsin L focused attention on this specific mechanism of control of cathepsin L activity in connection with the maturation of MHC class II molecules and processing of antigens [88]. p41 fragment inhibits cathepsin L (K_i = 1.7 pM) [55] and cruzipain (K_i = 58 pM) [89] more strongly than other cysteine proteinases. The amino acid sequence of p41 fragment shows no homology with those of the cystatin family of cysteine proteinase inhibitors [55]. It is, however, similar to a number of thyroglobulin type-1 domain inhibitors, called thyropins [39].

The recently determined crystal structure of cathepsin L – p41 fragment complex [57] reveals the features that enable the p41 fragment to act as a proteinase inhibitor, and to inhibit cathepsin L selectively in the presence of cathepsin S, thereby facilitating antigen processing.

The structure represents a novel fold of the cysteine proteinase inhibitors, consisting of two subdomains stabilized by three disulfide bridges. The first N-terminal subdomain starts with an α-helix-strand motif and the second subdomain main motif is an antiparallel-β structure. The CWCV sequence motif, which is conserved in the thyroglobulin type-1 domain, is the core of the second subdomain and is involved in two disulfide bridges. The wedge-shaped p41 fragment is bound to the active site of cathepsin L by its three bottom loops (Fig. 5). The overall arrangement of the interacting loops resembles the binding motif found in the cystatins and is an example of convergent evolution of cysteine proteinase inhibitors. The different fold of the p41 fragment, however, enables additional contacts with the enzyme which determine the selectivity of the inhibitor. This is very important because cathepsin S is essential for effective invariant chain proteolysis, which is necessary to render class II molecules competent for binding peptides [82]. Although it is still unclear at which step the p41 fragment becomes capable of specific inhibition of cathepsin L, evidence suggests that, at a certain step of Ii degradation and antigenic epitope formation, the potentially destructive role of cathepsin L needs to be prevented. It was shown recently that cystatin C, an intracellularly-located inhibitor of lysosomal

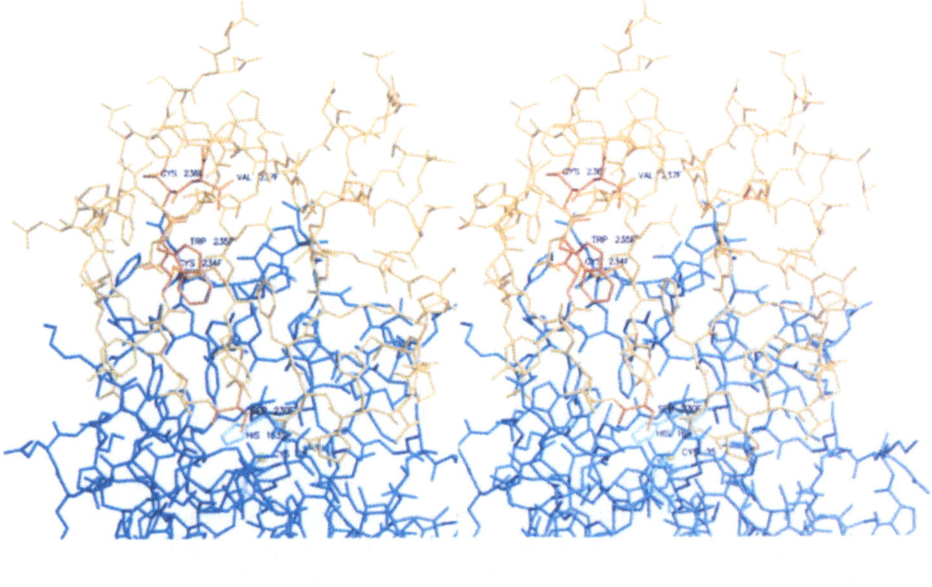

Figure 5. Stereo plot of the cathepsin L- p41 fragment complex. Blue sticks represent cathepsin L residues and yellow sticks invariant chain fragment. Catalytic residues Cys 25 and His 163 are highlighted in light blue color and the sulfur atom of the Cys 25 is shown as a yellow ball. In the fragment of the invariant chain the CWCV sequence motif and the Ser 230 residue, which forms the hydrogen bond with the catalytic Cys 25, are highlighted in orange. Figure was prepared using the program MOLSCRIPT [90] and rendered with the program RASTER 3D [91].

cysteine proteinases, helps to control the cathepsin S activity and thereby regulates maturation of MHC class II molecules in dendritic cells [85].

In conclusion, the importance of the increasing number of crystal structures is seen in providing a structural basis for some of the main steps in the control of antigen processing.

References

1 Rawling ND, Barret AJ (1994) Family of cysteine peptidases. *Meth Enzymol* 244: 461–486
2 Berti PJ, Storer AC (1995) Alignment/phylogeny of the papain superfamily of cysteine proteases. *J Mol Biol* 246: 273–283
3 Turk V, Bode W (1993) Lysosomal cysteine proteinases and their inhibitors cystatins. *In*: FX Avilés (ed.): *Innovations in Proteases and their Inhibitors*. Walter de Gruyter, Berlin, 161–178
4 Kirschke H, Wiederanders B (1994) Cathepsin S and related lysosomal endopeptidases. *Meth Enzymol* 244: 500–511
5 Kirschke H, Barrett AJ, Rawlings ND (1995) Proteinases 1. Lysosomal cysteine proteinases. *Protein Profile* 2: 1587–1643
6 Fox T, De Miguel E, Mort JS, Storer AC (1992) Potent slow-binding inhibition of cathepsin B by its propeptide. *Biochemistry* 31: 12571–12576
7 Carmona E, Dufour E, Plouffe C, Takebe S, Mason P, Mort JS, Ménard R (1996) Potency and selectivity of the cathepsin L propeptide as an inhibitor of cysteine proteinases. *Biochemistry* 35: 8149–8157
8 Tao K, Stearns NA, Dong J, Wu Q, Sahagian GG (1994) The proregion of cathepsin L is required for proper folding, stability and ER exit. *Arch Biochem Biophys* 311: 19–27
9 Dolenc I, Turk B, Pungerčič G, Ritonja A, Turk V (1995) Oligomeric structure and substrate induced inhibition of human cathepsin C. *J Biol Chem* 270: 21626–21631
10 Pariš A, Štrukelj B, Pungerčar J, Renko M, Dolenc I, Turk V (1995) Molecular cloning and sequence analysis of human preprocathepsin C. *FEBS Lett* 369: 326–330
11 Rao NV, Rao GV, Hoidal JR (1997) Human dipeptidyl-peptidase I. *J Biol Chem* 272: 10260–10265
12 Dolenc I, Turk B, Kos J, Turk V (1996) Interaction of human cathepsin C with chicken cystatin. *FEBS Lett* 392: 277–280
13 Velasco G, Ferrando AA, Puente XS, Sanchez LM, López-Otín C (1994) Human cathepsin O. *J Biol Chem* 269: 27136–27142
14 Shi GP, Chapman HA, Bhairi SM, De Leeuw C, Reddy VY, Weiss SJ (1995) Molecular cloning of human cathepsin O, a novel endoproteinase and homologue of rabbit OC2. *FEBS Lett* 357: 129–134
15 Inaoka T, Bilbe G, Ishibashi O, Tezuka K, Kumegawa M, Kokubo T (1995) Molecular cloning of human cDNA for cathepsin K: a novel cysteine proteinase predominantly expressed in bone. *Biochem Biophys Res Commun* 206: 89–96
16 Brömme D, Okamoto K, Wang BD, Biroc S (1996) Human cathepsin O_2, a matrix protein-degrading cysteine protease expressed in osteoclasts. *J Biol Chem* 271: 2126–2132
17 Drake FH, Dodds RA, James IE, Connor JR, Debouck C, Richardson S, Rykaczewsky EL, Coleman L, Rieman D, Barthlow R et al. (1996) Cathepsin K, but not cathepsins B, L or S, is abundantly expressed in human osteoclasts. *J Biol Chem* 271: 12511–12516
18 Linnevers C, Smeekens SP, Brömme D (1997) Human cathepsin W, a putative cysteine protease predominantly expressed in $CD8^+$ T-lymphocytes. *FEBS Lett* 405: 253–259
19 Vex T, Levy B, Smeekens SP, Ansorge S, Desnick RJ, Brömme D (1998) Genomic structure, chromosomal localization and expression of human cathepsin W. *Biochem Biophys Res Commun* 248: 255–261
20 Santamaria I, Velasco G, Pendás AM, Fueyo A, López-Otín C (1998) Cathepsin Z, a novel human cysteine proteinase with a short propeptide domain and a unique chromosomal location. *J Biol Chem* 273: 16816–16823
21 Drenth J, Jansonius R, Koekoek H, Swen M, Wolthers BG (1968) Structure of papain. *Nature* 218: 929–933
22 Musil D, Zučič D, Turk D, Engh RA, Mayr I, Huber R, Popovič T, Turk V, Towatari T, Katunuma N, Bode W (1991) The refined 2.15 Å X-ray crystal structure of human liver cathepsin B: the structural basis for its specificity. *EMBO J* 10: 2321–2330
23 Jia Z, Hasmain S, Hirama T, Lee X, Mort JS, To R, Huber CP (1995) Crystal structures of recombinant rat cathepsin B and a cathepsin B-inhibitor complex. *J Biol Chem* 270: 5527–5533
24 Fujishima A, Imai Y, Nomura T, Fujisawa Y, Yamamoto Y, Sugawara T (1997) The crystal structure of human cathepsin L complexed with E-64. *FEBS Lett* 407: 47–50
25 McGrath ME, Klaus JL, Barnes MG, Brömme D (1997) Crystal structure of human cathepsin K complexed with a potent inhibitor. *Nat Struct Biol* 4: 105–108
26 Zhao B, Janson CA, Amegadzie BY, D'Alessio K, Griffin C, Hanning CR, Jones C, Kurdyla J, McQueney M,

Qiu X et al. (1997) Crystal structure of human osteoclast cathepsin K complex with E-64. *Nat Struct Biol* 4: 109–111
27 Gunčar G, Podobnik M, Pungerčar J, Štrukelj B, Turk V, Turk D (1998) Crystal structure of porcine cathepsin H determined at 2.1 Å resolution: location of the mini-chain C-terminal carboxyl group defines cathepsin H aminopeptidase function. *Structure* 6: 51–61
28 McGrath ME, Palmer JT, Brömme D, Somoza JR (1998) Crystal structure of human cathepsin S. *Protein Sci* 7: 1294–1302
29 McGrath ME, Eakin AE, Engel JC, McKerrow JH, Craik CS, Fletterick RJ (1995) The crystal structure of cruzipain: a therapeutic target for Chaga's disease. *J Mol Biol* 247: 251–259
30 Brocklehurst K, Willenbrook K, Salih E (1987) Cysteine proteinases. *In*: A Neuberger, K Brocklehurst (eds): *Hydrolytic enzymes*. Elsevier, Amsterdam, 39–158
31 Dunn AD, Crutchfield HE, Dunn JT (1991) Thyroglobulin processing by thyroidal proteases. *J Biol Chem* 266: 20198–20204
32 Chapman HA (1998) Endosomal proteolysis and MHC class II function. *Curr Opin Immunol* 10: 93–102
33 Neurath H (1993) The regulation of protease action: an overview. *In*: FX Avilés (ed.): *Innovations in proteases and their inhibitors*. Walter de Gruyter, Berlin, 3–12
34 Twiming SS (1994) Regulation of proteolytic activity in tissues. *Crit Rev Biochem Mol Biol* 29: 315–383
35 Podobnik M, Kuhelj R, Turk V, Turk D (1997) Crystal structure of the wild-type human procathepsin B at 2.5 Å resolution reveals the native active site of a papain-like cysteine protease zymogen. *J Mol Biol* 271: 774–788
36 Turk D, Podobnik M, Kuhelj R, Dolinar M, Turk V (1996) Crystal structure of human procathepsin B at 3.2 and 3.3 Å resolution reveal an interaction motif between a papain-like cysteine protease and its propeptide. *FEBS Lett* 384: 211–214
37 Cygler M, Sivaraman J, Grochulski P, Coulombe R, Storer AC, Mort J (1996) Structure of rat procathepsin B: model for inhibition of cysteine protease activity by the proregion. *Structure* 4: 405–416
38 Turk V, Bode W (1991) The cystatins: protein inhibitors of cysteine proteinases. *FEBS Lett* 285: 213–219
39 Lenarčič B, Bevec T (1998) Thyropins – new structurally related proteinase inhibitors. *Biol Chem* 379: 105–111
40 Turk B, Dolenc I, Turk V, Bieth JG (1993) Kinetics of the pH-induced inactivation of human cathepsin L. *Biochemistry* 32: 375–380
41 Turk B, Dolenc I, Žerovnik E, Turk D, Gubenšek F, Turk V (1994) Human cathepsin B is a metastable enzyme stabilized by specific ionic interactions associated with the active site. *Biochemistry* 33: 14800–14806
42 Turk V, Podobnik M, Turk B, Turk D (1997) Regulation of cysteine proteinase activity by protein inhibitors, zymogen activation and pH. *In*: VK Hopsu-Havu, M Järvinen, H Kirschke (eds): *Proteolysis in cell function*. IOS Press, Amsterdam, 128–136
43 Barrett AJ, Rawlings ND, Woessner JF (eds) (1998) *Handbook of proteolytic enzymes*. Academic Press, San Diego
44 Shaw E (1990) Cysteinyl proteinases and their selective inactivation. *Adv Enzymol Relat Areas Mol Biol* 63: 271–347
45 Shaw E (1994) Peptidyl diazomethanes as inhibitors of cysteine and serine proteinases. *Meth Enzymol* 244: 649–656
46 Shaw E, Mohanty S, Čolič A, Stoka V, Turk V (1993) The affinity-labelling of cathepsin S with peptidyl diazomethyl ketones. Comparison with the inhibition of cathepsin L and calpain. *FEBS Lett* 334: 340–342
47 Hanada K, Tamai M, Yamagishi M, Ohmura S, Sawada J, Tanaka I (1978) Studies on thiol protease inhibitors. Part I. Isolation and characterization of E-64, a new thiol proteinase inhibitor. *Agric Biol Chem* 42: 523–528
48 Towatari T, Nikawa T, Murata M, Yokoo C, Tamai M, Hanada K, Katunuma N (1991) Novel epoxysuccinyl peptides. A selective inhibitor of cathepsin B, *in vivo*. *FEBS Lett* 280: 311–315
49 Barrett AJ, Rawlings ND, Davies ME, Machleidt W, Salvesen G, Turk V (1986) Cysteine proteinase inhibitors of cystatin superfamily. *In*: Barrett AJ, G Salvesen (eds): *Proteinase inhibitors*. Elsevier, Amsterdam, 515–569
50 Turk B, Turk V, Turk D (1997) Structural and functional aspects of papain-like cysteine proteinases and their protein inhibitors. *Biol Chem* 378: 141–150
51 Lenarčič B, Križaj I, Žunec P, Turk V (1996) Differences in specificity for the interactions of stefins A, B, and D with cysteine proteinases. *FEBS Lett* 395: 113–118
52 Turk B, Stoka V, Björk I, Boudier C, Johansson G, Dolenc I, ČolićA, Bieth JG, Turk V (1995) High-affinity binding of two molecules of cysteine proteinases to low-molecular-weight kininogen. *Protein Sci* 4: 1874–1880
53 Jerala R, Kroon-Žitko L, Popovič T, Turk V (1994) Elongation on the amino-terminal part of stefin B decreases inhibition of cathepsin H. *Eur J Biochem* 224: 797–802
54 Stubbs MT, Laber B, Bode W, Huber R, Jerala R, Lenarčič B, Turk V (1990) The refined 2.4 Å X-ray crystal structure of recombinant human stefin B in complex with the cysteine proteinase papain: a novel type of proteinase inhibitor interaction. *EMBO J* 9: 1939–1947
55 Bevec T, Stoka V, Pungerčič G, Dolenc I, Turk V (1996) Major histocompatibility complex class II-associated p41 invariant chain fragment is a strong inhibitor of lysosomal cythepsin L. *J Exp Med* 183: 1331–1338
56 Turk D, Gunčar G, Podobnik M, Turk B (1998) Revised definition of substrate binding sites of papain-like

cysteine proteases. *Biol Chem* 379: 137–147
57 Gunčar G, Pungerčič G, Klemenčič I, Turk V, Turk D (1999) Crystal structure of MHC class II-associated p41 Ii fragment bound to cathepsin L reveals the structural basis for differentiation between cathepsins L and S. *EMBO J* 18: 793–803
58 Turk D, Podobnik M, Popovič T, Katunuma N, Bode W, Huber R, Turk V (1995) Crystal structure of cathepsin B with CA030 at 2.0 Å resolution: A basis for the design of specific epoxysuccinyl inhibitors. *Biochemistry* 34: 4791–4797
59 Illy V, Quraishi O, Wang J, Purisima E, Vernet T, Mort JS (1997) Role of the occluding loop in cathepsin B activity. *J Biol Chem* 272: 1197–1202
60 Koga H, Mori N, Yamada H, Nishimura Y, Tokuda K, Kato K, Imoto T (1992) Endo- and aminopeptidase activities of rat cathepsin H. *Chem Pharm Bull* 40: 965–970
61 Abrahamson M, Mason RW, Hanson H, Buttle DJ, Grubb A, Ohlson K (1991) Human cystatin C. Role of the N-terminal segment in the inhibition of human cysteine proteinases and in its inactivation by leucocyte elastase. *Biochem J* 273: 621–626
62 Drenth J, Kalk KH, Swen HM (1976) Binding chloromethyl ketone substrate analogues to crystalline papain. *Biochemistry* 15: 3731–3738
63 Yamamoto A, Hara T, Tomoo K, Ishida T, Fujii T, Hata Y, Murata M, Kitamura K (1997) Binding mode of CA074, a specific irreversible inhibitor, to bovine cathepsin B as determined by X-ray crystal analysis of the complex. *J Biochem* 121: 974–977
64 Turk V, Bode W (1994) Human cysteine proteinases and their inhibitors, stefins and cystatins. *In*: N Katunama, K Suzuki, J Travis, H Fritz (eds): *Biological functions of proteases and inhibitors*. JSC Press, Tokyo, 47–59
65 Lenarčič B, Ritonja A, Štrukelj B, Turk B, Turk V (1997) Equistatin, a new inhibitor of cysteine proteinases from *Actinia equina*, is structurally related to thyroglobulin type-1 domain. *J Biol Chem* 272: 13899–13903
66 Lenarčič B, Ritonja A, Štrukelj B, Turk B, Turk V (1998) Equistatin, a new inhibitor of cysteine proteinases from *Actinia equina*, is structurally related to thyroglobulin type-1 domain. *J Biol Chem* 273: 12682–12683
67 Schechter I, Berger A (1967) On the size of the active site in proteases. I. Papain. *Biochem Biophys Res Commun* 27: 157–162
68 Mach L, Mort JS, Glössl J (1994) Maturation of human procathepsin B. *J Biol Chem* 269: 13030–13035
69 Chen Y, Plouffe C, Menard R, Storer AC (1996) Delineating functionally important regions and residues in the cathepsin B propeptide for inhibitory activity. *FEBS Lett* 393: 24–26
70 McIntyre GF, Godbold GD, Erickson AH (1994) The pH-dependent membrane association of procathepsin L is mediated by a 9-residue sequence within the propeptide. *J Biol Chem* 269: 567–572
71 Coulombe R, Grochulski P, Sivaraman J, Ménard R, Mort JS, Cygler M (1996) Structure of human procathepsin L reveals the molecular basis of inhibition by the prosegment *EMBO J* 15: 5492–5503
72 Groves MR, Taylor MAJ, Scott M, Cummings NJ, Pickersgill RW, Jenkins JA (1996) The prosequence of procaricain forms an α-helical domain that prevents access to the substrate binding cleft. *Structure* 4: 1193-1203
73 Mach L, Mort JS, Glössl J (1994) Noncovalent complexes between the lysosomal proteinase cathepsin B and its propeptide account for stable, extracellular, high molecular mass forms of the enzyme. *J Biol Chem* 269: 13036–13040
74 Jerala R, Žerovnik E, Kidrič J, Turk V (1998) pH induced conformational transitions of the propeptide of human cathepsin L. *J Biol Chem* 273: 11498–11504
75 Fox T, de Miguel E, Mort JS, Storer AC (1992) Potent slow binding inhibition of cathepsin B by its propeptide. *Biochemistry* 31: 12571–12576
76 Groves M, Coulombe R, Jenkins J, Cygler M (1998) Structural basis for specificity of papain-like cysteine protease proregions toward their cognate enzymes. *Proteins* 32: 504–514
77 Germain RN, Margulies DH (1993) The biochemistry and cell biology of antigen processing and presentation. *Annu Rev Immunol* 403–450
78 Wolf PR, Ploegh HL (1995) How MHC class II molecules aquire peptide cargo: biosynthesis and trafficking through the endocytic pathway. *Annu Rev Cell Dev Biol* 267–306
79 Weenink SM, Gantam AM (1997) Antigen presentation by MHC class II molecules. *Immunol Cell Biol* 75: 69–81
80 Fineschi B, Miller J (1997) Endosomal proteases and antigen processing. *Trends Biochem Sci* 22: 377–382
81 Deussing J, Roth W, Saftig P, Peters C, Ploegh HL, Villadangos JA (1998) Cathepsin B and D are dispensable for major histocompatibility complex class II-mediated antigen presentation. *Proc Natl Acad Sci USA* 95: 4516–4521
82 Riese RJ, Wolf PR, Brömme D, Natkin LR, Villadangos JA, Ploegh HL, Chapman HA (1996) Essential role of cathepsin S in MHC class II-associated invariant chain processing and peptide loading. *Immunity* 4: 357–366
83 Riese RJ, Mitchell RN, Villadangos JA, Shi GP, Palmer JT, Karp ER, DeSanctis GT, Ploegh HL, Chapman HA (1998) Cathepsin S activity regulates antigen presentation and immunity. *J Clin Invest* 101: 2351–2363
84 Fineschi B, Sakaguchi K, Appella E, Miller J (1996) The proteolytic environment involved in MHC class II-restricted antigen presentation can be modulated by the p41 form of invariant chain. *J Immunol* 157: 3211–3215
85 Pierre P, Mellman I (1998) Developmental regulation of invariant chain proteolysis controls MHC class II traf-

ficking in mouse dendritic cells. *Cell* 93: 1135–1145
86 Nakagawa T, Roth W, Wong P, Nelson A, Farr A, Denssing J, Villadangos JA, Ploegh H, Peters C, Rudensky AY (1998) Cathepsin L: Critical role in Ii degradation and $CD4^+$ T cell selection in the thymus. *Science* 280: 450–453
87 Cresswell P (1998) Proteases processing, and thymic selection. *Science* 280: 394–395
88 Ogrinc T, Dolenc Ritonja A, Turk V (1993) Purification of the complex of cathepsin L and the MHC class II-associated invariant chain fragment from human kidney. *FEBS Lett* 336: 555–559
89 Bevec T, Stoka V, Pungerčič G, Cazzulo JJ, Turk V (1997) A fragment of the major histocompatibility complex class II-associated p41 invariant chain inhibits cruzipain, the major cysteine proteinase from *Trypanosoma cruzi*. *FEBS Lett* 401: 259–261
90 Kraulis PJ (1991) MOLSCRIPT: A program to produce both detailed and schematic plots. of protein structures. *J Appl Cryst* 24: 946–950
91 Merritt EA, Bacon DJ (1997) Raster3D: Photorealistic Molecular Graphics. *Meth Enzymol* 277: 505–524
92 Carson M (1991) Ribbons 2.0. *J Appl Cryst* 24: 958–961

Subject index

ADAM 82 see also disintegrin and metallo-
protease domain
acidic environment 114
actin 188
activation 111, 116-118
alternative splicing product 167
Alzheimer's disease 149
aminopeptidase 138
amplification loop 111
amyloid β protein (Aβ) 149
amyloid β protein (Aβ)-degrading protease
153
β-amyloid protein 190
anaphylotoxin 25
antigen processing 54
apoptosis 185, 195
Arg-gingipain (gingipain-R, RGP) 175
Arg-gingipain, enzymatic properties 176
Arg-gingipain, pathophysiological role 179
Arg-gingipain, structural characterization
176
Arg-X-specific cysteine proteinase 175
Asp specificity 208
aspartic proteinase 59, 101
ATPase complex 51
ATP-hydrolyzing metalloprotease of mito-
chondria 77
atypical calpain 161
autocatalysis 116
autolysis 162

basement membrane 109, 112
basic carboxypeptidase 16
binary complex 20
black pigmentation 182
bone morphogenetic protein-1 80
bradykinin 25

Ca^{2+}-activated neutral protease (CANP) 159
CalpA 161
calpain 159, 190
calpain, 30K subunit of 159
calpain, 80K subunit of 159
calpain, ubiquitous 161
calpain, tissue-specific 161
μ-calpain 159
μ/M-calpain 164
calpain family 161
calpainolysis 190
calpastatin 163
carboxypeptidase 13, 138
carboxypeptidase, 3-D structure 17
carboxypeptidase, catalytic mechanism 17
carboxypeptidase, inhibitor 17, 18
carboxypeptidase, pancreatic 13
carboxypeptidase, regulatory 22
carboxypeptidase, substrate 18
carboxypeptidase A 15, 16
carboxypeptidase D 22, 26
carboxypeptidase H 23, 25

carboxypeptidase M 22, 23, 25
carboxypeptidase N 22, 25
carboxypeptidase U 15, 26
carboxypeptidase Z 22, 26
carcinoma 113, 115, 117, 118
caspase 187, 196
caspase recruitment domain (CARD) 200
cathepsin, biochemical properties 212
cathepsin, propeptide 212
cathepsin, specifity 213
cathepsin, subtrate specifity 216
cathepsin, three-dimensional structure 213
cathepsin B 113
cathepsin D (CD) 59
cathepsin E (CE) 59
cathepsin E (CE), biosynthesis 61
cathepsin E (CE), complex type 61
cathepsin E (CE), enzymatic properties 63
cathepsin E (CE), functional diversity 65
cathepsin E (CE), high mannose-type 61
cathepsin E (CE), processing 61
cathepsin E (CE), structural characterisation 61
cathepsin E (CE), transport 61
cathepsin H, mini-chain, octapeptide 215
cathepsin L – p41 fragment complex 216, 217
cathepsin L, inhibition by p41 fragment 219
cathepsin L, invariant chain degradation 219
cathepsin S, invariant chain degradation 219
ced-3 186
CED-3 195
cell cycle 189
cell surface interaction 116
cell surface receptor 109, 117, 118
chaperone 28
chick embryo chorioallantoic membrane (CAM) 113
chymosin 59
chymotrypsin 19
chymotrypsinogen B (CTGB) 20
chymotrypsinogen C (CTGC) 20
coagulation 132
cocoonase 135
complex type of oligosaccharides 61
connectin/titin 165
CrmA 188
CrmA, poxvirus 200
cuticle 137
cystatin 213
cysteine 4
cysteine peptidase 211
cysteine protease 113
cysteine proteinase 128, 212
cytotoxic T lymphocyte (CTL) 190

death effector domain (DED) 200
degradation of regulatory proteins 53
diagnostic tool 113
dichloroisocoumarin 191
digestion 126
dimerization of recombinant domain IV 164
disintegrin and metalloprotease domain, a (ADAM) 82
DNA-dependent protein kinase (DNA-PK) 188
domain II 160

domain IV 160

effect of various phospholipids 162
elastase 19
endogenous inhibitor 111
endopeptidase 2
endothelin-converting enzyme (ECE) 86
Entamoeba 95
Enzyme Commission (EC) 2, 3, 11
Enzyme Commission (EC) number 3
Enzyme Commission (EC) recommendation 4, 5
Enzyme Commission (EC) system 3
enzyme nomenclature 3
equistatin 216
exopeptidase 2
extracellular matrix 109

Fas 186
Fasciola 103
fat/fat mutation 26
fertilin/PH-30 84
fibroblast 112, 113, 118
fimbria 181
focal contact 111
focal contact site 112
fodrin 190
fragmentin 2 191

GAS 2 188
gastricsin 59
Giardia 95
gluzincin 73
gp180 22
granzyme 190
granzyme A (GraA) 190
granzyme B (GraB) 190, 199
growth factor 25

helminth 95
hemagglutination 177
heme accumulation 182
hemoglobin, adsorption and degradation 182
hemolymph 126
hen oviduct signal peptidase (HOSP) 39
high mannose-type oligosaccharide 61

ICE family protease 185
individual peptidase 10
inhibitor of apoptosis (IAP) 201
inhibitor 48, 114
insect proteinase 125
insulin 26
insulysin 75
integrin $\alpha_V\beta_3$ 117
γ-interferon 48, 52
International Union of Biochemistry and Molecular Biology (IUBMB) 2, 3
invadopodia 110, 111, 117
invasion 109, 112, 114
inverzincin 73

KELL 87
KGP-deficient mutant 180
kinin 25

kininogen 213
kpg gene 179
KUZ/ADAM10 83

lactacystin 189
lamin 188
leishmania 95
limb-girdle muscular dystrophy type 2A (LGMD2A) 165
limited proteolysis 19, 196
localization 110, 114, 115, 118
Lys-gingipain (gingipain-K, KGP) 175
Lys-gingipain, enzymatic properties 176
Lys-gingipain, pathophysiological role 179
Lys-gingipain, structural characterization 176
lysosomal cathepsin B 211
lysosomal cathepsin H 211
lysosomal cathepsin L 211
lysosomal cathepsin S 211
lysosome 27, 114
Lys-X-specific cysteine proteinase 175

α_2-macroglobulin 114
α_2-macroglobulin receptor/LDL (low density lipoprotein) receptor-related protein 111, 114
malignancy 117
mannose-6-phosphate receptor (MPR) 60
mast cell 22
mast cell carboxypeptidase A 26
matrix metalloproteinase (MMP) 116
m-calpain 159
membrane protein 35
membrane receptor 118
membrane-type matrix metalloproteinase (MT-MMP) 88
membrane-type matrix metalloproteinase (MT1-MMP) 116-118
membrane-type matrix metalloproteinase (MT2-MMP) 117
membrane-type matrix metalloproteinase (MT2-MMP) 118
meprin A 79
meprin B 79
MEROPS 2-4, 11
metallocarboxypeptidase 13
metalloprotease-disintegrin 82
metalloproteinase 100
metastasis 109, 118
metizincin 73
MHC class II, antigen presentation 219
MHC class II, invariant chain (Ii) 219
MHC class II, p41 219
microglia-type processing system 61
midgut 126
migration 109
mitochondrial intermediate peptidase 77
mitochondrial processing peptidase 77
multicatalytic endopeptidase 47

N-arginine dibasic convertase 76
natural killer (NK) cell 190
nCL-2(2') 161
nematode 104
neprilysin 84
neurolysin 74
N-tosyl-L-phenylalanyl chloromethyl ketone (TPCK) 185

N^{α}-tosyl-L-lysyl chloromethyl ketone (TLCK) 185

oxytocin 26

p34cdc2 191
p35 188, 189
p35, baculovirus 200
p94 161
palB 170
pancreas-specific protein 16
papain-stefin B complex 216
parasite 95
pepsin 59
peptidase 1, 2
peptidase, aspartic type 4
peptidase, catalytic type 2, 4
peptidase, classification 1, 3
peptidase, metallo type 4
peptidase, nomenclature 1
peptidase, recommended name 4
peptidase, terminology 2
peptidase, unknown catalytic type 4
peptidase clan 6
peptidase family 6
peptide hormone 25
perforin 190
periodontal disease 175
PEX 87
pH optimum 25
pH optimum, caspases 203
phosphorylation 53
plasma carboxypeptidase B 15
plasmin 112
plasmin-plasminogen system 110
plasminogen 110
plasminogen activator inhibitor 1 (PAI-1) 111
plasminogen activator inhibitor 2 (PAI-2) 111
Plasmodium falciparum 98
poly (ADP ribose) polymerase (PARP) 188
Porphyromonas gingivalis 175
post ischaemic activation of μ-calpain 164
precursor protein 36
prefimbrilin, processing 182
presenilin 154
procathepsin B 212, 218
procathepsin B, autoactivation 218
procathepsin L, propeptide 219
procollagen C-proteinase 80
procollagen N-proteinase 88
proenzyme 110, 111, 116
proenzyme activation 20
programmed cell death (PCD) 185, 195
propeptide 218
proprotein 50
proproteinase E (PPE) 20
pro-segment 17, 19
protease 2
protease inhibitor 185
proteasome 45, 102, 154, 189
26S proteasome 51
proteasome inhibitor 48, 189
75-kDa protein 182
protein degradation, ER-associated 54
protein degradation, ubiquitin-dependent 51
protein kinase Cδ (PKCδ) 188

proteinase inhibitor 129
proteinase 2, 95
proteolytic cascade 198
proteolytic enzyme 2, 13
protoheme 181
protozoa 95

racemization 155
receptor 115
recombinant cell-type processing system 63
recombinational rearrangement 179
regulation 109
11S regulator 52
19S regulatory complex 51
renin 59
rgpA gene 177
rgpA rgpB double mutant 181
rgpA rgpB kgp triple mutant 180
rgpB gene 177
RGP-deficient mutant 180

Schistosoma 193
α-secretase 149
β-secretase 151
γ-secretase 153
secretory pathway 23
secretory vesicle 23
serine 4
serine carboxypeptidase 26
serine protease 110
serine proteinase 100, 127
serine-carboxypeptidase 13
sheddase 149
signal peptidase, from yeast 40
signal peptidase, inhibition 40
signal peptide 19, 24, 36
signal peptide, removal of 37
signal peptide complex, microsomal 38
signal peptide gene, bacterial 38
Sm-calpain 170
Sol 161, 168
stefin 213
sterol regulatory element-binding protein (SREBP)-1 188
sterol regulatory element-binding protein (SREBP)-2 188
sterol-regulatory element-binding protein (SREBP) protease – S2P 78
structure of cathepsin B, human 212, 214
structure of cathepsin H, porcine 212
structure of cathepsin K, human 212, 214
structure of cathepsin L, human 212, 214
structure of cathepsin S, human 212, 214
structure of procaricain 217
structure of procathepsin B 217, 218
structure of procathepsin L 217, 218
substrate specifity 16

ternary complex 20
thimet oligopeptidase 74
threonine protease 49
thymocyte-type processing system 62
thyropin, equistatin 212
thyropin, thyroglobulin type-1 domain inhibitor 212
tissue inhibitor of metalloproteinases (TIMP) 116

tissue inhibitor of metalloproteinases 2 (TIMP2) 118
tissue-type plasminogen activator (tPA) 110
tolloid 80
Tra-3 161
TRA-3 169
Trichomona 95
trypanosomatid 95
trypsin 19
trypsin-like activity 175
tumor 112, 114, 115
tumor invasion 109, 117
tumor necrosis factor-α converting enzyme (TACE) 83
type I signal peptidase 35
typical calpain 161

U1 small nuclear ribonucleoprotein, 70kDa subunit of 188
ubiquitin 45, 154
ubiquitination 189
urokinase-type plasminogen activator (uPA) 110-112
urokinase-type plasminogen activator receptor (uPAR) 111-113

vacuolar enzyme 27
vasopressin 26
virulence factor of progressive periodontal disease 179

World Wide Web 3

zinc-binding motif 15
Z-Leu-Leu-leucinal (ZLLLal, MG-132) 189
zymogen 16
zymogen activation 197
zymogen activation, propeptide 217

EXS 85

D-Amino Acids in Sequences of Peptides of Multicellular Organisms

Jollès, P.,
Muséum National d'Histoire Naturelle, Paris, France (Ed.)

Life on earth almost exclusively uses laevorotatory or left-handed amino acids (L-enantiomers), rather than D-enantiomers. Nevertheless, with improved analytical methods, D-amino acids have been detected in a variety of peptides of multicellular organisms during recent years.

This book takes stock of our present knowledge in this rapidly expanding research area. In a series of chapters it discusses the characterization and analysis of D-amino acids, their occurrence and function in animal peptides and proteins, some possible biosynthetic pathways, and their appearance during ageing. Furthermore, one chapter approaches the puzzling question of homochirality and life.

Contents:

List of Contributors

Preface

Characterization and analysis of D-amino acids
Scaloni, A., Simmaco, M. and Bossa, F.:
Characterization and analysis of D-amino acids

Occurrence and function of D-amino acids-containing peptides and proteins
Mignogna, G., Simmaco, M. and Barra, D.:
Occurrence and function of D-amino acids-containing peptides and proteins: Antimicrobial peptides

Yasuda-Kamatani, Y.:
Molluscan neuropeptides
Amiche, M., Delfour, A. and Nicolas, P.:
Opioid peptides from frog skin

Huberman, A. and Aguilar, M. B.:
D-Amino acids in crustacean hyperglycemic neurohormones

Biosynthesis: A new family of isomerases
Volkmann, R. A. and Heck, S. D.:
Biosynthesis of D-amino acid-containing peptides: Exploring the role of peptide isomerases

Appearance of D-amino acids during aging
Fisher, G. H.:
Appearance of D-amino acids during aging: D-Amino acids in tumor proteins

Ingrosso, D. and Perna, A. F.:
D-Amino acids in aging erythrocytes

D-Amino acids hydrolysing enzymes
Yamada, R. and Kera, Y.:
D-Amino acid hydrolysing enzymes

Homochirality and life
Bonner, W.A.:
Homochirality and life

Subject index

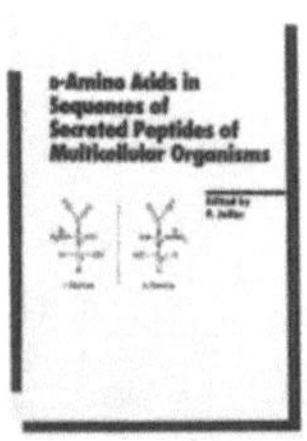

EXS 85
Jollès, P. (Ed.)
D-Amino Acids in Sequences of Peptides of Multicellular Organisms
1998. 200 pages. Hardcover
ISBN 3-7643-5814-9

http://www.birkhauser.ch

PIR
Progress in Inflammation Research

Medicinal Fatty Acids in Inflammation

Kremer, J.M.,
Albany Medical College, Albany, USA (Ed.)

This volume is a unique assembly of contributions focusing on the biochemical, immunological and clinical benefits of n-3 fatty acids in inflammation.

Leading clinical investigators from fields as diverse as rheumatology, dermatology, nephrology, gastroenterology and neurology have authored chapters. The basic scientific underpinnings of their findings are elucidated as well.

The work is a highly accessible, one-of-a-kind source which will well serve lipid researchers, graduate students, dieticians and members of the food industry.

Contents

List of contributors

Preface

Calder, P. C.:
n-3 Polyunsaturated fatty acids and mononuclear phagocyte function

Zurier, R. B.:
Gammalinolenic acid treatment of rheumatoid arthritis

Ziboh, V. A.:
The role of n-3 fatty acids in psoriasis

Horrobin, D. F.:
n-6 Fatty acids and nervous system diorders

Fernandes, G.:
n-3 Fatty acids on autoimmune disease and apoptosis

Belluzzi, A. and Miglio, F.:
n-3 Fatty acids in the treatment of Crohn's disease

Rodgers, J. B.:
n-3 Fatty acids in the treatment of ulcerative colitis

Geusens, P. P.:
n-3 Fatty acids in the treatment of rheumatoid arthritis

Grande, J. P. and Donadio, J. V.:
n-3 Polyunsaturated fatty acids in the treatment of patients with IgA nephropathy

Subject index

PIR – Progress in Inflammation Research
Kremer, J.M., (Ed.)
Medicinal Fatty Acids in Inflammation
1998. 162 pages. Hardcover
ISBN 3-7643-5854-8

BioSciences with Birkhäuser

For orders originating from all over the world except USA and Canada:

Birkhäuser Verlag AG
P.O. Box 133
CH-4010 Basel / Switzerland
Fax: +41 / 61 / 205 07 92
e-mail: orders@birkhauser.ch

For orders originating in the USA and Canada:

Birkhäuser Boston, Inc.
333 Meadowland Parkway
USA-Secaucus, NJ 07094-2491
Fax: +1 / 201 348 4033
e-mail: orders@birkhauser.com

Birkhäuser

MIX
Papier aus verantwortungsvollen Quellen
Paper from responsible sources
FSC® C105338

If you have any concerns about our products,
you can contact us on
ProductSafety@springernature.com

In case Publisher is established outside the EU,
the EU authorized representative is:
Springer Nature Customer Service Center GmbH
Europaplatz 3, 69115 Heidelberg, Germany

Printed by Libri Plureos GmbH
in Hamburg, Germany